Water Reuse

POTENTIAL FOR EXPANDING THE NATION'S WATER SUPPLY THROUGH REUSE OF MUNICIPAL WASTEWATER

Committee on the Assessment of Water Reuse as an Approach
for Meeting Future Water Supply Needs

Water Science and Technology Board

Division on Earth and Life Studies

NATIONAL RESEARCH COUNCIL
OF THE NATIONAL ACADEMIES

THE NATIONAL ACADEMIES PRESS
Washington, D.C.
www.nap.edu

THE NATIONAL ACADEMIES PRESS   **500 Fifth Street, NW**   **Washington, DC 20001**

NOTICE: The project that is the subject of this report was approved by the Governing Board of the National Research Council, whose members are drawn from the councils of the National Academy of Sciences, the National Academy of Engineering, and the Institute of Medicine. The members of the panel responsible for the report were chosen for their special competences and with regard for appropriate balance.

Support for this study was provided by the Environmental Protection Agency under contract number EP-C-09-003: TO#7, the National Science Foundation under grant number CBET-0924454, the National Water Research Institute under grant number 08-KM-006, the U.S. Bureau of Reclamation under grant number R11AP81325, the Water Research Foundation under agreement 04276:PF, and the Monterey Regional Water Pollution Control Agency. Any opinions, findings, conclusions, or recommendations expressed in this publication are those of the author(s) and do not necessarily reflect the views of the organizations or agencies that provided support for the project.

International Standard Book Number-13:   978-0-309-25749-7
International Standard Book Number-10:   0-309-25749-2
Library of Congress Control Number: 2012936028

Additional copies of this report are available from the National Academies Press, 500 Fifth Street, NW, Keck 360, Washington, DC 20001; (800) 624-6242 or (202) 334-3313; http://www.nap.edu.

THE NATIONAL ACADEMIES
Advisers to the Nation on Science, Engineering, and Medicine

The **National Academy of Sciences** is a private, nonprofit, self-perpetuating society of distinguished scholars engaged in scientific and engineering research, dedicated to the furtherance of science and technology and to their use for the general welfare. Upon the authority of the charter granted to it by the Congress in 1863, the Academy has a mandate that requires it to advise the federal government on scientific and technical matters. Dr. Ralph J. Cicerone is president of the National Academy of Sciences.

The **National Academy of Engineering** was established in 1964, under the charter of the National Academy of Sciences, as a parallel organization of outstanding engineers. It is autonomous in its administration and in the selection of its members, sharing with the National Academy of Sciences the responsibility for advising the federal government. The National Academy of Engineering also sponsors engineering programs aimed at meeting national needs, encourages education and research, and recognizes the superior achievements of engineers. Dr. Charles M. Vest is president of the National Academy of Engineering.

The **Institute of Medicine** was established in 1970 by the National Academy of Sciences to secure the services of eminent members of appropriate professions in the examination of policy matters pertaining to the health of the public. The Institute acts under the responsibility given to the National Academy of Sciences by its congressional charter to be an adviser to the federal government and, upon its own initiative, to identify issues of medical care, research, and education. Dr. Harvey V. Fineberg is president of the Institute of Medicine.

The **National Research Council** was organized by the National Academy of Sciences in 1916 to associate the broad community of science and technology with the Academy's purposes of furthering knowledge and advising the federal government. Functioning in accordance with general policies determined by the Academy, the Council has become the principal operating agency of both the National Academy of Sciences and the National Academy of Engineering in providing services to the government, the public, and the scientific and engineering communities. The Council is administered jointly by both Academies and the Institute of Medicine. Dr. Ralph J. Cicerone and Dr. Charles M. Vest are chair and vice chair, respectively, of the National Research Council.

www.national-academies.org

Preface

Starting in the late 19th and through most of the 20th century, the United States built a substantial infrastructure to capture fresh water and bring it to our farms and cities. Although efforts to add to that infrastructure continue, by most measures the amount of water delivered has not materially increased in the past 30 years, but the U.S. population has continued to climb. The National Research Council (NRC, 2001) said, "In this new century, the United States will be challenged to provide sufficient quantities of high-quality water to its growing population." This report is part of an ongoing effort by the NRC to understand the tools the nation has available to address the challenge identified in that statement—in this case, the role water reuse might play in the nation's water future.

The committee formed by the NRC's Water Science and Technology Board performed a critical assessment of water reuse as an approach to meet future water supply needs. The report presents a brief summary of the nation's recent history in water use and shows that, although reuse is not a panacea, the amount of wastewater discharged to the environment is of such quantity that it could play a significant role in the overall water resource picture and complement other strategies, such as water conservation. The report also identifies a research agenda designed to help the nation progress in making the most appropriate use of the resource.

For each of us, our most precious resource is our time. This project was a substantial project, involving eight meetings. I want to thank the members of this committee for their most generous contribution of their personal time to this project. That time is especially valuable because of the unique individual expertise and intellect each of member brought to the task. Once again, as it does so well, the NRC assembled a collection of the nation's best minds from a broad spectrum of disciplines and assigned them to work together to address an issue important to the nation's future. Once again, the process worked beautifully and, in a collaborative spirit, these individuals worked together to produce many insights none of us had as individuals when we walked into our first meeting and a report that the committee should be proud of.

Those who have been on an NRC committee know that staff play a critical role in the success of the project. Our study director, Stephanie Johnson, is an amazing woman—organized, disciplined, persistent, able to cope with great detail, and a fabulous technical writer. She was in constant communication with all of us; reminding us of our assignments, providing us with critical comments, personally writing some sections of the report, and thoroughly editing our myriad styles to produce a document that speaks with a single voice. This report would not have happened were it not for her effort. The committee is also grateful for the assistance provided by Stephen Russell and Sarah Brennan, project assistants, who handled administrative details of the meetings, did supporting research, and aided in report preparation.

Thanks are also due to the sponsors who provided support for the study. This report was undertaken with support from a myriad of sponsors. More than half of the study funding was provided by the Environmental Protection Agency, with the remaining funding from the U.S. Bureau of Reclamation, the National Science Foundation, the National Water Research Institute, the Centers for Disease Control and Prevention, the Water Research Foundation, Orange County Water District, Orange County Sanitation District, Los Angeles Department of Water and Power, Irvine Ranch Water District, West Basin Water District, Inland Empire Utilities Agency, Metropolitan Water District of Southern California, Los Angeles County Sanitation Districts, and the Monterey Regional Water Pollution Control Agency.

The committee held meetings at several locations, including California, Florida, Colorado, Texas, and Washington D.C. In particular the committee would like to thank the individuals and agencies who gave presentations and provided tours to help the committee in its deliberations (see Acknowledgments).

In draft form the report was reviewed by individuals chosen for their breadth of perspective and technical expertise in accordance with the procedures approved by the National Academies' Report Review Committee. The purpose of this independent review was to provide candid and critical comments to assist the NRC in ensuring that the final report is scientifically credible and that it meets NRC standards for objectivity, evidence, and responsiveness to the study charge. The reviewer comments and the draft manuscript remain confidential to protect the deliberative process. We thank the following reviewers for their criticisms, advice, and insight, all of which were considered and many of which were wholly or partly incorporated in the final report:

Bryan Brooks, Baylor University; Charles Gerba, University of Arizona; Jerome Gilbert, Engineering Perfection, PLLC; Robert Hultquist, California Department of Public Health; Anna Hurlimann, The University of Melbourne; Blanca Jimenez, Instituto de Ingenieria UNAM; Stuart Khan, University of New South Wales; Margaret Nellor, Nellor Environmental Associates, Inc.; Larry Roesner, Colorado State University; Dan Tarlock, Chicago Kent College of Law; George Tchobanoglous, University of California, Davis (emeritus); Michael Wehner, Orange County Water District; and Paul Westerhoff, Arizona State University.

Although reviewers were asked to, and did, provide constructive comments and suggestions, they were not asked to endorse the conclusions and recommendations nor did they see the final draft of the report before its release. The review of this report was overseen by Edward Bouwer, Johns Hopkins University, and Michael Kavanaugh, Geosyntec Consultants. Appointed by the NRC, they were responsible for making certain that an independent examination of this report was carried out in accordance with NRC procedures and that all review comments received full consideration. Responsibility for the final content of this report rests entirely with the authoring committee and the NRC.

R. Rhodes Trussell, *Chair*
Committee on the Assessment of
Water Reuse as an Approach for
Meeting Future Water Supply Needs

Acknowledgments

Many individuals assisted the committee and the National Research Council staff in their task to create this report. We would like to express our appreciation to the following people who have provided presentations to the committee and served as guides during the field trips:

Richard Atwater, Inland Empire Utilities Agency
Jared Bales, U.S. Geological Survey
Robert Bastian, U.S. Environmental Protection Agency
Curt Brown, U.S. Bureau of Reclamation
Shonnie Cline, AWWA Research Foundation
Glenn Clingenpeel, Trinity River Authority
Betsy Cody, Congressional Research Service
Phil Cross, Conserv II
James Dobrowolski, U.S. Department of Agriculture
Mark Elsner, Southwest Florida Water Management District
Chris Ferraro, Florida Department of Environmental Protection
James Franckiewicz, U.S. Agency for International Development
Bertha Goldenberg, Miami-Dade Water and Sewer Department
Brian Good, Denver Water
Bruce Hamilton, National Science Foundation
Larry Honeybourne, Orange County Health Care Agency
Martin Jekel, Technical University of Berlin, Germany

Josh Johnson, Senate Committee on Energy and Natural Resources
Rai Kookana, CSIRO Land and Water, Australia
Mark LeChevallier, American Water
Audrey Levine, U.S. Environmental Protection Agency
Mong Hoo Lim, Public Utilities Board, Singapore
Dean Marrone, U.S. Bureau of Reclamation
James McDaniel, Los Angeles Department of Water and Power
Mark Millan, Data Instincts
Wade Miller, WateReuse Foundation
David Moore, Southwest Florida Water Management District
John Morris, Metropolitan Water District of Southern California
Jeff Mosher, National Water Research Institute
Lynn Orphan, Clean Water Coalition
Pankaj Parekh, Los Angeles Department of Water and Power
Larry Parsons, University of Florida
Mark Pifher, Aurora Water
Robert Quint, U.S. Bureau of Reclamation
Mark Sees, Orlando Easterly Wetlands
Peter Silva, U.S. Environmental Protection Agency
Mark Squillace, University of Colorado Law School
Marsi Steirer, City of San Diego Department of Water
Frank Stephens, Gwinnett County Water Resources
Ray Tremblay, Los Angeles County Sanitation Districts

Bob Vincent, Florida Department of Health
Joe Waters, West Basin Municipal Water District
Michael Wehner, Orange County Water District
Ron Wildermuth, Orange County Water District
Hal Wilkening, Southwest Florida Water
    Management District

Dan Woltering, Water Environment Research
    Foundation
Max Zarate-Bermudez, U.S. Center for Disease
    Control

We would also like to thank Sangam Tiwari, Trussell
Technologies, Inc. for her detailed verification of the
risk exemplar.

Contents

Summary

As the world enters the 21st century, the human community finds itself searching for new paradigms for water supply and management. As communities face water supply challenges amidst continued population growth and climate change, water reuse, or the use of highly treated wastewater effluent (also called reclaimed water) for either potable or nonpotable purposes, is attracting increasing attention. Many communities have implemented inexpensive water reuse projects, such as irrigating golf courses and parks or providing industrial cooling water in locations near the wastewater reclamation plant. In the process, these communities have become familiar with the advantages of water reuse, such as improved reliability and drought resistance of the water supply. However, increased use of reclaimed water typically poses greater financial, technical, and institutional challenges than traditional sources and some citizens are concerned about the safety of using reclaimed water for domestic purposes. These challenges have limited the application of water reuse in the United States.

The National Research Council's (NRC's) Committee on Assessment of Water Reuse as an Approach for Meeting Future Water Supply Needs was formed to conduct a comprehensive study of the potential for water reclamation and reuse of municipal wastewater to expand and enhance the nation's available water supply alternatives (see Box S-1 for the statement of task). The study is sponsored by the Environmental Protection Agency, the Bureau of Reclamation, the National Science Foundation, the National Water Research Institute, the Centers for Disease Control and Prevention, the Water Research Foundation, the Orange County Water District, the Orange County Sanitation District, the Los Angeles Department of Water and Power, the Irvine Ranch Water District, the West Basin Water District, the Inland Empire Utilities Agency, the Metropolitan Water District of Southern California, the Los Angeles County Sanitation Districts, and the Monterey Regional Water Pollution Control Agency.

In this report, the committee analyzes technical, economic, institutional, and social issues associated with increased adoption of water reuse and provides an updated perspective since the NRC's last report, Issues in Potable Reuse (NRC, 1998). This report considers a wide range of reuse applications, including drinking water, nonpotable urban uses, irrigation, industrial process water, groundwater recharge, and ecological enhancement.

CONTEXT AND POTENTIAL FOR WATER REUSE

Municipal wastewater reuse offers the potential to significantly increase the nation's total available water resources. Approximately 12 billion gallons of municipal wastewater effluent is discharged each day to an ocean or estuary out of the 32 billion gallons per day discharged nationwide. Reusing these coastal discharges would directly augment available water resources (equivalent to 6 percent of the estimated total U.S. water use or 27 percent of public supply).[1] When reclaimed water is used for nonconsumptive

---

[1] See Chapter 1 for details on how the committee calculated this discharge total and the percentages.

BOX S-1
Statement of Task

A National Research Council committee, convened by the Water Science and Technology Board, conducted a comprehensive study of the potential for water reclamation and reuse of municipal wastewater to expand and enhance the nation's available water supply alternatives. The committee was tasked to address the following issues and questions:

1. Contributing to the nation's water supplies. What are the potential benefits of expanded water reuse and reclamation? How much municipal wastewater effluent is produced in the United States, what is its quality, and where is it currently discharged? What is the suitability—in terms of water quality and quantity—of processed wastewaters for various purposes, including drinking water, nonpotable urban uses, irrigation, industrial processes, groundwater recharge, and environmental restoration?

2. Assessing the state of technology. What is the current state of the technology in wastewater treatment and production of reclaimed water? How do available treatment technologies compare in terms of treatment performance (e.g., nutrient control, contaminant control, pathogen removal), cost, energy use, and environmental impacts? What are the current technology challenges and limitations? What are the infrastructure requirements of water reuse for various purposes?

3. Assessing risks. What are the human health risks of using reclaimed water for various purposes, including indirect potable reuse? What are the risks of using reclaimed water for environmental purposes? How effective are monitoring, control systems, and the existing regulatory framework in assuring the safety and reliability of wastewater reclamation practices?

4. Costs. How do the costs (including environmental costs, such as energy use and greenhouse gas emissions) and benefits of water reclamation and reuse generally compare with other supply alternatives, such as seawater desalination and nontechnical options such as water conservation or market transfers of water?

5. Barriers to implementation. What implementation issues (e.g., public acceptance, regulatory, financial, institutional, water rights) limit the applicability of water reuse to help meet the nation's water needs and what, if appropriate, are means to overcome these challenges? Based on a consideration of case studies, what are the key social and technical factors associated with successful water reuse projects and favorable public attitudes toward water reuse? Conversely, what are the key factors that have led to the rejection of some water reuse projects?

6. Research needs. What research is needed to advance the nation's safe, reliable, and cost-effective reuse of municipal wastewater where traditional sources of water are inadequate? What are appropriate roles for governmental and nongovernmental entities?

---

uses, the water supply benefit of water reuse could be even greater if the water can again be captured and reused. Inland effluent discharges may also be available for water reuse, although extensive reuse has the potential to affect the water supply of downstream users and ecosystems in water-limited settings. Water reuse alone cannot address all of the nation's water supply challenges, and the potential contributions of water reuse will vary by region. However, water reuse could offer significant untapped water supplies, particularly in coastal areas facing water shortages.

Water reuse is a common practice in the United States. Numerous approaches are available for reusing wastewater effluent to provide water for industry, irrigation, and potable supply, among other applications, although limited estimates of water reuse suggest that it accounts for a small part (<1 percent) of U.S. water use. Water reclamation for nonpotable applications is well established, with system designs and treatment technologies that are generally accepted by communi-

ties, practitioners, and regulatory authorities. The use of reclaimed water to augment potable water supplies has significant potential for helping to meet future needs, but planned potable water reuse only accounts for a small fraction of the volume of water currently being reused. However, potable reuse becomes more significant to the nation's current water supply portfolio if de facto (or unplanned) water reuse[2] is included. **The de facto reuse of wastewater effluent as a water supply is common in many of the nation's water systems, with**

---

[2] *De facto reuse* is defined by the committee as a drinking water supply that contains a significant fraction of wastewater effluent, typically from upstream wastewater discharges, although the water supply has not been permitted as a water reuse project. There is no specific cutoff for how much effluent in a water source is considered de facto reuse, because water quality is affected by the extent of instream contaminant attenuation processes and travel time. However, water supplies where effluent accounts for more than a few percent of the overall flow are usually considered to be undergoing de facto reuse. For a detailed discussion of the extent of effluent contributions to water supplies, see Chapter 2.

some drinking water treatment plants using waters from which a large fraction originated as wastewater effluent from upstream communities, especially under low-flow conditions.

An analysis of the extent of de facto potable water reuse should be conducted to quantify the number of people currently exposed to wastewater contaminants and their likely concentrations. A systematic analysis of the extent of effluent contributions to potable water supplies has not been made in the United States for over 30 years. Such an analysis would help water resource planners and public health agencies understand the extent and importance of de facto water reuse.

WATER QUALITY AND WASTEWATER RECLAMATION TECHNOLOGY

The very nature of water reuse suggests that nearly any substance used or excreted by humans has the potential to be present at some concentration in the treated product. Modern analytical technology allows detection of chemical and biological contaminants at levels that may be far below human and environmental health relevance. Therefore, if wastewater becomes part of a reuse scheme (including de facto reuse), the impacts of wastewater constituents on intended applications should be considered in the design of the treatment systems. Some constituents, such as salinity, sodium, and boron, have the potential to affect agricultural and landscape irrigation practices if they are present at concentrations or ratios that exceed specific thresholds. Some constituents, such as microbial pathogens and trace organic chemicals, have the potential to affect human health, depending on their concentration and the routes and duration of exposure (see Chapter 6). Additionally, not only are the constituents themselves important to consider but also the substances into which they may transform during treatment. Pathogenic microorganisms are a particular focus of water reuse treatment processes because of their acute human health effects, and viruses necessitate special attention based on their low infectious dose, small size, and resistance to disinfection.

A portfolio of treatment options, including engineered and managed natural treatment processes, exists to mitigate microbial and chemical contaminants in reclaimed water, facilitating a multitude of process combinations that can be tailored to meet specific water quality objectives. Advanced treatment processes are also capable of addressing contemporary water quality issues related to potable reuse involving emerging pathogens or trace organic chemicals. Advances in membrane filtration have made membrane-based processes particularly attractive for water reuse applications. However, limited cost-effective concentrate disposal alternatives hinder the application of membrane technologies for water reuse in inland communities.

Natural systems are employed in most potable water reuse systems to provide an environmental buffer. However, it cannot be demonstrated that such "natural" barriers provide any public health protection that is not also available by other engineered processes (e.g., advanced treatment processes, reservoir storage). Environmental buffers in potable reuse projects may fulfill some or all of three design elements: (1) provision of retention time, (2) attenuation of contaminants, and (3) blending (or dilution). However, the extent of these three factors varies widely across different environmental buffers under differing hydrogeological and climatic conditions. In some cases engineered natural systems, which are generally perceived as beneficial to public acceptance, can be substituted for engineered unit processes, although the science required to design for uniform protection from one environmental buffer to the next is not available. The lack of clear and standardized guidance for design and operation of engineered natural systems is the biggest deterrent to their expanded use, in particular for potable reuse applications.

QUALITY ASSURANCE

Reuse systems should be designed with treatment trains that include reliability and robustness. Redundancy strengthens the reliability of contaminant removal, particularly important for contaminants with acute affects, while robustness employs combinations of technologies that address a broad variety of contaminants. Reuse systems designed for applications with possible human contact should include redundant barriers for pathogens that cause waterborne diseases. Potable reuse systems should employ diverse processes that can function as barriers for many types of chemicals, considering the wide range of physiochemical properties of chemical contaminants.

Reclamation facilities should develop monitoring and operational plans to respond to variability, equipment malfunctions, and operator error to ensure that reclaimed water released meets the appropriate quality standards for its use. Redundancy and quality reliability assessments, including process control, water quality monitoring, and the capacity to divert water that does not meet predetermined quality targets, are essential components of all reuse systems. A key aspect involves the identification of easily measureable performance criteria (e.g., surrogates), which are used for operational control and as a trigger for corrective action.

Monitoring, contaminant attenuation processes, post-treatment retention time, and blending can be effective tools for achieving quality assurance in both nonpotable and potable reuse schemes. Today most projects find it necessary to employ all these elements, and different configurations of unit processes can achieve similar levels of water quality and reliability. **In the future, as new technologies improve capabilities for both monitoring and attenuation, it is expected that retention and blending requirements currently imposed on many potable reuse projects will become less significant in quality assurance.**

The potable reuse of highly treated reclaimed water without an environmental buffer is worthy of consideration, if adequate protection is engineered within the system. Historically, the practice of adding reclaimed water directly to the water supply without an environmental buffer—a practice referred to as direct potable reuse—has been rejected by water utilities, by regulatory agencies in the United States, and by previous NRC committees. However, research during the past decade on the performance of several full-scale advanced water treatment operations indicates that some engineered systems can perform equally well or better than some existing environmental buffers in diluting and attenuating contaminants, and the proper use of indicators and surrogates in the design of reuse systems offers the potential to address many concerns regarding quality assurance. Environmental buffers can be useful elements of design that should be considered along with other processes and management actions in formulating the composition of

potable water reuse projects. However, environmental buffers are not essential elements to achieve quality assurance in potable reuse projects. Additionally, the classification of potable reuse projects as indirect (i.e., includes an environmental buffer) and direct (i.e., does not include an environmental buffer) is not productive from a technical perspective because the terms are not linked to product water quality.

UNDERSTANDING THE RISKS

Health risks remain difficult to fully characterize and quantify through epidemiological or toxicological studies, but well-established principles and processes exist for estimating the risks of various water reuse applications. Absolute safety is a laudable goal of society; however, in the evaluation of safety, some degree of risk must be considered acceptable (NAS, 1975; NRC, 1977). To evaluate these risks, the principles of hazard identification, exposure assessment, dose-response assessment, and risk characterization can be used, as outlined in Chapter 6. Risk assessment screening methods enable estimates of potential human health effects for circumstances where dose-response data are lacking. Although risk assessment will be an important input in decision making, it only forms one of several such inputs, and risk management decisions incorporate a variety of other factors, such as cost, equitability, social, legal and regulatory factors, and qualitative public preferences.

The occurrence of a contaminant at a detectable level does not necessarily pose a significant risk. Instead, only by using dose-response assessments can a determination be made of the significance of a detectable and quantifiable concentration.

A better understanding and a database of the performance of treatment processes and distribution systems are needed to quantify the uncertainty in risk assessments of potable and nonpotable water reuse projects. Failures in reliability of a water reuse treatment and distribution system may cause a short-term risk to those exposed, particularly for acute contaminants (e.g., pathogens) where a single exposure is needed to produce an effect. To assess the overall risks of a system, the performance (variability and uncertainty) of each of the steps needs to be understood. Although a good understanding of the typical performance of different treatment processes exists, an improved understanding of the duration and extent of any variations in performance at removing contaminants is needed.

When assessing risks associated with reclaimed water, the potential for unintended or inappropriate uses should be assessed and mitigated. If the risk is then deemed unacceptable, some combination of more stringent treatment barriers or more stringent controls against inappropriate uses would be necessary if the project is to proceed. Inadvertent cross connection of potable and nonpotable water lines represent one type of unintended outcome that poses significant human health risks from exposure to pathogens. To significantly reduce the risks associated with cross connections, particularly from exposure to pathogens, nonpotable reclaimed water distributed to communities via dual distribution systems should be disinfected to reduce microbial pathogens to low or undetectable levels. Enhanced surveillance during installation of reclaimed water pipelines may be necessary for nonpotable reuse projects that distribute reclaimed water that has not received a high degree of treatment and disinfection.

EVALUATING THE RISKS OF POTABLE REUSE IN CONTEXT

It is appropriate to compare the risk of water produced by potable reuse projects with the risk associated with the water supplies that are presently in use. In Chapter 7, the committee presents the results of an original comparative analysis of potential health risks of potable reuse in the context of the risks of a conventional drinking water supply derived from a surface water that receives a small percentage of treated wastewater. By means of this analysis, termed a risk exemplar, the committee compares the estimated risks of a common drinking water source generally perceived as safe (i.e., de facto potable reuse) against the estimated risks of two other potable reuse scenarios.

The committee's analysis suggests that the risk from 24 selected chemical contaminants in the two potable reuse scenarios does not exceed the risk in common existing water supplies. The results are helpful in providing perspective on the relative importance of different groups of chemicals in drinking water.

For example, disinfection byproducts, in particular nitrosodimethylamine (NDMA), and perfluorinated chemicals deserve special attention in water reuse projects because they represent a more serious human health risk than do pharmaceuticals and personal care products. Despite uncertainties inherent in the analysis, these results demonstrate that following proper diligence and employing tailored advanced treatment trains and/or natural engineered treatment, potable reuse systems can provide protection from trace organic contaminants comparable to what the public experiences in many drinking water supplies today.

With respect to pathogens, although there is a great degree of uncertainty, the committee's analysis suggests the risk from potable reuse does not appear to be any higher, and may be orders of magnitude lower, than currently experienced in at least some current (and approved) drinking water treatment systems (i.e., de facto reuse). State-of-the-art water treatment trains for potable reuse should be adequate to address the concerns of microbial contamination if finished water is protected from recontamination during storage and transport and if multiple barriers and quality assurance strategies are in place to ensure reliability of the treatment processes. The committee's analysis is presented as an exemplar (see Appendix A for details and assumptions made) and should not be used to endorse certain treatment schemes or determine the risk at any particular site without site-specific analyses.

ECOLOGICAL APPLICATIONS OF WATER REUSE

Currently, few studies have documented the environmental risks associated with the purposeful use of reclaimed water for ecological enhancement. Water reuse for the purpose of ecological enhancement is a relatively new and promising area of investigation, but few projects have been completed and the committee was unable to find any published research in the peer-reviewed literature investigating potential ecological effects at these sites. As environmental enhancement projects with reclaimed water increase in number and scope, the amount of research conducted with respect to ecological risk should also increase, so that the potential benefits and any issues associated with the reuse application can be identified.

The ecological risk issues and stressors in ecological enhancement projects are not expected to exceed those encountered with the normal surface water discharge of municipal wastewater. Further, the presence of contaminants and potential ecological impacts may be lower if additional levels of treatment are applied. The most probable ecological stressors include nutrients and trace organic chemicals, although stressors could also include temperature and salinity under some circumstances. For some of these potential stressors (e.g., nutrients), there is quite a bit known about potential ecological impacts associated with exposure. Less is known about the ecological effects of trace organic chemicals, including pharmaceuticals and personal care products, even though aquatic organisms can be more sensitive to these chemicals than humans. Sensitive ecosystems may necessitate more rigorous analysis of ecological risks before proceeding with ecological enhancement projects with reclaimed water.

COSTS

Financial costs of water reuse are widely variable because they are dependent on site-specific factors. Financial costs are influenced by size, location, incoming water quality, expectations and/or regulatory requirements for product water quality, treatment train, method of concentrate disposal, extent of transmission lines and pumping requirements, timing and storage requirements, costs of energy, interest rates, subsidies, and the complexity of the permitting and approval process. Capital costs in particular are site specific and can vary markedly from one community to another. Data on reuse costs are limited in the published literature, although Chapter 9 provides reported capital and operations and maintenance costs for nine utilities (representing 13 facilities) that responded to a committee questionnaire.

Distribution system costs can be the most significant component of costs for nonpotable reuse systems. Projects that minimize those costs and use effluent from existing wastewater treatment plants are frequently cost-effective because of the minimal additional treatment needed for most nonpotable applications beyond typical wastewater disposal requirements. When large nonpotable reuse customers are located far from the water reclamation plant, the total costs of nonpotable projects can be significantly greater than potable reuse projects, which do not require separate distribution lines.

Although each project's costs are site specific, comparative cost analyses suggest that reuse projects tend to be more expensive than most water conservation options and less expensive than seawater desalination. The costs of reuse can be higher or lower than brackish water desalination, depending on concentrate disposal and distribution costs. Water reuse costs are typically much higher than those for existing water sources. The comparative costs of new water storage alternatives, including groundwater storage, are widely variable but can be less than those for reuse.

To determine the most socially, environmentally, and economically feasible alternative, water managers and planners should consider nonmonetized costs and benefits of reuse projects in their comparative cost analyses of water supply alternatives. Water reuse projects offer numerous benefits that are frequently not monetized in the assessment of project costs. For example, water reuse systems used in conjunction with a water conservation program can be effective in reducing seasonal peak demands on the potable system, which reduces capital and operating costs and prolongs existing drinking water resources. Water reuse projects can also offer improved reliability, especially in drought, and can reduce dependence on imported water supplies. Depending on the specific designs and pumping requirements, reuse projects may have a larger or smaller carbon footprint than existing supply alternatives. They can also reduce water flows to downstream users and ecosystems.

Current reclaimed water rates do not typically return the full cost of treating and delivering reclaimed water to customers. Nonpotable water reuse customers are often required to pay for the connection to the reclaimed water lines; therefore, some cost incentive is needed to attract customers for a product that is perceived to be of lower quality based on its origin. Frequently, other revenue streams, including fees, drinking water programs, and subsidies, are used to offset the low rates. As the need for new water supplies in water-limited regions becomes the driving motivation for water reuse, reclaimed water rates are likely to climb so that reclaimed water resources are used as efficiently as the potable water supplies they are designed to augment.

SOCIAL, LEGAL, AND REGULATORY FACTORS

Water rights laws, which vary by state, affect the ability of water authorities to reuse wastewater. States are continuing to refine the relationship between wastewater reuse and the interests of downstream entities. Regardless of how rights are defined or assigned, projects can proceed through the acquisition of water rights after water rights have been clarified. The right to use aquifers for storage can be clarified by states through legislation or court decision. The clarification of these legal issues can provide a clearer path for project proponents.

Scientifically supportable risk-based federal regulations for nonpotable water reuse would provide uniform nationwide minimum acceptable standards of health protection and could facilitate broader implementation of nonpotable water reuse projects. Existing state regulations for nonpotable reuse are developed at the state level and are not uniform across the country. Further, no state water reuse regulations or guidelines for nonpotable reuse are based on rigorous risk assessment methodology that can be used to determine and manage risks. The U.S. Environmental Protection Agency (EPA) has published suggested guidelines for nonpotable reuse that are based, in part, on a review and evaluation of existing state regulations and guidelines and are not based on rigorous risk assessment methodology. Federal regulations would not only provide a uniform minimum standard of protection, but would also increase public confidence that a water reuse project does not compromise public health. If nonpotable reuse regulations were developed at the federal level through new enabling legislation, this process should be informed by extensive scientific research to address the wide range of potential nonpotable reuse applications and practices, which would require resources beyond the reach of most states. A more detailed discussion of the advantages and disadvantages of federal reuse regulations is provided in Chapter 10. EPA should fully consider the advantages and disadvantages of

federal reuse regulations on the future application of water reuse to address the nation's water needs while appropriately protecting public health.

Modifications to the structure or implementation of the Safe Drinking Water Act (SDWA) would increase public confidence in the potable water supply and ensure the presence of appropriate controls in potable reuse projects. Although there is no evidence that the current regulatory framework fails to protect public health when planned or de facto reuse occurs, federal efforts to address potential exposure to wastewater-derived contaminants will become increasingly important as planned and de facto potable reuse account for a larger share of potable supplies. The SDWA was designed to protect the health of consumers who obtain potable water from supplies subject to many different sources of contaminants but does not include specific requirements for treatment or monitoring when source water consists mainly of municipal wastewater effluent. Presently, many potable reuse projects include additional controls (e.g., advanced treatment and increased monitoring) in response to concerns raised by state or local regulators or the recommendations of expert advisory panels. Adjustment of the SDWA to consider such requirements when planned or de facto potable reuse is practiced could serve as a mechanism for achieving a high level of reliability and public health protection and nationwide consistency in the regulation of potable reuse. In the process, public confidence in the federal regulatory process and the safety of potable reuse would be enhanced.

Application of the legislative tools afforded by the Clean Water Act (CWA) and SDWA to effluent-impacted water supplies could improve the protection of public health. Increasingly, we live in a world where municipal effluents make up a significant part of the water drawn for many water supplies, but this is not always openly and transparently recognized. Recognition of this reality necessitates increased consideration of ways to apply both the CWA and SDWA toward improved drinking water quality and public health. For example, the CWA allows states to list public water supply as a designated use of surface waters. Through this mechanism, some states have set up requirements on discharge of contaminants that could adversely affect downstream water supplies.

Updates to the National Pretreatment Program's list of priority pollutants would help ensure that water reuse facilities and de facto reuse operations are protected from potentially hazardous contaminants. The National Pretreatment Program has led to significant reductions in the concentrations of toxic chemicals in wastewater and the environment. However, the list of 129 priority pollutants presently regulated by the National Pretreatment Program has not been updated since its development more than three decades ago, even though the nation's inventory of manufactured chemicals has expanded considerably since that time, as has our understanding of their significance. Updates to the National Pretreatment Program's priority pollutant list can be accomplished through existing rulemaking processes. Until this can be accomplished, EPA guidance on priority chemicals to include in local pretreatment programs would assist utilities implementing potable reuse.

Enhanced public knowledge of water supply and treatment are important to informed decision making. The public, decision makers, and decision influencers (e.g., members of the media) need access to credible scientific and technical materials on water reuse to help them evaluate proposals and frame the issues. A general investment in water knowledge, including improved public understanding of a region's available water supplies and the full costs and benefits associated with water supply alternatives, could lead to more efficient processes that evaluate specific projects. Public debate on water reuse is evolving and maturing as more projects are implemented and records of implementation are becoming available.

RESEARCH NEEDS

The committee identified 14 water reuse research priorities that are not currently being addressed in a major way. These research priorities in the areas of health, social, and environmental issues and performance and quality assurance (detailed in Chapter 11,

Box 11-1) hold significant potential to advance the safe, reliable, and cost-effective reuse of municipal wastewater where traditional sources are inadequate.

Improved coordination among federal and nonfederal entities is important for addressing the long-term research needs related to water reuse. Addressing the research needs identified by the committee will require the involvement of several federal agencies as well as support from nongovernmental research organizations. If the federal government decides to develop national regulations for water reuse, a more robust research effort will be needed to support that initiative with enhanced coordination among federal and nonfederal entities. Such an effort would benefit from the leadership of a single federal agency, which could serve as the primary entity for coordination of research and for information dissemination.

*   *   *

Solutions to the nation's water challenges will require an array of approaches, involving conservation, supplemented as needed by alternative water supply technologies, such as reuse. Both potable and nonpotable reuse can increase the nation's water supply, although nonpotable reuse can be more expensive in existing communities that are not already equipped with dual water distribution systems. With recent advances in technology and treatment design, potable reuse can reduce the concentrations of chemical and microbial contaminants to levels comparable to or lower than those present in many drinking water supplies. Adjustments to the federal regulatory framework, including scientifically supportable risk-based regulations for nonpotable reuse and modifications to the structure or implementation of the SDWA for potable reuse projects, would ensure a high level of public health protection for both planned and de facto reuse and increase public confidence in water reuse. Additionally, improved coordination among federal and nonfederal entities could more effectively address key research needs.

1

A New Era of Water Management

As the world enters the 21st century, the human community finds itself searching for new paradigms for water supply and management in light of expanding populations, sprawling development, climate change, and the limits of existing conventional supplies. This introductory chapter explores the context for this new era of water management, within which water reuse is attracting increasing attention.

POPULATION GROWTH AND WATER SUPPLY

In the year 1900, the population of the world was between 1.6 and 1.8 billion persons (U.S. Census, 2010e). By the end of the 20th century, it was just short of 6.1 billion persons (U.S. Census, 2010d), an increase of approximately 270 percent. The United States finds itself in the same situation. Between 1900 and 2000, the population of the United States grew from 76 million persons to 282 million persons, an increase of 240 percent (U.S. Census 2010c). Along with this increase in population has come an increase in the demand for water.

To address the water supply needs of this expanding population in the United States, the 20th century was a time for building major water infrastructure, particularly dams (Figure 1-1) and aqueducts (Morgan, 2004). In the southwestern United States, ambitious projects built on the Colorado River, the Central Valley of California, and in central Arizona provided water and power that supported rapid population growth and increases in irrigated agriculture. Smaller projects in

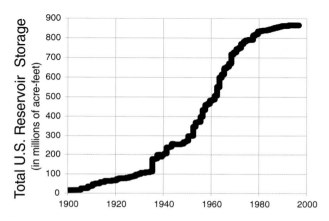

FIGURE 1-1 Reservoir capacity in the continental United States from 1900 to 1996.
SOURCE: Data from Graf (1999).

Texas, Florida, Colorado, and Georgia also expanded the nation's water supply capacity as population growth accelerated. Although a limited number of water supply and storage projects are still being built, the rate of construction of water supply infrastructure has dropped off significantly in recent decades (Graf, 1999; Gleick, 2003).

This decline in construction of new capacity has occurred in spite of continuing projections for increased demand, suggesting that the strategy of fulfilling increased water demand by building large dams and aqueducts to capture water from freshwater streams is reaching its limit. This change is attributable to a number of causes, among them: (1) a diminishing number of rivers whose flow is not already claimed by other users, (2) increased concern about adverse impacts of impoundments on stream ecology, and (3) a better understanding of water quality problems caused by irrigated agriculture (NRC, 1989).

Regional development and migration have placed further stress on our water sources. Large populations have migrated to warmer climates in California, Nevada, Arizona, Texas, and Florida, causing growth rates of 85 percent to more than 400 percent between 1970 and 2009 in those states while the national population has increased by less than 50 percent (Figure 1-2). In some places, these changes have necessitated infrastructure to collect and move water on a grand scale (e.g., the infrastructure on the Colorado River, the California State Water Project, and the Central Arizona Project).

An even broader perspective on this migration is provided in the U.S. county-level population projections through 2030 prepared by the U.S. Global Change Research Program (Figure 1-3). Continued development of these population centers in the southwest and arid west and continued migration from population centers in the eastern and midwestern United States will require substantial transformation in the way water is procured and used by the people who live and work in these geographies.

The shift in population and associated water demand is further complicated by potential impacts of climate change on the water cycle. Increases in evapotranspiration due to higher temperatures will increase water use for irrigated agriculture and landscaping while changes in precipitation patterns (see Figure 1-4) may diminish the ability of existing water infrastructure to capture water. This is particularly important in the

western United States where shifts in the timing and location of precipitation and decreases in snowfall are expected (NRC, 2007).

Considerable uncertainty remains about the impacts of climate change on water supplies. Improvements in models and the collection of additional data are likely to reduce the uncertainties associated with these estimates in coming decades. However, the pressures placed on water supplies by the combination of population growth and the likely impacts of climate change necessitate a reexamination of the ways in which water is acquired and used, before all of the questions about climate change impacts on the hydrological cycle are resolved (NRC, 2011a).

NEW APPROACHES TO WATER MANAGEMENT

The increase in population coupled with the decreased rate of construction of reservoirs, dams, and other types of conventional water supply infrastructure is leading to a new era in water management in the United States. The pressures on water supplies are changing virtually every aspect of municipal, industrial, and agricultural water practice. These changes in water management strategies take two principal forms: reducing water consumption through water conservation and technological change and seeking new sources of water.

Reducing Water Consumption

Improvements in water efficiency and programs for water conservation have begun to change our national water use habits, reducing per capita water consumption. More changes of this kind are likely in the future across many sectors. In Table 1-1, selected data on water use collected by the U.S. Geological Survey (Kenny et al., 2009) are summarized, where changes in water use by both agriculture and industry are clearly evident.

While the U.S. population grew from roughly 150 million to 300 million persons during the 60-year period, industrial water use—an application that was once the third highest use of water in the United States—grew only modestly between 1950 and 1970 and has been on the decline for 45 years now. These decreases are due to increased efficiency, higher prices for water and energy, and a shift away from water-

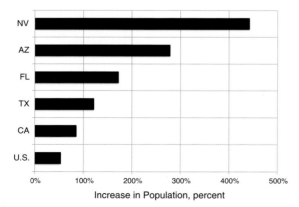

FIGURE 1-2 Population growth in selected states between 1970 and 2009.
SOURCE: Data from U.S. Census (2010b).

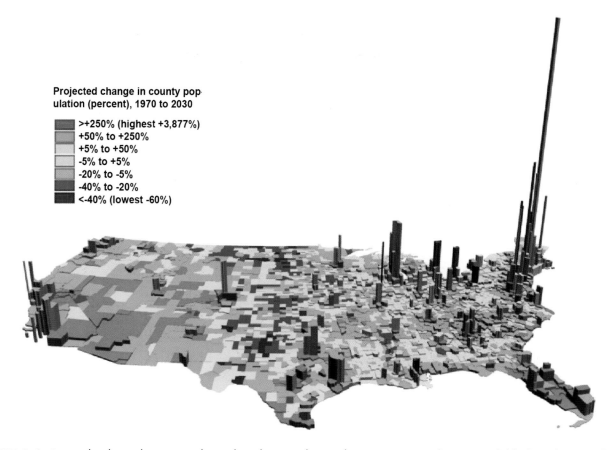

Projected change in county population (percent), 1970 to 2030

>+250% (highest +3,877%)
+50% to +250%
+5% to +50%
-5% to +5%
-20% to -5%
-40% to -20%
<-40% (lowest -60%)

FIGURE 1-3 County-level population growth trends in the United States between 1970 and 2030. Each block on the map illustrates one county in the United States. The height of each block is proportional to that county's population density in the year 2000, and so the volume of the block is proportional to the county's total population. The color of each block shows the county's projected change in population between 1970 and 2030, with shades of orange denoting increases and blue denoting decreases.
SOURCE: USGCRP (2000).

intensive manufacturing. More recently transfer of manufacturing outside the United States may also have been important.

Water use for irrigation peaked in 1980 and has now declined below 1970 levels. New technologies have been developed in irrigation practice (Gleick, 2003) and indications are that these technologies, if more widely adopted, could result in significant additional improvement (Postel and Richter, 2003). Water exchanges between municipal and agricultural entities are also taking place with increasing frequency. Agreements with agricultural interests by both the Metropolitan Water District of Southern California and the San Diego Water Authority are examples. This practice puts further pressure on agriculture to get value for the water it uses.

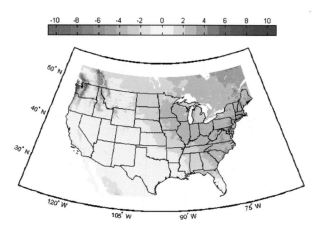

FIGURE 1-4 Downscaled climate projections showing the change in 30-year mean annual precipitation between 1971–2000 and 2041–2070, in centimeters per year. The median difference is based on 112 projections.
SOURCE: Brekke et al. (2009).

TABLE 1-1   Summary of Water Use (billion gallons per day) in the United States, 1950–2005

Year	Public Supply	Self-Supplied Domestic	Irrigation	Livestock, Aquaculture	Thermoelectric Power Use	Other Industrial Use	Total (Excluding Power Use)
1950	14.0	2.1	89	1.5	40	37	144
1955	17.0	2.1	110	1.5	72	39	170
1960	21.0	2.0	110	1.6	100	38	173
1965	24.0	2.3	120	1.7	130	46	194
1970	27.0	2.6	130	1.9	170	47	209
1975	29.0	2.8	140	2.1	200	45	219
1980	33.0	3.4	150	2.2	210	45	234
1985	36.4	3.3	135	4.5	187	30.5	210
1990	38.8	3.4	134	4.5	194	29.9	211
1995	40.2	3.4	130	5.5	190	29.1	208
2000	43.2	3.6	129	6.0	195	23.2	205
2005	44.2	3.8	128	10.9	201	22.2	209

NOTE: Includes both freshwater and saline water sources.
SOURCE: Data from Kenny et al. (2009).

Thermoelectric power use also peaked in 1980, but this use is misleading because a large fraction consists of "once-through" cooling water, which is primarily a nonconsumptive use (Kenny et al., 2009). Thus, reduction of use of this water would not necessarily provide new water resources, although it may have other environmental benefits. Furthermore, plants employing freshwater once-through cooling are often located in areas with ample water resources where water demands are not growing rapidly.

Whereas the total consumption for industry and irrigation have both decreased in recent decades, water use for primarily public supply continues to rise. During the period between 1950 and 2005, water used for public supply more than tripled as the nation's population doubled. Much of the increase in per capita consumption of water during this period (most notably between 1950 and 1985) can be tied to increased water use for landscaping, especially in arid climates. Consequently, there is significant potential for water conservation in the public supply sector.

Overall, U.S. water use (excluding thermoelectric power uses) has been stable at approximately 210 billion gallons per day (BGD; 795 million cubic meters per day [m³/d]) since 1985. This flat water-use trend corresponds with the slowdown in construction of new impoundments in the United States (Figure 1-1).

When these water use data are combined with population data from the U.S. Census Bureau and examined on a per capita basis, it becomes clear that irrigation and nonpower industrial use are now on the

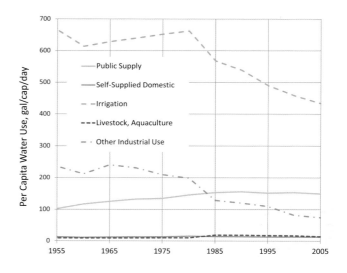

FIGURE 1-5 Past trends in water use in the United States, expressed on a per capita basis.
SOURCE: Data from Kenny et al. (2009).

decline (Figure 1-5). Per capita industrial water use has been on the decline since 1965; per capita agricultural use was flat between 1955 and 1980 and has been declining since then. Municipal use (referred to as public water supply in Kenny et al., 2009) continued to grow until 1990, but even this sector has begun to see the effects of water conservation in recent years. It is reasonable to expect that conservation will continue to play an increasingly important role in the nation's water management in the decades ahead, thereby reducing the demand for new water supplies. Including all sectors (except thermoelectric power), per capita water

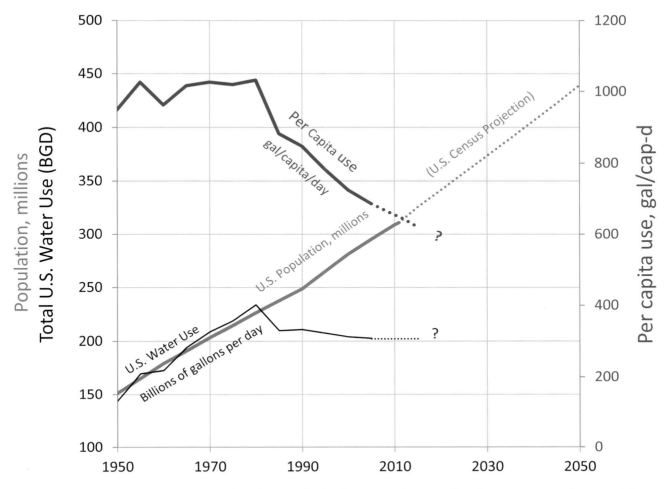

FIGURE 1-6 Changes in U.S. water use and implications for the future. Population and total U.S. water use shown on left axis; per capita water use on right axis. Per capita water use includes all water uses except thermoelectric power, which is dominated by once-through cooling.
SOURCE: Data from Kenny et al. (2009) and U.S. Census Bureau (2008).

use was relatively stable between 1950 and 1980 but has dropped precipitously since that time (Figure 1-5).

The U.S. Census Bureau predicts that the nation's population will increase by over 50 percent between 2010 and 2060. This population growth is displayed in Figure 1-6 along with the history of total water use and the history of per capita water use as well. If the U.S. Census estimates are correct, then, barring the development of major new water sources, per capita use must decline further. Both more efficient water use and the development of new sources of water beyond those the nation has traditionally used may be necessary in areas with limited existing water supplies.

Searching for New Water Sources

In addition to conservation efforts, the other major emphasis in the new era of water management involves a search for untapped water sources. These sources include the desalination of seawater and brackish groundwater, the recovery of groundwater impaired by previous anthropogenic activity, off-stream or underground storage of seasonal surpluses from existing impoundments, the recovery of rainwater and stormwater runoff, on-site greywater[1] reuse, and the reuse of

---

[1] Greywater is water from bathing or washing that does not contain concentrated food or human waste.

municipal wastewater effluent. The role of each of these approaches in the nation's future water supply portfolio is likely to be dictated by considerations related to public health, economics, impacts on the environment, and institutional considerations. The NRC recently published studies on desalination (NRC, 2008b), stormwater management (NRC, 2009c) and underground storage (NRC, 2008c). In this new water era, the reuse of municipal effluent for beneficial purposes may also be important. This topic—herein termed water reuse—is the focus of this report. See Box 1-1 for additional reuse terminology.

BOX 1-1
REUSE TERMINOLOGY

The terminology associated with treating municipal wastewater and reusing it for beneficial purposes differs within the United States and globally. For instance, although the terms are synonymous, some states and countries use the term *reclaimed water* and others use the term *recycled water*. Similarly, the terms *water recycling*, and *water reuse*, have the same meaning. In this report, the terms *reclaimed water* and *water reuse* are used. Definitions for these and other terms are provided below.

Reclaimed water: Municipal wastewater that has been treated to meet specific water quality criteria with the intent of being used for beneficial purposes. The term *recycled water* is synonymous with reclaimed water.

Water reclamation: The act of treating municipal wastewater to make it acceptable for beneficial reuse.

Water reuse: The use of treated wastewater (reclaimed water) for a beneficial purpose. Synonymous with the term *wastewater reuse*.

Potable reuse: Augmentation of a drinking water supply with reclaimed water.

Nonpotable reuse: All water reuse applications that do not involve potable reuse (e.g., industrial applications, irrigation; see Chapter 2 for details).

De facto reuse: a situation where reuse of treated wastewater is in fact practiced, but is not officially recognized (e.g., a drinking water supply intake located downstream from a wastewater treatment plant discharge point).

SOURCE: These definitions are taken from Crook, 2010.

Water Reuse

During the past several decades, treated wastewater (also called reclaimed water) has been reused to accomplish two primary purposes: (1) to create a new water supply and thereby reduce demands on limited traditional water supplies and (2) to prevent ecological impacts that can occur when nutrient-rich effluent is discharged into sensitive environments.[2] Increasingly, the basic need for additional water supply is becoming the central motivator for water reuse. In addition to growing water demands, the further adoption of water reuse will be affected by a variety of issues, including water rights, environmental concerns, cost, and public acceptance.

The context for water reuse and common reuse applications for nonpotable reuse (e.g., water reuse for irrigation or industrial purposes) and potable water reuse (e.g., returning reclaimed water to a public water supply) are described in detail in Chapter 2. Potable reuse is commonly broken into two categories: indirect potable reuse and direct potable reuse. This classification considered potable reuse to be "indirect" when the reclaimed water spent time in the environment after treatment but before it reached the consumer. Inherent in this distinction was the idea that the natural environment (or environmental buffer, discussed in Chapter 2) provided a type of treatment that did not occur in engineered treatment systems. An example of these definitions can be found in the NRC (1998) report, *Issues in Potable Reuse*. The committee has chosen not to use these terms but rather to speak about the project elements required to protect public health when potable reuse is contemplated and to try to understand the attributes of the protection provided by an environmental buffer (see Chapters 2, 4, and 5).

In NRC (1998) a distinction was also made between "planned" and "unplanned" potable water reuse. For this report, the committee has chosen not to use these terms, because they presume that water managers are unaware of the integrated nature of the nation's

---

[2] For example, the water reuse program in St. Petersburg, Florida, was started in response to state legislation in 1972 (the Wilson-Grizzle Act) requiring all wastewater treatment plants discharging to Tampa Bay to either upgrade to include advanced wastewater treatment (including nutrient removal) or to cease discharging to Tampa Bay (Crook, 2004).

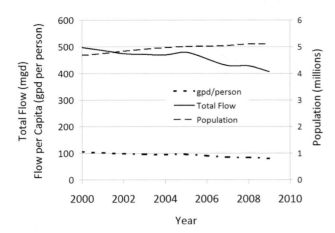

FIGURE 1-7 Reduction in per capita flow to the Los Angeles County Joint Outfall during the beginning of the 21st century (2000–2007).
SOURCE: Data from S. Highter, Los Angeles County Sanitation District, personal communication, 2010.

water system (e.g., when downstream drinking water systems use surface waters that receive upstream wastewater discharges). In the committee's view, the use of effluent-impacted water supplies is reuse in fact, if not reuse in name. Therefore, the committee will refer to the less carefully scrutinized practice of using effluent-impacted water supplies for potable water sources as "de facto" reuse, rather than the term unplanned reuse (see Chapter 2 for more discussion of de facto reuse).

Municipal wastewater effluent is produced from households, offices, hospitals, and commercial and industrial facilities and conveyed through a collection system to a wastewater treatment plant. In 2004, over 16,000 publicly owned wastewater treatment plants were in operation in the United States, receiving over 33 BGD (120 million m^3/d) of influent flow (EPA, 2008b). These publicly owned wastewater plants serve approximately 222 million Americans, or 75 percent of the population. Thus, the total discharge averages approximately 150 gallons (0.56 m^3) per day per person.[3] Recently, however, per capita wastewater flows have been decreasing, largely because of conservation practices (see Figure 1-7 for one example). Thus, water conservation and water reuse are linked, and projections of water available for reuse based on today's wastewater

flows need to take some allowance for reductions in wastewater production due to conservation and reduced sewer flows during future periods of water restriction.

Although a map depicting the location of all of the effluent discharges in the country is not available, the distribution of wastewater discharges should roughly track the population distribution, assuming similar per capita domestic and industrial wastewater generation rates occur across the country (Figure 1-8). Figure 1-8 illustrates that much of the nation's wastewater is discharged to inland waterways. As a result, de facto reuse of wastewater is already an important part of the current water supply portfolio. The ongoing practice of de facto reuse and the likelihood that all of the reclaimed water will not be returned to the water supply also means that increased water reuse will not necessarily increase the nation's net water resource by an equal amount. In fact in many western U.S. jurisdictions, downstream users possess a water right that could prevent or inhibit municipal reuse (see Chapter 10).

Based on data provided by the U.S. Environmental Protection Agency (EPA, 2008c), the committee calculated that approximately 12 BGD (45 million m^3/d) of U.S. municipal wastewater was discharged directly into or just upstream of an ocean or estuary in 2008 out of 32 BGD (120 million m^3/d) discharged nationwide (38 percent).[4] Because there are no downstream cities that rely on these discharges to augment their water supplies, reuse of coastal discharges could directly augment the nation's overall water resource. If all of these coastal discharges were reused, the additional water available would represent approximately 6 percent of estimated U.S. total water use or about 27 percent of municipal use in 2005 (Kenney et al., 2009). However, not all of the water available for reuse is located in areas where it is needed. Additionally, the health of some coastal estuaries may be dependent on the freshwater inflows provided by coastal wastewater discharges, particularly in water-scarce regions. Thus, the extent of availability

---

[3] Calculated from 33 BGD divided by 222 million people. Thus, this per capita discharge includes all discharges to wastewater treatment plants, not just residential discharges.

[4] The raw data of the wastewater treatment plants along the continental U.S. coastline is from EPA's Clean Watersheds Needs Survey: 2008 Data and Reports. The cited numbers are the sum of the outflow from wastewater treatment plants that discharge into watersheds having a fourth-level hydrologic unit code–defined area that directly borders or is immediately upstream of a major estuary or ocean, such that the wastewater discharge is unlikely to be part of the water supply of any downstream users.

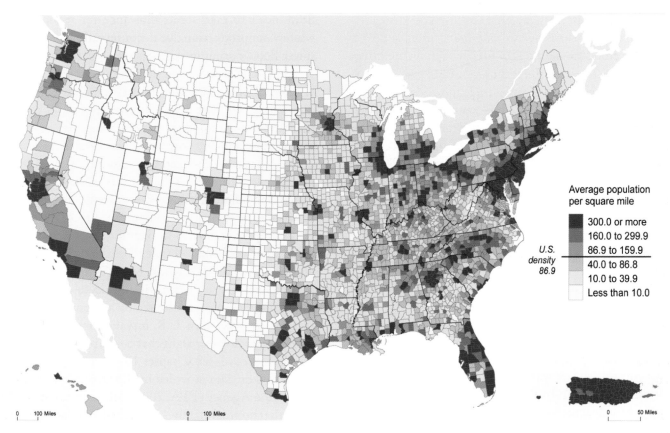

FIGURE 1-8 Distribution of the U.S. population in 2009, which can be used to approximate discharge volumes of municipal wastewater effluent.
SOURCE: U.S. Census Bureau (http://www.census.gov/popest/gallery/maps/PopDensity_09.pdf).

of these coastal discharges for reuse would be dependent on site-specific analysis.

If reclaimed water was used largely for nonconsumptive uses, the water supply benefit of water reuse could be even greater because, in many cases, the wastewater can be again captured and reused. It is also evident that many inland discharges could be productively used as well, suggesting the potential for an even larger impact from water reuse on the nation's water supplies.

CURRENT CHALLENGES

Important challenges remain that must be addressed before the potential of municipal water reuse can be fully harnessed. These challenges are discussed in this section and explored in more depth in the remainder of the report.

It is important to recognize that many communities currently practicing water reuse have already "picked the low-hanging fruit," through practices such as irrigating golf courses, landscapes, municipally owned parks, and medians near wastewater treatment plants or by converting industrial applications that are less sensitive to water quality (e.g., cooling) to reclaimed water. Where these projects have been implemented, communities have become familiar with the advantages of reuse, particularly improved reliability and drought resistance of the water supply and reduced nutrient loading to sensitive downstream ecosystems. On the other hand, while many of these initial types of water reuse projects were inexpensive and relatively simple to implement, many future water reclamation projects are likely to pose greater challenges.

In addition, utilities will have to consider public skepticism about the health risks associated with reuse projects, and the public decision-making process can be a difficult one, particularly for projects with a potable reuse component. People have been trained for generations to provide separation in both time and space between their wastes and their water supplies, and therefore the public is concerned about the safety of using wastewater effluent for domestic purposes. At the same time, several high-profile reports detailing the presence of pharmaceuticals and personal care products in water supplies (e.g., Kolpin et al., 2002; Benotti et al., 2009) have increased awareness of the common practice of de facto water reuse, which has increased with population growth. Today, many U.S. communities rely on drinking water sources that are exposed to wastewater discharges. Nevertheless, the quality of U.S. drinking water continues to improve, largely because of improvements in treatment technology. Perhaps the question is not whether reuse should be considered; rather the question should be how reuse can be planned so that it better incorporates appropriate engineered barriers. In many cases the alternative to building new, engineered water reuse systems is increased reliance on de facto water reuse, with fewer engineered controls and monitoring.

A century ago, circumstances as well as best professional judgment supported policies in which water was considered to be potable after it spent a certain period of time in the natural environment. This is illustrated by an official policy of the state of Massachusetts allowing sewage (untreated wastewater) discharges to rivers serving as a drinking water supply provided the outfall was located more than 20 miles (32 km) upstream of the drinking water intake (Hazen, 1909; Sedgwick, 1914; Tarr, 1979). Today, we increasingly rely on the application of treatment technologies and sophisticated monitoring to ensure that safe drinking water conditions are achieved. In recent decades, advances in the capability of water treatment systems have been substantial, and these systems are now able to routinely achieve a level of protection that exceeds anything imaginable in the middle of the 20th century. Despite this progress, how do we determine when treated wastewater has reached the point where it has become suitable for potable supply? How can this decision be made in a way that engenders public confidence? What monitoring tools are needed to provide assurance that promised performance is being delivered on a continuous basis?

Every treatment technique takes advantage of the specific properties of each contaminant in order to remove it, and no one treatment technique or combination of treatment techniques can be relied upon to reduce all possible contaminants to levels below the limits of detection. Robust analytical methods will continue to be developed that will detect organic compounds and pathogens at increasingly lower levels. Thus, water managers are faced with the challenge of knowing a contaminant is present at low levels without knowing if its presence at those levels is significant.

In the decades since the NRC published its groundbreaking report *Risk Assessment in the Federal Government: Managing the Process* (NRC, 1983), the nation has developed a sophisticated infrastructure for assessing the risk of anthropogenic chemicals in the environment and a significant cadre of experts trained in its application. Significant progress also has been made in the assessment of risks from waterborne pathogens. Whereas this infrastructure is well suited for the support of national regulations designed to manage risk and also for application to the assessment of important regional decisions, it is not as well suited to facilitate the decisions of individual communities comparing the costs, risks, and benefits of planned reuse with other water supply alternatives. Thus, communities face challenges in finding adequate technical support for complex water management decisions.

STATEMENT OF COMMITTEE TASK AND REPORT OVERVIEW

The challenges discussed in the previous section have limited the application of water reuse in the United States. In 2008, the NRC's Committee on Assessment of Water Reuse as an Approach for Meeting Future Water Supply Needs was formed to conduct a comprehensive study of the potential for water reclamation and reuse of municipal wastewater to expand and enhance the nation's available water supply alternatives. Effluent reuse has long been a topic of discussion and the NRC has issued several reports on the subject in the past (see Box 1-2).

This broad study considers a wide range of uses, including drinking water, nonpotable urban uses, irrigation, industrial process water, groundwater recharge, and water for environmental purposes. The study also considers technical, economic, institutional, and social challenges to increased adoption of water reuse to pro-

---

BOX 1-2
NRC Reports Relating to Water Reuse

At least seven NRC reports over the last 30 years have addressed water reuse or related technologies:

- *Quality Criteria for Water Reuse* (NRC, 1982) provided advice for assessing the suitability of water from impaired sources such as wastewater. The report addressed chemical and microbiological contaminants in reclaimed water, health effects testing for reclaimed water, sample concentration methods, and monitoring strategies. It also contained an assessment and criteria for potable water reuse.
- *The Potomac Estuary Experimental Water Treatment Plant* (NRC, 1984) assessed the U.S. Army Corps of Engineers' operation, maintenance, and performance of the experimental water treatment plant using an impaired water source containing treated wastewater. The report praised the Corps for development of a database of microbiological contaminants and toxicological indicators and for demonstrating the reliability of advanced treatment processes. The report, however, questioned whether there was enough data to ensure protected public health and concluded that failure to detect viruses cannot be accepted as an indication that they are absent.
- *Ground Water Recharge Using Waters of Impaired Quality* (NRC, 1994a) addressed issues concerning identification of potentially toxic chemicals and the limits of natural constituent removal mechanisms. Public health was the principal concern of the committee, and constant monitoring as well as federal leadership were identified as crucial steps if groundwater recharge using impaired waters is to be used. The committee recommended significant further research in both epidemiology and toxicology to assess appropriate risk limits and to identify emerging contaminants.
- *Use of Reclaimed Water and Sludge in Food Crop Production* (NRC, 1996) examined the safety and practicality of using treated municipal wastewater and sewage sludge for production of crops for human consumption. The report concluded that risks from organic compounds were negligible, and Class A water standards appeared to be adequate to protect human health. The committee's concerns were primarily demand-side; acceptance from farmers and consumers was expected to be a much larger hurdle for significant use of reclaimed water in food crops.
- *Issues in Potable Reuse* (NRC, 1998) provided technical and policy guidance regarding use of treated municipal wastewater as a potable water supply source. The committee recommended the most protected source be targeted first for use, combined with nonpotable reuse, conservation, and demand management. While direct potable reuse is not yet viable, indirect potable reuse may be viable when careful, thorough, project-specific assessments are completed, including monitoring, health and safety testing, and system reliability evaluation.
- *Prospects for Managed Underground Storage* (NRC, 2008c) identified research, education needs, and priorities in managed underground storage technology and implementation. The report concluded that better knowledge of contaminants in water and chemical constituents in the subsurface and a systematic way to deal with emerging contaminants are needed. The report stated that technologies such as ultraviolet, ozone, and membranes can be made more efficient, and new surrogates or indicators may be needed to monitor for a wider suite of contaminants.
- *Desalination: A National Perspective* (NRC, 2008b) assessed the state of the art in desalination technologies and addressed cost and implementation challenges. Several of the technologies discussed in the report, such as reverse osmosis and concentrate disposal, are also relevant to water reuse.

---

vide practical guidance to decision makers evaluating their water supply alternatives. The study is sponsored by the EPA, the Bureau of Reclamation, the National Science Foundation, the National Water Research Institute, the Centers for Disease Control and Prevention, the Water Research Foundation, the Orange County Water District, the Orange County Sanitation District, the Los Angeles Department of Water and Power, the Irvine Ranch Water District, the West Basin Water District, the Inland Empire Utilities Agency, the Metropolitan Water District of Southern California, the Los Angeles County Sanitation District, and the Monterey Regional Water Pollution Control Agency.

The committee was specifically tasked to address the following questions:

1. Contributing to the nation's water supplies. What are the potential benefits of expanded water reuse and reclamation? How much municipal wastewater effluent is produced in the United States, what is its quality, and where is it currently discharged? What is the suitability—in terms of water quality and quantity—of processed wastewaters for various purposes, including drinking water, nonpotable urban uses, irrigation, industrial processes, groundwater recharge, and environmental restoration?

2. Assessing the state of technology. What is the current state of the technology in wastewater treatment and production of reclaimed water? How do available treatment technologies compare in terms of treatment performance (e.g., nutrient control, contaminant control, pathogen removal), cost, energy use, and environmental impacts? What are the current technology challenges and limitations? What are the infrastructure requirements of water reuse for various purposes?

3. Assessing risks. What are the human health risks of using reclaimed water for various purposes, including indirect potable reuse? What are the risks of using reclaimed water for environmental purposes? How effective are monitoring, control systems, and the existing regulatory framework in assuring the safety and reliability of wastewater reclamation practices?

4. Costs. How do the costs (including environmental costs, such as energy use and greenhouse gas emissions) and benefits of water reclamation and reuse generally compare with other supply alternatives, such as seawater desalination and nontechnical options such as water conservation or market transfers of water?

5. Barriers to implementation. What implementation issues (e.g., public acceptance, regulatory, financial, institutional, water rights) limit the applicability of water reuse to help meet the nation's water needs and what, if appropriate, are means to overcome these challenges? Based on a consideration of case studies, what are the key social and technical factors associated with successful water reuse projects and favorable public attitudes toward water reuse? Conversely, what are the key factors that have led to the rejection of some water reuse projects?

6. Research needs. What research is needed to advance the nation's safe, reliable, and cost-effective reuse of municipal wastewater where traditional sources of water are inadequate? What are appropriate roles for governmental and nongovernmental entities?

The committee's report and its conclusions and recommendations are based on a review of relevant technical literature, briefings, and discussions at its eight meetings, field trips to water reuse facilities, and the experience and knowledge of the committee members in their fields of expertise.

Following this brief introduction, the statement of task is addressed in nine subsequent chapters of this report:

- Chapter 2 provides context for this report by describing the history of reuse, common reuse applications, and the use of reuse technologies in the United States and globally.
- Chapter 3 discusses water quality and contaminants of concern in wastewater effluent.
- Chapter 4 provides an overview of the state of the science in water reuse with respect to treatment technology.
- Chapter 5 examines design and operational strategies to ensure reclaimed water quality.
- Chapter 6 discusses the risk assessment framework as it applies to water reuse.
- Chapter 7 explores the risks of reuse in context by evaluating the relative risks of various reuse practices to human health compared with de facto reuse practices that are generally perceived as safe.
- Chapter 8 discusses applications of water reuse for ecological enhancement.
- Chapter 9 examines the financial and economic circumstances surrounding reuse and examines the benefits of reuse.
- Chapter 10 describes the social and institutional factors, including regulatory concerns, legal considerations, and public perception.
- Chapter 11 discusses actions needed to advance the capacity to use reuse to address water demands, including research needs and the roles of federal and nonfederal agencies.

Note that this report covers all types of reuse, but not all chapters include equal coverage of all reuse applications. The committee has chosen to focus more intensely on applications for which there are specific unresolved issues that may be limiting the ability of communities and local decision makers to make wise choices about their future water supply options; thus, the reader will find greater discussion on potable reuse relative to nonpotable reuse. Additionally, on the basis of the statement of task, the committee focused its efforts on the reuse of municipal wastewater effluent. The issues discussed in the report have applicability to both large and small municipal wastewater treatment plants.

However, the committee does not discuss building-scale reuse or greywater reuse in depth in this report.

CONCLUSION

As populations are increasing, particularly in water-limited regions, water managers are looking toward sustainable water management solutions to address shortfalls in supply from conventional water sources. Efforts to increase the efficiency of water use through enhanced conservation and improved technologies and the development of new sources of water may both be necessary to address future water demand in areas facing extreme water shortfalls. Potable and nonpotable reuse are attracting increasing attention in the search for untapped water supply sources. Out of the 32 BGD (121 million m^3/d) of municipal wastewater effluent discharged nationwide, approximately 12 BGD (45 million m^3/d) is discharged to an ocean or estuary (equivalent to 6 percent of the estimated total U.S. water use or 27 percent of public supply). Reuse of these coastal discharges, where feasible, in water-limited regions could directly augment available water resources. When reclaimed water is used for nonconsumptive uses, the water supply benefit of water reuse could be even greater if the water can again be captured and reused. Inland effluent discharges may also be available for water reuse, although extensive reuse has the potential to affect the water supply of downstream users and ecosystems (e.g., in-stream habitats, coastal estuaries) in water-limited settings. **Municipal wastewater reuse, therefore, offers the potential to significantly increase the nation's total available water resources.** However, reuse alone cannot address all of the nation's water supply challenges, and the potential contributions of water reuse will vary by region.

2

Current State of Water Reuse

This chapter provides the background needed to understand the role of water reuse in the nation's water supply. After presenting a brief overview of how sewage collection and treatment developed during the 19th and 20th centuries, the chapter describes the ways in which reclaimed water has been used for industrial applications, agriculture, landscaping, habitat restoration, and water supply. Through descriptions of current practices and case studies of important water reclamation projects, the chapter provides a means of understanding the potential for expansion of different types of water reuse and identifies factors that could limit future applications.

CONTEXT FOR WATER REUSE

To understand the potential role of water reuse in the nation's water supply, it is important to consider the infrastructure that has been developed to enable the collection, treatment, and disposal of municipal wastewater because these systems serve as the source of reclaimed water. By understanding the ways in which wastewater collection and treatment systems developed and are currently operated, it is possible to gain insight into many of the technical issues discussed in later sections of the report. In particular, this section describes the practice of unplanned, or de facto, water reuse (see Box 1-1), which is an important but underappreciated part of our current water supply, as well as the different types of systems that have been developed as part of planned water reclamation projects.

Historical Perspectives on Sewage and Municipal Wastewater Treatment

Prior to the installation of piped water supplies, most cities did not have sewers or centralized systems for disposing of liquid waste. Feces and urine were collected in privy vaults or cesspools (Billings, 1885). When the vaults were filled, wastes were removed and applied to agricultural fields, dumped in watercourses outside of the city, or the vault was abandoned (Tarr et al., 1984). Other liquid wastes, from cooking or clothes washing, were discharged to gutters or unlined dry wells. Sewers were only employed to a limited extent in densely populated areas to prevent flooding by conveying runoff to nearby rivers. In many cities, it was illegal to discharge human wastes to sewers (Billings, 1885).

Emergence of Sewer Collection Systems

With the advent of pressurized potable water, per capita urban water use increased from approximately 5 gal/d (20 L/d) to over 105 gal/d (400 L/d; Tarr et al., 1984). When ample freshwater supplies became available, the popularity of the flush toilet grew and the resulting large volumes of liquid waste overwhelmed the capacity of privy vaults, cesspools, and gutters. The public health and aesthetic problems associated with the liquid wastes led to the widespread construction of sewer systems in populated areas. During the initial phase of sewer system construction, in the late 1800s, most cities in the United States built combined sewers to convey sewage and stormwater runoff from the city to nearby waterways (Tarr, 1979). Separate sanitary sewers (that conveyed mainly waste from homes and businesses) were built in several dozen cities because they were less expensive and the concentrated wastes could be used as fertilizers (Tarr, 1979). By 1890, approximately 70 percent of the urban population lived in areas that were served by one of the two types of sewer systems (Figure 2-1).

Throughout this period, the wastes conveyed by combined sewer systems were usually discharged to surface waters without any treatment because the available treatment methods (e.g., chemical precipitation) were considered to be too expensive (Billings, 1885). As a result of the rapid growth of cities and the relatively large volumes of water discharged by sewers, drinking water supplies of cities employing sewers and their downstream neighbors were compromised by waterborne pathogens, resulting in increased mortality due to waterborne diseases (Tarr et al., 1984). For example, severe outbreaks of typhoid fever in Lowell and Lawrence, Massachusetts, in 1890 and 1891, in which over 200 people died, were traced back to the discharge of sewage by communities located approximately 12 miles (20 km) upstream of Lawrence (Sedgwick, 1914).

In cities with separate sanitary sewers, treatment was more common because of the smaller volumes and

consistent quality of the waste. In some communities, sewage was applied directly to orchards or farms (in a practice known as sewage farming (Anonymous, 1893; see Box 2-1). Sewage farming led to high crop yields, especially in locations where water was limited. The nutrients in the sewage made sewage farming attractive to farmers, but the practice eventually died out in the 1920s as public health officials expressed concerns about exposure to pathogens in fruits and vegetables grown on sewage farms.

As downstream communities became aware of the impact that upstream communities were having on their water supplies, there were debates about the obligations of communities to remove contaminants from sewage prior to discharge. Leading engineers, such as Allen Hazen, advocated for downstream cities to install drinking water treatment systems (Hazen, 1909) while public health scientists, like William Sedgwick (1914), advocated a requirement for cities to treat sewage. Many sanitary engineers supported their assertion that wastewater treatment was unnecessary by a belief that flowing water undergoes a process of self-purification. They asserted that as long as a water supply was located at a sufficient distance downstream of the sewage discharge, the water would be safe to drink. In fact, this concept was instrumental in the state of Massachusetts' policy of allowing sewage discharges to rivers if the outfall was located more than 20 miles (32 km) from a drinking water intake (Hazen, 1909; Sedgwick, 1914; Tarr, 1979). As a result of these debates, downstream communities often took the responsibility for ensuring the safety of their own water supply by building drinking water treatment plants or relocating their water supplies to protected watersheds.

Emergence of Wastewater Treatment

In 1900, less than 5 percent of the municipal wastewater in the United States was treated in any way prior to discharge (Figure 2-1). However, increases in population density, especially in cities, coupled with the growth of the progressive movement, which created a greater awareness of natural resources, led to increased construction of wastewater treatment systems (Burian et al., 2000). Coincident with these trends was the development of more cost-effective methods of biological wastewater treatment, such as activated

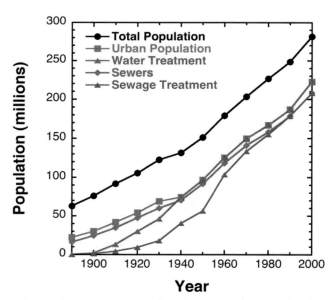

FIGURE 2-1 Comparison of total U.S. population with urban population, population served by sewers, population served by water treatment plants, and population served by wastewater treatment plants.
SOURCES: Tarr et al. (1984), (EPA, 2008b).

BOX 2-1
Sewage Farming

Throughout history, farmers have recognized the potential benefits of applying human wastes to agricultural land. With the widespread popularity of the water closet (i.e., the flush toilet) in the latter part of the 19th century, the water content of wastes increased and the traditional system for transporting waste to agricultural fields became impractical. To obtain the benefits of land application of wastes, scientists in Europe began evaluating the potential for using pipelines to transport sewage to farms where the water and nutrients could be used to grow plants. Eventually, large sewage farms were built and operated in Edinburgh, Paris, and Berlin where they produced fodder for cattle, fruits, and vegetables (Hamlin, 1980). At the turn of the century, the majority of the sewage produced in Paris was being treated on sewage farms (Reid, 1991).

In the United States, sewage farming was especially popular in arid western states because water supplies were limited (see figure below). For example, in California the practice of irrigating food crops with raw sewage reached a peak in 1923 with 70 municipalities applying their sewage to food crops (Reinke, 1934). In some locations, chemical treatment followed by settling was used prior to irrigation (Tarr, 1979). Eventually sewage farming became less prevalent as cities expanded, fertilizers became less expensive, and modern wastewater treatment plants provided an alternative means of sewage disposal. Sewage farming continued in France and Germany until the second half of the 20th century. Despite the public health risks associated with potential exposure to pathogens in raw sewage, almost all of the wastewater produced in Mexico City is sent to sewage farms (Jiménez and Chavez, 2004).

A sewer farm near Salt Lake City, Utah.
SOURCE: Utah Historical Society, circa 1908.

---

sludge. By 1940, 55 percent of the urban population of the United States was served by wastewater treatment plants (EPA, 2008b). Concerns associated with raw sewage discharges increased during the postwar period, with the passage of the Water Pollution Control Acts of 1948 and 1956, which provided federal funding for wastewater treatment plant construction (Everts and Dahl, 1957; Melosi, 2000). By 1968, 96.5 percent of the urban population of the United States lived in areas where wastewater was treated prior to discharge (EPA, 2008b), but the extent of treatment varied considerably, with many plants only removing suspended solids through primary treatment.

Concerns associated with sewage pollution grew during the 1960s and culminated with the allocation of $24.6 billion in construction and research grants for wastewater treatment plants as part of the Clean Water Act of 1972 (Burian et al., 2000). Most of the municipal wastewater treatment plants built in the United States during the late 1960s and early 1970s were equipped with primary and secondary treatment (see Box 2-2 and Chapter 4), which are capable of removing from wastewater over 90 percent of the total suspended solids and both oxygen-demanding organic wastes (i.e., biochemical oxygen demand [BOD] and chemical oxygen demand [COD]). By 2004, only 40 of more than 16,000 publicly owned wastewater treatment plants in the United States reported less than secondary treatment (see Table 2-1; EPA, 2008b).

The increased number of wastewater treatment

BOX 2-2
Stages of Wastewater Treatment

Primary — Removal of a portion of the suspended solids and organic matter form the wastewater.

Secondary — Biological treatment to remove biodegradable organic matter and suspended solids. Disinfection is typically, but not universally, included in secondary treatment.

Advanced treatment — Nutrient removal, filtration, disinfection, further removal of biodegradable organics and suspended solids, removal of dissolved solids and/or trace constituents as required for specific water reuse applications.

SOURCE: Adapted from Asano et al. (2007).

---

sources, ammonia concentrations often reached levels that were toxic to aquatic organisms. In other locations, wastewater effluent discharges caused excessive growth of algae and aquatic macrophytes due to the elevated concentrations of nutrients (i.e., nitrogen and phosphorus) in the effluent. To address these issues, treatment plants were often retrofitted or new treatment plants were built with technologies for removing nutrients (see Chapter 4 for detailed descriptions). These nutrient removal processes, which are sometimes referred to as tertiary treatment processes, became increasingly popular in the 1970s.

To protect downstream recreational users, wastewater effluent is often disinfected before discharge. The most common means of disinfection in the United States is effluent chlorination, a process in which a small amount of dissolved chlorine gas or hypochlorite (i.e., bleach) is added to the effluent prior to discharge. However, concerns about potential hazards associated with handling of chlorine coupled with the need to minimize the formation of disinfection byproducts that are toxic to humans and aquatic organisms have caused some utilities to switch to other means of effluent disinfection (Sedlak and von Gunten, 2011). In particular, disinfection with ultraviolet light has become more common as the technology has become less expensive. Ozone also is being used for effluent disinfection in some locations because it also oxidizes trace organic

plants built during the postwar period had immediate and readily apparent impacts on the aesthetics of surface waters and the integrity of aquatic ecosystems. However, effluent from wastewater treatment plants sometimes caused problems. In locations where effluent was insufficiently diluted with water from other

TABLE 2-1 Treatment Provided at U.S. Publicly Owned Wastewater Treatment Plants

Level of Treatment	Treatment Facilities in Operation in 2004[a]				
	Number of Facilities	Existing Flow (MGD)	Present Design Capacity	Number of People Served	Percent of U.S. Population
Less than Secondary[b]	40	441	570	3,306,921	1.1
Secondary	9,221	14,622	19,894	96,469,710	32.4
Greater than Secondary	4,916	16,522	23,046	108,506,467	36.5
No Discharge[c]	2,188	1,565	2,296	14,557,817	4.9
Partial Treatment[d]	218	507	632	—	—
Total[e]	16,583	33,657	46,438	222,840,915	74.9

[a]Alaska, American Samoa, Guam, the Northern Mariana Islands and the Virgin Islands did not participate in the CWNS 2004. Arizona, California, Georgia, Massachusetts, Michigan, Minnesota, North Dakota, and South Dakota did not have the resources to complete the updating of their data. All other states, the District of Columbia, and Puerto Rico completed more than 97 percent of the data entry or had fewer than 10 facilities that were not updated.

[b]Less-than-secondary facilities include facilities granted or pending section 301(h) waivers from secondary treatment for discharges to marine waters.

[c]No-discharge facilities do not discharge treated wastewater to the Nation's waterways. These facilities dispose of wastewater via methods such as industrial reuse, irrigation, or evaporation.

[d]These facilities provide some treatment to wastewater and discharge their effluents to other wastewater facilities for further treatment and discharge. The population associated with these facilities is omitted from this table to avoid double accounting.

[e]Totals include best available information from states and territories that did not have the resources to complete the updating of the data or did not participate in the CWNS 2004 in order to maintain continuity with previous reports to Congress. Forty operational and 43 projected treatment plants were excluded from this table because the data related to population, flow, and effluent levels were not complete.

SOURCE: EPA (2008b).

contaminants (see Chapter 4 for details). It is worth noting that effluent disinfection is not practiced at all wastewater treatment plants because of variations in local regulations.

Increasing Importance of De Facto Water Reuse

Irrespective of the treatment process employed, municipal wastewater effluent that is not directly reused is discharged to the aquatic environment where it reenters the hydrological cycle. As a result, almost every municipal wastewater treatment plant, with the exception of coastal facilities, practices a form of water reuse, because the discharged treated wastewater is made available for reuse by downstream users. In many cases, effluent-impacted surface water is employed for nonpotable applications, such as irrigation. However, there are numerous locations where wastewater effluent accounts for a substantial fraction of a potable water supply (Swayne et al., 1980). This form of reuse, which is also referred to as de facto reuse (Asano et al., 2007), is important to the evaluation of water reuse projects and may be a useful source of data on potential public health risks. In many cases, the degree of treatment that this municipal wastewater receives prior to entering the potable water supply is less than that applied in planned reuse projects.

Rivers and lakes that receive wastewater effluent discharges are sometimes referred to as effluent-impacted waters.[1] Box 2-3 describes an example of a watershed where wastewater effluent accounts for about half of the water in a drinking water reservoir. The concentration of wastewater-derived contaminants in a drinking water treatment plant water intake from an effluent-impacted source water depends upon the wastewater treatment plant, the extent of dilution, residence time in the surface water, and the characteristics of the surface water (including depth and temperature, which affect the rates of natural contaminant attenuation processes). Although it is currently difficult to estimate the total contribution of de facto reuse to the

nation's potable water supply, monitoring efforts (e.g., the U.S. Geological Survey [USGS] Toxic Substances Hydrology Program) have documented the presence of wastewater-derived contaminants in watersheds throughout the country (Kolpin et al., 2002). In a recent study of drinking water supplies, one or more prescription drugs was detected in approximately 25 percent of samples collected at the intakes of drinking water treatment plants in 25 states and Puerto Rico (Focazio et al., 2008).

Although detection of wastewater-derived organic compounds demonstrates the occurrence of de facto reuse, making precise estimates of the contribution of effluent to a water supply is more challenging. Aside from anecdotal reports from watersheds such as the Trinity River (Box 2-3), it is challenging to find good estimates of effluent contributions to water supplies. Attempts to quantify the fraction of the overall flow of a river that was derived from wastewater effluent require detailed information about the hydrology of the watershed and the quantity of effluent discharged. In 1980, EPA conducted a scoping study to characterize the contribution of wastewater effluent to drinking water supplies (see Box 2-4). Results indicated that more than 24 major water utilities used rivers from which effluent accounted for over 50 percent of the flow under low-flow conditions (Swayne et al., 1980).

Since that time, the urban population of the United States has increased by over 35 percent (U.S. Census, 2010c, 2011), with much of the growth occurring in the southeastern and western regions. As a result, it is likely that the contribution of wastewater effluent to water supplies has increased since the 1980 EPA scoping study. In 1991, data from EPA indicated that 23 percent of all permitted wastewater discharges were made into surface waters that consisted of at least 10 percent wastewater effluent under base-flow conditions. More recently, Brooks et al. (2006) estimated that 60 percent of the surface waters that received effluent discharges in EPA Region 6 (i.e., Arkansas, Louisiana, New Mexico, Oklahoma, and Texas) consisted of at least 10 percent wastewater effluent under low-flow conditions.[2]

---

[1] Effluent-impacted surface waters can also discharge to groundwater. As a result, groundwater wells located proximate to effluent-impacted surface waters can be a route for de facto potable water reuse. The number of people who acquire their drinking water from wells under the influence of effluent-dominated waters that are not intentionally operated as potable water reuse systems is unknown.

[2] The committee recognizes that temporal variations in dilution flows will affect surface water quality, but it was beyond the committee's charge to assess specific flow criteria (e.g., average flow, 7Q10 [average low-flow over 7 consecutive days with a 10-year return frequency]) that should be used to evaluate the extent and

BOX 2-3
De Facto Reuse in the Trinity River Basin

The Trinity River in Texas is an example of an effluent-dominated surface water system where de facto potable water reuse occurs. The section of the river south of Dallas/Forth Worth consists almost entirely of wastewater effluent under base flow conditions (Fono et al., 2006; TRA, 2010). In response to concerns about nutrients, the wastewater treatment plants in Dallas/Fort Worth that collectively discharge about 500 million gallons per day (MGD; 2 million m³/d) of effluent employ nutrient removal processes (Fono et al., 2006). Little dilution of the effluent-dominated waters occurs as the water travels from Dallas/Fort Worth to Lake Livingston, which is one of the main drinking water reservoirs for Houston (see figure below). Once the water reaches Lake Livingston, it is subjected to conventional drinking water treatment prior to delivery to consumers in Houston.

Results from hydrological models and contaminant monitoring indicate that contaminant attenuation takes place in the river and reservoir. During the estimated 2-week travel time between Dallas/Fort Worth and Lake Livingston, many of the trace organic contaminants undergo transformation by microbial and photochemical processes (Fono et al., 2006). Additional contaminant attenuation and pathogen inactivation also may occur during the water's residence time in the reservoir. On an annual basis, about half of the water flowing into Lake Livingston is derived from precipitation. Therefore, water entering the drinking water treatment plant consists of approximately 50 percent wastewater effluent that has spent approximately 2 weeks in the Trinity River and up to a year in the reservoir before it becomes a potable water supply. The potable water from the Trinity River meets all of the Environmental Protection Agency's water quality regulations and this de facto potable reuse system is an important element in the region's water resource planning.

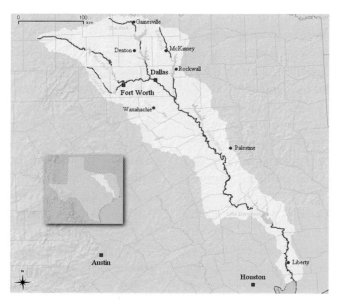

Trinity River Basin, showing Dallas/Fort Worth in the headwaters of the water supply for the city of Houston.
SOURCE: http://wapedia.mobi/en/File:Trinity_Watershed.png.

Improved integration of hydrological data and better watershed models make it possible to estimate the fraction of wastewater effluent in surface waters under a

significance of de facto reuse. The existing regulatory structure for drinking water addresses this issue through requirements for periodic monitoring. For chemicals where the risk is based on lifetime exposure, average concentrations of contaminants are used. For pathogens and chemicals where risks are based on shorter exposures, low-flow measures might be appropriate, although it is beyond the committee's charge to evaluate.

range of conditions. For example, Andrew Johnson and Richard Williams (Centre for Ecology and Hydrology, personal communication, 2009) used readily available data on river flows and volumes of wastewater effluent discharged by individual treatment plants to develop a hydrological model that predicts the fraction of wastewater effluent in different surface waters in and around Cambridge, UK, under base-flow conditions (Figure 2-2). Such hydrological data are available in

---

BOX 2-4
The Presence of Wastewater in Drinking Water Supplies Circa 1980

A survey of wastewater discharges upstream of drinking water intakes was conducted on behalf of EPA, reflecting water systems that collectively served 76 million persons (Swayne, et al., 1980). Data are shown in the below figure for average flow conditions and low flow (i.e., 7-day, 10-year low flow) conditions. Utilities serving 32 million people (of the 76 million total reflected in the survey) reported that no wastewater was discharged upstream of the water intakes. However, of the remaining 44 million people served by the utilities surveyed, more than 20 million relied upon source water with a wastewater content of 1 percent or more under average flow conditions, and a similar number relied on source water with a wastewater content of 10 percent or more during low-flow conditions. No comparable more recent data are available, but these percentages have likely increased significantly since the EPA data were collected, given the population growth and increasing water use over the last 30 years. Although some of the supplies represented by the data on the right side of the figure below are controversial, most of these urban water supplies are considered safe, conventional water supplies by the public.

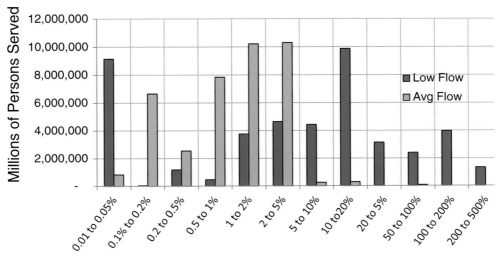

Persons served by a water supply with wastewater content according to EPA's 1980 survey of wastewater discharged upstream of drinking water intakes.
SOURCE: Data from Swayne et al. (1980).

---

the United States through the EPA's Better Assessment Science Integrating Point and Nonpoint Sources (BASINS) system[3] and have been adapted by scientists working for the pharmaceutical industry to make such calculations for 11 watersheds serving as drinking water supplies for 14 percent of the U.S. population (Anderson et al., 2004). Maps that show the contribution of wastewater under current and future scenarios could be extremely useful to water resource planners and public health experts as part of efforts to manage the nation's water resources in a safe and reliable manner.

USGS maintains stream gauging stations and has an active research and monitoring program for wastewater-derived contaminants. EPA has considerable experience in the development and application of surface water quality models. Through a collaborative effort drawing upon the expertise of both agencies, agency scientists could provide water resource planners with a better understanding of the extent of de facto reuse in their catchment and provide data useful to estimating contaminant attenuation between effluent discharge and potable water intakes (e.g., residence time, water quality, depth).

[3] See http://water.epa.gov/scitech/datait/models/basins/index.cfm.

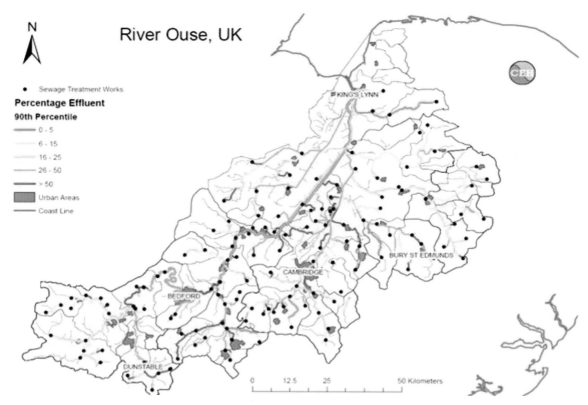

FIGURE 2-2 Estimated Contribution of wastewater effluent to overall river flow in the River Ouse (UK).
SOURCE: Andrew Johnson and Richard J. Williams, CEH, personal communication, 2009.

PLANNED NONPOTABLE WATER REUSE APPLICATIONS

As an alternative to releasing wastewater effluent into the environment, reclaimed wastewater can be reused for a variety of purposes (Table 2-2). Currently, most reclaimed water is used for nonpotable applications, such as agricultural and landscape irrigation. (Data on the extent of various reuse applications in several states is presented toward the end of this chapter.) The following section discusses a variety of nonpotable reuse applications and associated technical and water quality considerations. Economics, the regulatory framework, and public acceptance also influence planning decisions about nonpotable reuse, and these factors are examined in Chapters 9 and 10.

Urban Reuse Applications

A wide array of uses for nonpotable reclaimed water have been identified in urban areas. Urban water reuse systems currently provide reclaimed water for landscape irrigation, decorative water features, toilet and urinal flushing, fire protection, cooling water for air conditioners, commercial uses (e.g., car washes, laundries), dust suppression, and street washing, among others. For example, in Florida, urban nonpotable applications (i.e., industrial uses, public access irrigation) represented at least 68 percent of total reclaimed water use by flow volume in 2010 (FDEP, 2011). Industrial and landscape irrigation reuse applications are discussed in more detail below, along with dual distribution systems that enable these applications.

Landscape Irrigation

Landscape irrigation is the most widely used application of reclaimed water in urban environments and typically involves the spray irrigation of golf courses, parks, cemeteries, school grounds, freeway medians, residential lawns, and similar areas. Because public contact with the applied water presents potential health

TABLE 2-2 Uses of Reclaimed Water

Category of Use	Specific Types of Use	Limitations
Landscape irrigation	Parks, playgrounds, cemeteries, golf courses, roadway rights-of-way, school grounds, greenbelts, residential and other lawns	• Dual distribution system costs • Uneven seasonal demand • High–total dissolved solids (TDS) reclaimed water can adversely affect plant health
Agricultural irrigation	Food crops, fodder crops, fiber crops, seed crops, nurseries, sod farms, silviculture, frost protection	• Use and source are often some distance apart • Dual distribution system costs • Uneven seasonal demand • High-TDS reclaimed water can adversely affect plant health
Nonpotable urban uses (other than irrigation)	Toilet and urinal flushing, fire protection, air conditioner chiller water, commercial laundries, vehicle washing, street cleaning, decorative fountains and other water features	• Dual distribution system costs • Building-level dual plumbing may be required • Greater burden on cross-connection control
Industrial uses	Cooling, boiler feed, stack scrubbing, process water	• Dual distribution system cost to industrial sites varies based on proximity • Treatment required depends on end use
Impoundments	Ornamental, recreational (including full-body contact)	• Dual distribution system costs • Nutrient removal required to prevent algal growth • Potential ecological impacts depending on reclaimed water quality and sensitivity of species
Environmental uses	Stream augmentation, marshes, wetlands	• Nutrient and ammonia removal may be required. • Potential ecological impacts depending on reclaimed water quality and sensitivity of species
Groundwater recharge	Aquifer storage and recovery, seawater intrusion control, ground subsidence control	• Appropriate hydrogeological conditions needed • High level of treatment may be required • Potential for water quality degradation in subsurface
Potable water supply augmentation	Water supply treatment	• Very high level of treatment required • Requires post-treatment storage • Can be energy intensive
Miscellaneous	Aquaculture, snow making, soil compaction, dust control, equipment washdown, livestock watering	

SOURCE: Adapted from Washington State Department of Health (2007).

risks if microbial pathogens are present in the water, reclaimed water typically is subjected to high doses of disinfectants. Chemical contaminants usually are not a major concern in landscape irrigation projects. When used for landscape irrigation, reclaimed water usually does not have adverse impacts on plants, although in some cases high levels of salts or constituents such as boron can adversely affect vegetation (see Chapter 8). Furthermore, the potential for ingestion of irrigation water is limited.

Depending on the area being irrigated, its location relative to populated areas, and the extent of public access or use of the grounds, the microbiological requirements and operational controls placed on the system may differ. Irrigation of areas not subject to public access (e.g., highway medians) have limited potential for creating public health problems, whereas

microbiological requirements become more restrictive as the expected level of human contact with reclaimed water increases (e.g., parks, golf courses, schoolyards). Operational considerations include limiting aerosol formation and dispersal, managing application rates to avoid ponding and runoff, and maintaining proper disinfection (EPA, 2004).

Landscape irrigation with reclaimed water is well accepted and widely practiced in the United States. For example, in 2005 there were more than 200 water reclamation facilities that provided reclaimed water to more than 1,600 individual park, playground, or schoolyard sites for irrigation (Crook, 2005b). The majority of the sites were in California and Florida. Irrigation of golf courses is one of the most common uses of reclaimed water, and 525 golf courses in Florida alone used reclaimed water for irrigation in 2010 (FDEP, 2011).

Industrial Applications

Effluent from conventional wastewater treatment plants is of adequate quality for many industrial applications. Major industrial uses of reclaimed water include cooling, process water, stack scrubbing, boiler feed, washing, transport of material, and as an ingredient in industrial products (MCES, 2007). When used for these applications, reclaimed water has the important advantage of being a reliable supply. This is particularly advantageous for industries located near populated areas that generate large volumes of wastewater effluent.

Cooling Water. The predominant application of reclaimed water by industry is for cooling water. There are more than 40 power plants in the United States that use municipal wastewater as plant makeup water (Veil, 2007). Examples of a steam electric generating plant and a nuclear plant that use reclaimed water for cooling are provided in Boxes 2-5 and 2-6. In general, the major problems experienced by power plants employing reclaimed water for cooling are scale formation, biological growth, and corrosion.

Power plants often use disinfected secondary effluent for cooling, but in recirculating cooling systems, additional treatment, such as filtration, chemical precipitation, ion exchange or reverse osmosis, is often necessary. In some cases, only additional chemical treatment is necessary (e.g., antifoaming agents, polyphosphates to control corrosion, polyacrylates to disperse suspended solids, chlorine to control of biological growth; see EPA, 2004).

Boiler Feedwater. When used as feedwater in boilers, reclaimed water requires extensive treatment with quality requirements that increase with the operating pressure of the boiler. Typically, both potable and reclaimed water need to be treated to remove inorganic constituents that can damage the boilers (EPA, 2004). For example, calcium, magnesium, silica, and aluminum contribute to scale formation in boilers, while excessive alkalinity and high concentrations of potassium and sodium can cause foaming (WPCF, 1989). Bicarbonate alkalinity can lead to the release of carbon dioxide, which can increase the acidity in the steam and corrode the equipment. Because of the relatively small quantities of makeup water and extensive treatment required, reclaimed water is typically a poor candidate for boiler feed. However, reclaimed water is used at a few facilities that provide additional treatment (e.g., reverse osmosis).

Process Water. The acceptability of reclaimed water for industrial process water depends on the specific application. Whereas secondary treatment effluent may be acceptable for some applications (e.g., concrete manufacturing), advanced treatment is needed for applications such as carpet dyeing because water used in textile manufacturing must be nonstaining and the iron, manganese, and organic matter in secondary effluent could compromise the quality of the final product. Divalent metal cations cause problems in some of the dyeing processes that use soap, and nitrates and nitrites may also cause problems (WPCF, 1989). Exceptionally high-quality water is required for some other industrial process uses (e.g., water used to wash circuit boards in the electronics industry often requires reverse osmosis treatment to remove salts).

Reclaimed water is used in the paper and pulp industry, although higher quality paper products are more sensitive to water quality. Certain metal ions, such as iron and manganese, can cause discoloration of the paper, microorganisms can affect its texture and uniformity, and suspended solids may affect its brightness (Rommelmann et al., 2004). The use of reclaimed water in the manufacture of paper products used as food wrap or beverage containers is prohibited in some states (e.g., Florida) to prevent the possibility of contaminants that pose health risks leaching into consumable products.

In the chemical industry, water requirements vary widely depending on the processes involved. In general, water that is in the neutral pH range (6.2 to 8.3), moderately soft (i.e., low calcium and magnesium), and relatively low in silica, suspended solids, and color is required (WPCF, 1989). Total dissolved solids and chloride content generally are not critical.

Dual Distribution and Distributed Systems for Urban Water Reuse

Increasing use of reclaimed water in urban areas has resulted in the development of large dual-water systems in several communities that distribute two

---

BOX 2-5
Xcel Energy Cherokee Station, Denver, Colorado

The Xcel Energy Cherokee Station (pictured below) is a coal-fired, steam electric generating station with four operating units that can produce 717 MW of electricity. The plant, located just north of downtown Denver, Colorado, also is capable of burning natural gas as fuel. The power plant uses 7.1–9.0 MGD (27,000 to 34,000 m^3/d) of water for cooling towers. Historically, all cooling tower feedwater originated from ditch systems that provided raw water to the plant. The Xcel Energy Cherokee Station began using reclaimed water from Denver's Water Recycling Plant as one of its sources of cooling water in 2004 to reduce the plant's freshwater consumption. The Cherokee Station is the largest customer of Denver Water's Recycling Plant, using up to 4.7 MGD (18,000 m^3/d) of reclaimed water. Raw water and reclaimed water are brought to the site and mixed in a large reservoir before feeding the cooling towers. The blend of reclaimed and raw water is also used onsite for ash silo washdown and fire protection. The major benefit of reclaimed water to the power plant is the availability of a new water source and an overall increased water supply to ensure that Xcel Energy will be able to obtain needed water even in dry or drought years.

Denver Water's Recycling Plant, which currently has a treatment capacity of 30 MGD (110,000 m^3/d) and is designed for expansion to 45 MGD (170,000 m^3/d), receives secondary effluent from the Metro Wastewater Treatment Plant. Treatment at the Water Recycling Plant, which is located in close proximity to the Cherokee Station, includes the following

- Nitrification with biologically aerated filters
- Coagulation with aluminum sulfate for phosphorus reduction
- Flocculation and high rate sedimentation
- Filtration with deep-bed anthracite filters
- Chlorine disinfection with free chlorine or chloramines depending on season and need

The cooling towers typically run four to five cycles, and sodium hypochlorite is used as a biocide. Blowdown from the cooling towers is treated with lime and ferric chloride to ensure discharge permit compliance before it is discharged into the South Platte River.

The Xcel Energy Cherokee Station.
SOURCE: Photo courtesy of Xcel Energy (www.XcelEnergy.com)

---

grades of water to the same service area: potable water and nonpotable reclaimed water. The nonpotable reclaimed water can be used for residential irrigation, toilet flushing, and fire protection, among other applications (see Table 2-2). To minimize microbial health risks associated with inadvertent contact or ingestion of reclaimed water (see also Chapter 6), dual-water systems generally provide filtered, disinfected effluent where significant portions of the population could be exposed to the reclaimed water.

Dual-water distribution systems vary considerably in aerial extent, reclaimed water uses, volumes, and complexity of the systems. Infrastructure requirements vary but often include storage facilities, pumping facilities, transmission and distribution pipelines, valves and meters, and cross-connection control devices. There

BOX 2-6
Palo Verde Nuclear Generating Station

The Palo Verde Nuclear Generating Station (pictured below) is the largest nuclear power plant in the nation. The plant is located in the desert, approximately 55 miles (89 km) west of Phoenix, Arizona. The facility uses reclaimed water for cooling purposes and has zero discharge. The sources of the cooling water are two secondary wastewater treatment plants, located in Phoenix and Tolleson, Arizona. The plant used 22 billion gallons (83 million m^3) of reclaimed water in 2008, which is about 61 MGD (230,000 m^3/d) as an average. It has a capacity to treat and use 90 MGD (340,000 m^3/d) of reclaimed water, which receives additional treatment by trickling filters to reduce ammonia, lime/soda ash softening to reduce scale- and corrosion-causing constituents, and filtration to reduce suspended solids. The filtered water is stored in two water storage reservoirs to supply cooling to the steam turbines. Water is routed through condensers and cooling towers an average of 25 cycles until the TDS approaches 30,000 mg/L. About 200 million pounds (91 megagrams) of TDS are sent to the evaporation ponds. Currently, three evaporation ponds that total 650 acres (263 hectares) are used to evaporate liquid waste from blowdown. New evaporation ponds are constructed as needed, and the residual in the ponds will not be sent offsite for disposal until the plant is decommissioned.

SOURCE: Day and Conway (2009).

Palo Verde Nuclear Generating Station.
SOURCE: Photo courtesy of Henry Day.

are economic and other advantages to installing dual distribution systems in new communities as they are being developed as opposed to retrofitting reclaimed water distribution lines in established areas. Installing dual distribution pipelines and appurtenances in existing urban areas often results in considerably higher construction and operational costs than installing a system in a new or developing community (see also Chapter 9).

Operation and management of a dual-water system is similar to that for a potable water system. However,

because the distributed water is nonpotable reclaimed water, special attention needs to be given to public health protection. This includes using color-coded (e.g., purple) pipe for reclaimed water lines, conducting routine water quality monitoring, and periodically testing the system to protect against inadvertent cross-connections with the potable water system (see Box 6-4).

The oldest dual-water system in the United States is located in Grand Canyon Village, Arizona, where less than 1 MGD (3,800 m^3/d) of disinfected advanced effluent is used for landscape irrigation, toilet flushing, cooling water makeup, vehicle washing, and construction uses when needed (Fleming, 1990; Okun, 1996). The original system began operation in 1926. In contrast, in the late 1970s, large systems were implemented in St. Petersburg, Florida (see Box 2-7) and at the Irvine Ranch Water District in Orange County, California, that provided large volumes of reclaimed water for multiple uses within those communities. These pioneering communities helped develop many of the practices that are necessary to ensure the safe and efficient operation of dual distribution systems as documented in a recent manual published by the American Water Works Association (AWWA, 2009).

In areas where local governments have imposed sewer moratoriums or sewer-capacity restrictions, onsite wastewater reclamation and reuse systems have been used successfully in schools and office buildings. More than 30 individual onsite wastewater treatment systems in the United States provide reclaimed water for outside irrigation or for toilet and urinal flushing in office buildings, schools, shopping centers, and manufacturing plants. Because the committee was specifically charged to address municipal wastewater effluent, this report does not discuss onsite reuse systems in detail.

Agricultural

In many parts of the United States, the demand for irrigation water is nearing or exceeds the supply of fresh water. Reclaimed water provides a constant and reliable source of water, even during drought conditions. Agricultural irrigation currently represents the largest use of reclaimed water both in the United States and worldwide (Jiménez and Asano, 2008). Crops irrigated vary from grazing pastures to food crops eaten raw, although irrigation of produce and other food crops eaten raw is prohibited in some states (see also Chapter 10 for state regulation of water reuse). Because agricultural irrigation with reclaimed water has a long history, the technology and suitability of the practice are relatively well understood and do not need to be repeated here. The chemical composition of reclaimed water that has received secondary or higher levels of treatment normally meets existing guidelines for irrigation water (NRC, 1998). Regulatory controls directed at ensuring an adequate level of health protection address reclaimed

water treatment and quality, method of irrigation, type of crops to be irrigated, and operation and management of the distribution system and use area and are described in detail in the EPA Guidelines for Water Reuse (EPA, 2004).

Nitrogen, phosphorus, and potassium in reclaimed waters contribute valuable nutrients to plants and reduce the need for fertilizers, which can result in considerable cost savings; however, excessive nitrogen stimulates vegetative growth in most crops and may also delay maturity and reduce crop quality and quantity. Excessive nitrate in forages can cause an imbalance of nitrogen, potassium, and magnesium in grazing animals if forage is used as a primary feed source for livestock (EPA, 2004). The cost of reclaimed water is often less than the real cost of subsidized agricultural irrigation water or the cost of potable water used for irrigation.

There are numerous examples of agricultural irrigation water reuse projects in the United States. For example, Bakersfield, California, has used its effluent for irrigation since 1912 (Crook and Okun, 1993). During the early years, first raw sewage and then primary effluent were used for irrigation. Today, secondary wastewater effluent from Bakersfield is used to irrigate corn, alfalfa, cotton, barley, and sugar beets. Secondary effluent from the city of Lubbock, Texas, has been used to irrigate cotton, grain sorghum, and wheat on a local farm since 1938 (Crook, 1999). In Orange County, Florida, a project known as Water CONSERV II has been supplying reclaimed water for citrus irrigation since 1986. After disinfection and advanced treatment, reclaimed water has been used to irrigate produce and other food crops eaten raw in Monterey County, California, since 1998 following extensive research conducted to demonstrate its safety (see Box 2-8).

Seawater Intrusion Barrier

In aquifers in which groundwater withdrawals exceed rates of recharge, seawater migrates inland. This process, often referred to as seawater intrusion, can result in high concentrations of salts (mainly sodium and chloride) that prevent use of the groundwater for potable, industrial, and agricultural water supply applications. The only long-term solution is to bring supply and demand in balance, but seawater intrusion can be

BOX 2-7
Dual Distribution in St. Petersburg, Florida

The city of St. Petersburg, Florida, with a population of about 255,000, is a residential community located on the west coast of Florida. In the early 1970s, the city relied upon municipal wells to satisfy a growing population, but St. Petersburg needed additional water. At roughly the same time, the Florida Legislature passed a bill to address water quality issues in Tampa Bay, which required all surrounding communities to stop discharging wastewater to Tampa Bay or to remove nutrients via advanced wastewater treatment prior to discharge. The city of St. Petersburg subsequently decided to upgrade its wastewater treatment plants to secondary treatment and eliminate wastewater discharge to surface waters by implementing a water reuse and deep-well injection program.

Reclaimed water was initially provided to sites with large irrigation requirements, such as golf courses, parks, schools, and large commercial areas, beginning in 1977. A few years later, the reclaimed water distribution system was expanded to include irrigation of residential property.

In FY 2009, the total average flow from the four water reclamation plants was about 33 MGD (125,000 m^3/d), of which an average of 17 MGD (64,000 m^3/d) was used for nonpotable reuse applications. Excess reclaimed water and treated wastewater that does not meet reuse water quality requirements is disposed of via deep well injection. The reclaimed water satisfies about 40 percent of the city's total water demand. The dual-water system serves more than 10,500 customers, including about 10,250 residential customers for landscape irrigation. Reclaimed water also is used for irrigation at 96 parks, 62 schools, 6 golf courses, and about 343 commercial sites (see figure below). The water also is used for fire protection via reclaimed water hydrants throughout the system and for cooling water at 13 sites.

Prior to distribution, reclaimed water is pumped to covered storage tanks at all four reclamation plants. The transmission mains from the four treatment plants are interconnected so that water flow and pressure can be maintained to all customers if one plant needs to be taken out of service. In all areas where dual-distribution lines provide reclaimed water, the potable water supplies are protected with cross-connection control backflow assembly devices, including double check-valve assemblies at residences that use reclaimed water for irrigation.

St. Petersburg residents that want to be connected to the nonpotable distribution system are required to pay the connection costs, which typically ranges from $500 to $1,200 per customer. Reclaimed water costs $15.62/month for the first acre (0.40 hectares) to be irrigated and $8.95/month for each additional acre or portion thereof. The flat-fee rate structure does not encourage water conservation, and most residents use more reclaimed water than is necessary for proper irrigation. The reclaimed water rate for commercial customers who have metered service is $0.45/1,000 gallons ($0.45/3.785 m^3). The current annual operating cost is $5.3 million. System revenue is $2.6 million; the remaining $2.7 million is subsidized by the city's water and wastewater utilities, each of which pays half of that cost. For additional discussion on the costs of water reuse, see Chapter 9.

SOURCE: Crook, 2005a, Bowen, E., St. Petersburg Water Resources Department, personal communication, 2010.

Landscape irrigation with reclaimed water in St. Petersburg.
SOURCE: Dennis MacDonald/World of Stock.

slowed or reversed by injection of water between the supply wells and the ocean. In densely populated areas, seawater intrusion barriers typically consist of a network of wells arrayed parallel to the shoreline to form a hydrostatic barrier to seawater intrusion (Figure 2-3).

In several cases, including four seawater intrusion barriers in Southern California (Figure 2-4), reclaimed water has been used to create the groundwater barrier. In 2007, a similar project was built near Barcelona, Spain (Mujeriego et al., 2008), as a means of protect-

BOX 2-8
Monterey County Water Reuse Project, California

As far back as 1975, the Monterey Regional Water Pollution Agency identified the potential for using reclaimed water to stem seawater intrusion from Monterey Bay, caused by overdrafting of underlying aquifers. A demonstration study began in 1976, with the goal of determining the safety of reclaimed water use on edible crops, including those eaten raw. The study tested traditional well water versus two treatment trains of reclaimed water, reclaimed water with advanced treatment that included chemical coagulation and clarification processes and reclaimed water with advanced treatment using direct filtration. Study results indicted that advanced treatment using direct filtration was acceptable for irrigation of food crops eaten raw (*Engineering-Science, 1987*).

Design of the treatment plant facilities, collectively named the Salinas Valley Reclamation Project, was completed in 1994 along with design of the distribution system, known as the Castroville Seawater Intrusion Project. The 30-MGD (110,000-m^3/d) Salinas Valley Reclamation Project began distributing 20 MGD (76,000 m^3/d) of irrigation water in 1998 to local farmers, covering 222 parcels of farmland in the 12,000-acre (4,900-ha) service area (see figure below). Reclaimed water is used to irrigate various crops, including lettuce, celery, broccoli, cauliflower, artichokes, and strawberries. The system has experienced only minor problems including flushing of construction debris from the system, excessive sand in the extracted water of some wells, and a few pipeline breaks.

The Recycled Water Food Safety Study was conducted prior to startup to determine if any viable pathogenic organisms of concern to food safety were present in reclaimed water (Jaques et al., 1999). Sampling began in 1997 and continues to the present. No *Escherichia coli* 0157:H7, *Salmonella*, helminth ova, *Shigella*, *Legionella*, or culturable natural (in situ) viruses were detected in any of the samples. An extremely low number of *Cyclospora* (one instance), *Giardia* with internal structure (one instance), and *Cryptosporidia* (in seven instances) were detected in the reclaimed water. The use of reclaimed water for agricultural irrigation in this region is expected to reduce the volume of seawater intrusion by 40 to 50 percent (Crook, 2004).

The Salinas Valley Reclamation Project in Monterey, California, which provides reclaimed water to area farms, thereby reducing seawater intrusion caused by overpumping the region's aquifers.
SOURCE: Monterey Regional Water Pollution Control Agency.

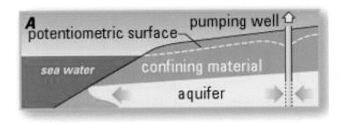

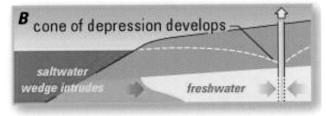

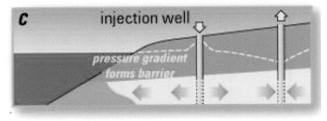

FIGURE 2-4 Locations of the four major Southern California seawater barriers employing reclaimed water. These barriers range in length from 2 miles (Alamitos Gap) to 9 miles (West Coast Barrier).

FIGURE 2-3 Effects of groundwater withdrawal on saltwater intrusion and the role of a seawater intrusion barrier. Image A depicts a normal coastal aquifer with a water table high enough to resist seawater intrusion. Image B depicts an aquifer that is being overpumped and is beginning to experience seawater intrusion. Image C shows the same aquifer after the installation of an injection well to form a hydrostatic barrier, protecting the aquifer.
SOURCE: Modified from Johnson (2007).

ing an aquifer that is important for urban water supply and agricultural production. In cases where some of the reclaimed water from the seawater barrier reaches wells used for drinking water supply, the practice is considered potable water reuse.

Impoundments

Reclaimed water impoundments, which are often used for system or seasonal storage, fall into two categories—aesthetic or recreational. Fishing, boating, or any other activity that may involve human contact with the reclaimed water is not allowed in aesthetic impoundments, which are also called landscape impoundments. Recreational impoundments can be subdivided into either non–body contact or body contact impoundments (or restricted and nonrestricted recreational im-

poundments, respectively). Non–body contact includes activities such as boating and fishing where there is only incidental contact with the reclaimed water, while body contact impoundments allow swimming. There are several recreational impoundments in the United States that allow fishing and boating, and one of the first of which was the Santee Recreational Lakes in San Diego County, California (see Box 2-9). At present there are no reclaimed water recreational impoundments in the United States that are used for full-body-contact activities, although such use is allowed in some states.

Regulatory guidelines for recreational impoundments are predicated on the assumption that the water should not contain chemical substances that are toxic following ingestion or irritating to the eyes or skin, and should be safe from a microbiological standpoint. Other concerns are temperature, pH, chemical composition, algal growth, and clarity. Clarity is important for several reasons, including safety, visual appeal, and recreational enjoyment. Recreational lakes composed entirely of reclaimed water are prone to eutrophication. The nutrients in the wastewater can cause excessive growth of algae, and nutrient removal may be necessary prior to reclaimed water discharge. Phosphorus is generally the limiting nutrient and can serve as a means of controlling algae in freshwater impoundments. Before fish, shellfish, or plants are harvested for human consumption from recreational impoundments containing reclaimed water, regulatory guidelines presume that both the microbiological and chemical quality of the

BOX 2-9
Santee Recreational Lakes

Reclaimed water has been used as a source of supply to recreational lakes in Santee, California, since 1961 (see figure below). The activities were limited initially to picnicking and boating, and progressed to a "fish for fun" program, and finally to a normal fishing program. In the early 1970s, a 3.8-MGD (14,000-m^3/d) activated sludge treatment plant replaced a pond system. The water was percolated through 400 ft (120 m) of sand and gravel and disinfected prior to discharge to the lake system. Because of the high nutrient levels in the reclaimed water, there was considerable algal growth in the lakes, which average 1,000 ft (300 m) in length and 2–10 ft (0.6-3 m) in depth. Algae control in the lakes via chemicals and mechanical harvesting was practiced. Flow has increased through the years and now includes a advanced treatment system consisting of a 1.9-MGD (7,200-m^3/d) Bardenpho (multistage biological treatment) plant followed by coagulation and flocculation using alum, a lamella settler for turbidity and excess phosphorus removal, a denitrification filter, and chlorine disinfection. The reclaimed water is dechlorinated prior to discharge to the lake system, which consists of seven lakes, which have a total surface area of about 60 acres (24 ha). The lakes are part of an extensive recreational area widely used by the local populace (Asano et al., 2007).

Santee Recreational Lakes.
SOURCE: http://Santeelakes.com.

source water will be thoroughly assessed for possible bioaccumulation of toxic contaminants through the food chain.

Habitat Restoration

In locations where surface water has been diverted for agriculture, industrial, or urban uses, decreases in water availability have had adverse impacts on aquatic habitat (NRC, 2004). The discharge of wastewater effluent can restore, and in some cases, create aquatic habitat. Most documented projects in which water

reclamation has resulted in the restoration or creation of aquatic habitat originally were designed either for the disposal of wastewater effluent or as an inexpensive means of improving water quality prior to surface water discharge. Nevertheless, the use of wastewater effluent for habitat restoration or creation is a potentially important application of reclaimed water, especially in rapidly growing regions with limited availability of surface water.

The most common restoration projects are engineered treatment wetlands, which often are built adjacent to wastewater treatment plants as a means of removing nitrate or phosphate (Kadlec and Knight, 1996). Engineered wetlands are typically not used for removal of ammonium—the other main form of nitrogen present in wastewater effluent—because ammonium is toxic to fish, which are important to the control of mosquitoes and other vectors. The wetlands typically consist of emergent vegetation (e.g., cattails) and shallow ponds that provide excellent habitat for waterfowl, birds, and species of fish that are adapted to shallow water. Although some treatment wetlands have been designed to receive secondary effluent (EPA, 1993a), good aquatic habitat is difficult to establish if the effluent contains ammonia, which is toxic to most aquatic organisms. Therefore, to provide acceptable habitat, wetlands are usually supplied with wastewater effluent that has been subjected to additional treatment to remove ammonia (see Chapter 4). Examples of engineered wetlands that provide wildlife habitat and associated recreational benefits (e.g., wildlife viewing, hunting) include the Easterly Wetlands in Orlando, Florida (see Box 2-10); the Prado Wetlands in Riverside County, California; Tres Rios Wetlands in Phoenix, Arizona; and the Tarrant Regional Wetlands near Dallas, Texas.

It is also possible to use reclaimed water to enhance surface water habitats, especially in arid regions where the original sources of water have been diverted for other uses. For example, San Luis Obispo Creek, which is located in California's Central Coast region, lost a considerable fraction of its overall flow when the nearby wastewater treatment plant began using its effluent for landscape irrigation. To maintain aquatic habitat in the creek, the utility discharges approximately 1.1 MGD (4,200 m³/d) of reclaimed water directly to the creek (Asano et al., 2007). To ensure that the water quality is cold enough for native species, the reclaimed water is passed through a cooling tower prior to discharge. The reclaimed water accounts for the majority of the flow during the dry summer season.

Wastewater effluent also has been used to create or restore habitat in coastal marshes (Day et al., 2004) and woodlands (Rohnke and Yahner, 2008). Although such systems are less common than treatment wetlands, there is evidence that the nutrients and added water supplied by the reclaimed water can create or restore a variety of habitat types.

Although wetlands and terrestrial systems that depend on wastewater effluent often support rich ecological communities, it is important to recognize that the restored or created systems may not be similar to those that were present prior to development. For example, a surface water wetland fed with wastewater effluent will not result in the same ecosystem as the nutrient-poor ephemeral stream that was present prior to development. Therefore, decisions about the type of treatment needed prior to using reclaimed water for habitat restoration need to be made in recognition of the needs of the specific type of ecosystem. These and other issues related to environmental applications of reclaimed water are discussed in more detail in Chapter 8.

POTABLE WATER REUSE

Potable reuse projects have been operated in the United States for almost 50 years. During this period, the treatment technologies employed in the advanced treatment systems have evolved considerably, with a gradual shift from reliance on physical processes, such as lime clarification and adsorption of contaminants on activated carbon (Table 2-3), to membrane filtration and advanced oxidation (see Chapter 4 for descriptions of treatment technologies). In 2010, approximately 355 MGD (1,350 m³/d) of reclaimed water was used for planned potable reuse projects in the United States. Although this accounts for only about 0.1 percent of the municipal wastewater undergoing treatment, reclaimed water can account for the majority of the drinking water supply in some areas.

The use of reclaimed water for drinking water supplies has historically been divided into two categories: indirect potable reuse (IPR) and direct potable reuse. Both employ a sequence of treatment processes after conventional wastewater treatment (detailed in Chapter 4). However, IPR projects were distinguished from direct potable reuse projects by the presence of an environmental buffer between the wastewater effluent and the potable water supply. An environmental buffer is a water body or aquifer, perceived by the public as natural, which serves to sever the connection between the water and its history. The buffer may also (a) decrease the concentration of contaminants through various attenuation processes, (b) provide an opportunity to blend or dilute the reclaimed water, and (c) increase the amount of time between when the reclaimed water

---

BOX 2-10
The Easterly Wetlands Project

The Easterly Wetlands Project (see figure below) was constructed approximately 30 miles (48 km) east of Orlando, Florida, in 1993. The 1,650-acre (670-ha) wetland was built by constructing 18 miles (29 km) of berms and importing wetland plants to create a series of wetland cells on a property that had been used as a cattle ranch after the natural wetland had been drained in the 1850s. Between approximately 20 and 35 MGD (76,000 to 130,000 m³/d) of wastewater effluent flows through the wetland before being discharged to the St. Johns River.

The wetland system reduces the concentrations of nutrients discharged to the sensitive St. Johns River. Phosphate is mainly removed by settling and plant uptake while much of the nitrogen is denitrified (i.e., released from the wetlands as nitrogen gas). Data collected over the first 3 years of the project indicated reductions of total phosphorus and total nitrogen of over 97 percent and over 90 percent, respectively (Mark Sees, Orlando Easterly Wetlands, personal communication, 2009).

The Easterly Wetlands also acts as a habitat for birds, such as the locally endangered Everglades snail kite, and various species of mammals, amphibians, and reptiles. The wetland facility has an educational center that regularly attracts visitors from local schools and bird watchers.

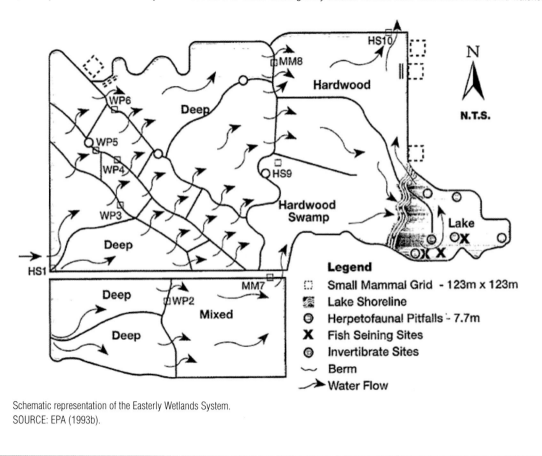

Schematic representation of the Easterly Wetlands System.
SOURCE: EPA (1993b).

---

is produced and when it is introduced into the water supply. Although the latter three functions of environmental buffers have potentially important implications for public health, performance standards for buffers have never been defined. The committee is unaware of any situation in which the time delay provided by a buffer has been used to respond to an unforeseen upset, and the residence time of reclaimed water in some environmental buffers (e.g., rivers, small lakes, and reservoirs) is short (e.g., hours or days) relative to the time needed to detect and respond to all but the most obvious system failures.

It was largely the passage of water through a natural system and its role in increasing public acceptance of the subsequent use of the water in potable supplies that led to the perception that environmental buffers

TABLE 2-3 Examples of Potable Reuse Schemes and Employed Treatment Technologies in the United States

Project Location	Type of Reuse	Project Size MGD (m³/d)	First Installation Year
Montebello Forebay, County Sanitation Districts of Los Angeles County, CA	Groundwater recharge via soil-aquifer treatment	44 (165)	1962
Water Factory 21, Orange County, CA	Groundwater recharge via seawater barrier	16 (60)	1976
Upper Occoquan Service Authority, VA	Surface water augmentation	54 (204)	1978
Hueco Bolson Recharge Project, El Paso Water Utilities, TX	Groundwater recharge via direct injection	10 (38)	1985
Clayton County Water Authority, GA	Surface water augmentation	18 (66)	1985
West Basin Water Recycling Plant, CA	Groundwater recharge via direct injection	12.5 (47)	1993
Gwinnett County, GA	Surface water augmentation	60 (227)	1999
Scottsdale Water Campus, AZ	Groundwater recharge via direct injection	14 (53)	1999
Los Alimitos Barrier Water Replenishment District of So. CA	Groundwater recharge via direct injection	2.7 (10)	2005
Chino Basin Groundwater Recharge Project, Inland Empire Utility Agency, Chico, CA	Groundwater recharge via soil-aquifer treatment	18 (69)	2007
Groundwater Replenishment System, Orange County, CA	Groundwater recharge via direct injection and spreading basins	70 (265)	2008
Arapahoe County/Cottonwood, CO	Groundwater recharge via spreading operation	9 (34)	2009
Cloudcroft, NM	Spring water augmentation	0.1 (0.38)	2009
Prairie Waters Project, Aurora, CO	Groundwater recharge via riverbank filtration	50 (190)	2010
Permian Basin, Colorado River Municipal Water District, TX	Surface water augmentation	2.5 (9.4)	2012
Dominguez Gap Barrier, City of Los Angeles	Groundwater recharge via direct injection	2.5	2012

SOURCE: Adapted from Drewes and Khan (2010)

were essential to potable water reuse projects. For the community, environmental buffers have been crucial to acceptance because they break the perceived historical connection between the ultimate water source (i.e., sewage) and the reclaimed water supply. The notion that potable water suppliers should avoid the use of effluent-impacted source waters was supported by outbreaks of waterborne disease that were common prior to the widespread installation of drinking water and wastewater treatment plants during the twentieth century, when consumers were exposed to untreated water supplies that were subjected to discharges of raw sewage. Given the improvements in treatment, such outbreaks are much less likely in systems where treated wastewater and drinking water undergo disinfection. However, the public's notion that water sources should be separated from waste discharges is a well-established precedent.

The committee recognizes that community acceptance is important to potable reuse projects (see Chapter 10) and this factor alone may motivate utilities to include buffers in potable reuse projects. However,

Current Status	Treatment Technologies				
	Suspended Solids	Organic Compounds	Residual Nutrients	Residual Salts	Pathogens
Ongoing	Media filtration	Soil-aquifer treatment	Soil-aquifer treatment	None	Chlorination, soil-aquifer treatment
Terminated 2004	Lime clarification	GAC filtration; Reverse osmosis; UV/AOP	Air stripping; reverse osmosis	Reverse osmosis	Lime clarification, chlorination, UV
Ongoing	Lime clarification, media filtration	GAC filtration	Ion exchange (optional)	None	Chlorination
Ongoing	Lime clarification, media filtration	Ozonation, GAC filtration	PAC augmented activated sludge system	None	Ozonation, chlorination
Ongoing	Land application system and wetlands	Land application system; wetlands	Land application system; wetlands	None	Chlorination, UV
Ongoing	Microfiltration	Reverse osmosis; UV/AOP	Reverse osmosis	Reverse osmosis	Microfiltration chloramination, UV
Ongoing	Ultrafiltration	Pozonation; GAC filtration	Chem. P-removal	None	Ultrafiltration, Ozone
Ongoing	Media filtration, microfiltration	Reverse osmosis	Reverse osmosis	Reverse osmosis	Microfiltration, Chlorination
Ongoing	Microfiltration	Reverse osmosis, UV	Reverse osmosis	Reverse osmosis	Microfiltration, UV
Ongoing	Media filtration	Soil-aquifer treatment	Soil-aquifer treatment	None	Chlorination
Ongoing	Microfiltration	Reverse osmosis, UV/AOP	Reverse osmosis	Reverse osmosis	Microfiltration; UV
Ongoing	Media filtration	Reverse osmosis, UV/AOP	Reverse osmosis	Reverse osmosis	Chlorination
Ongoing	Microfiltration; ultrafiltration	Reverse osmosis, UV/AOP	Reverse osmosis	Reverse osmosis	Chlorination
Ongoing	Riverbank filtration	Riverbank filtration, UV/AOP, BAC, GAC	Riverbank filtration; artificial recharge and recovery	Precipitative softening	Riverbank filtration, UV, chlorination
Under construction	Ultrafiltration	Reverse osmosis, UV-AOP	Reverse osmosis	Reverse osmosis	Chlorination
Ongoing	Microfiltration	Reverse osmosis	Reverse osmosis	Reverse osmosis	Microfiltration

the role of the environmental buffer in providing public health protection under the conditions encountered in planned potable reuse systems has not always been well documented. This is particularly important because each environmental buffer will have different attributes that affect the removal of contaminants, the amount of dilution, or the residence time (see also Chapter 4). For example, greater removal of contaminants by photochemical processes will occur in shallow, clear streams than in deep lakes or turbid rivers (Fono et al. 2006). As a result, it would be inappropriate to as-sume that contaminant attenuation by photochemical processes occurs at the same rates in these two types of systems. Without good data on site-specific character-istics, there will be considerable uncertainty about the ability of environmental buffers to remove contami-nants. Because of the limited and variable data on the performance of environmental buffers (see Chapter 4), the committee has chosen in this report to emphasize the key processes and attributes necessary for potable reuse, rather than specific design elements implied by the terms *direct* or *indirect potable reuse*. Thus, these terms are mainly used in this report in the context of historical or planned reuse projects, in recognition of the widespread practice of classifying potable reuse projects as *direct* or *indirect*, but these distinctions are deemphasized in the remainder of the report.

The overview of potable reuse projects in the following section is intended to provide representative examples of potable reuse projects, to illustrate the role of environmental buffers, and to describe current trends in potable water reuse. The performance of environmental buffers is discussed in detail in Chapter 4, and public perception is discussed in Chapter 10.

Surface Water Augmentation

Approximately two-thirds of the potable water delivered by public water systems in the United States comes from surface water sources, including rivers, lakes, and reservoirs (Hutson et al., 2000). In some cases, the entire surface water source is located in a protected watershed. Such systems usually provide water of high quality that can be delivered to consumers after disinfection (NRC, 2000). However, most surface water supplies are at least partially located in unprotected watersheds, where they may receive contaminants from upstream sources including agricultural and urban runoff, industrial process water, and municipal wastewater effluent. For example, wastewater effluent accounts for approximately half of the water entering one of the main water supply reservoirs for Houston (see Box 2-3). In recognition of the potential contributions of these sources of contamination, drinking water treatment plants that handle water from unprotected water sources often employ more sophisticated treatment technologies (see also Chapter 4).

Augmentation of surface waters with reclaimed water represents the addition of another source of water to the system. Surface water augmentation involves discharge of reclaimed water directly to a water supply reservoir, a lake, or a short stretch of river followed by capture in a reservoir or to a wetland adjacent to a river. Most reservoir systems receive a considerable fraction of their overall flow from other sources and as a result, reclaimed water undergoes substantial dilution. Furthermore, the relatively long hydraulic retention time in large reservoirs affords considerable opportunities for contaminant attenuation, although if nutrients are

not removed prior to discharge, the reclaimed water can result in excessive algal growth and water quality degradation.

As discussed in Chapter 3, the concentration of contaminants in reclaimed water depends on the source of the sewage and the treatment processes used. For example, wastewater reclamation plants using advanced treatment produce reclaimed water that contains lower concentrations of contaminants than what is commonly observed in surface waters subject to upstream discharges of typical wastewater effluent, urban runoff, and agricultural drainage. Thus, surface water augmentation may contribute better quality water to a drinking water treatment plant than other sources in the watershed. Assessments of surface water augmentation projects should therefore be viewed in the broader context of the water quality that already exists in the water body. Assessments of the public health risks associated with potable reuse projects also need to consider the potential for attenuation of contaminants to occur between the location where the reclaimed water enters the system and the consumer's tap (for a detailed discussion of risk, see Chapters 6 and 7).

The first permanent[4] surface water augmentation project in the United States was installed in Fairfax County, Virginia, in 1978. As part of the augmentation project, the Upper Occoquan Service Authority (UOSA) discharges approximately 54 MGD (204,000 m^3/day) of effluent from an advanced treatment plant into a water supply reservoir. In a typical year, the wastewater effluent accounts for less than 10 percent of the water flowing into the reservoir. However, during a drought in the early 1980s, reclaimed water accounted for more than 80 percent of the water entering the reservoir (AWWA/WEF, 1998). Using data on the size of the reservoir and the contribution of reclaimed water, the hydraulic retention time of the reclaimed water in the reservoir is estimated to vary from a few days to more than 6 months.

In 1982, a water utility near Atlanta, Georgia, began augmenting one of its reservoirs by using sprinklers to apply effluent from a conventional wastewater treatment plant to forestland adjacent to a water supply

---

[4] A reservoir supplying water for the City of Chanute, Kansas, was augmented with secondary wastewater effluent between 1956 and 1957 (Metzler et al., 1958).

reservoir. After passing through the soil, the reclaimed water flowed into the reservoir. As the water needs of the Clayton County Water Authority expanded, the land application system was replaced by a series of constructed wetlands that do not require as much land. The first set of engineered wetlands was installed in 2003 and was expanded to cover over 500 acres (202 ha) in subsequent years. Available estimates suggest that during droughts, wastewater effluent may contribute up to 50 percent of the flow into the reservoir (Guy Pihera, Water Production Manager, Clayton County Water Authority, personal communication, 2010).

Recent developments related to surface water augmentation in the Trinity River watershed of Texas (see Box 2-3) are noteworthy with respect to their design and water rights issues. As part of the region's integrative water planning efforts in anticipation of projected rapid population growth in the Dallas/Forth Worth area, regional water utilities have acknowledged the principle that wastewater effluent is a resource rather than a disposal problem. The first of several planned indirect potable water reuse projects in the watershed was initiated in 2002, when water from an effluent-dominated section of the Trinity River was diverted into a series of engineered treatment wetlands located approximately 50 miles (80 km) south of Dallas. The river water passes through the wetlands over a period of approximately 8 days prior to being discharged into a water supply reservoir. The Tarrant Regional Water District is currently permitted to discharge an average of 56 MGD (210,000 m³/d) of Trinity River water into the Richland Chambers Reservoir. The Trinity River water accounts for up to approximately 30 percent of the water entering the reservoir (D. Andrews, Tarrant Regional Water District, personal communication, 2010). A similar project is planned for the Cedar Creek Reservoir, which is located approximately the same distance downstream of Dallas on the other side of the river starting in 2018.

Two additional surface water augmentation projects under development in the Trinity River Basin will send reclaimed water directly into water supply reservoirs. By trading reclaimed water produced in different parts of the watershed, utilities in the basin can minimize capital costs for construction of pipelines as well as the costs associated with pumping water to different elevations. For example, the North Texas Met-ropolitan Water Authority is planning to discharge 30 MGD (110,000 m³/d) of reclaimed water into the City of Dallas's Lake Lewisville Reservoir in exchange for Dallas discharging the same volume of reclaimed water into North Dallas's Ray Hubbard Reservoir (Glenn Clingenpeel, Trinity River Authority, personal communication, 2010). Trades involving reclaimed water, or trades in which the discharge of reclaimed water to a river is used to offset the use of surface water from another location are useful to water resource planners and may lead to more surface water augmentation projects in the future.

Groundwater Recharge

Approximately one-third of the potable water provided by public water supplies in the United States is from groundwater sources (Hutson et al., 2000). In locations with high water demand and low precipitation, groundwater oversubscription can result in seawater intrusion, land subsidence, and exhaustion of wells (NRC, 2008c). The depletion of aquifers can be exacerbated in urbanized areas, where impervious surfaces (e.g., pavement) reduce groundwater recharge. Groundwater also is an important means of water storage, especially in areas where the construction of new surface water reservoirs is difficult due to the lack of available land or concerns about the environmental damage caused by reservoirs. In response to concerns about groundwater overdrafts, reclaimed water can be used to recharge aquifers.

The most common ways reclaimed water is introduced into groundwater are surface spreading basins and direct injection (UNEP, 2005). Riverbank filtration with effluent-dominated surface waters also has been used as a means of augmenting groundwater supplies. Each of these approaches has different requirements with respect to pretreatment. As a result, the concentrations of contaminants in recharged waters and the extent of attenuation occurring in the subsurface will vary among the different approaches. When an aquifer is used as the environmental buffer in a potable water reuse project, the extent of contaminant attenuation will be dictated by the pretreatment process, the degree of contact with surface soils (e.g., infiltration versus injection), the hydrogeology of the aquifer, and the amount of time that the water remains in the subsurface prior to abstraction.

The composition of reclaimed water and geology of the aquifer are important considerations in groundwater recharge projects. Highly treated reclaimed water is often depleted with respect to calcium, magnesium, and other common ions. As a result, minerals in the aquifer may dissolve as the reclaimed water is recharged. Alternatively, elevated concentrations of certain ions could lead to the formation of new mineral phases in aquifers. Over time, these processes can alter the permeability of the aquifer or result in the release of toxic trace elements, such as arsenic and chromium. To prevent such changes, post-treatment processes are frequently employed before introducing reclaimed water into an aquifer. However, the long-term responses of an aquifer to reclaimed water are not always completely understood when a project is initiated.

Surface Spreading Via Recharge Basins. Surface spreading is a method of groundwater recharge in which reclaimed water moves from the land surface to the aquifer, usually through unsaturated surface soils. Generally, surface spreading is accomplished in large bermed basins with sand or permeable soil above an unconfined aquifer where reclaimed water can percolate into the subsurface (see Figure 2-5). This practice is also called soil aquifer treatment or rapid infiltration.

In terms of water quality and contaminant attenuation, the process of infiltration provides opportunities

FIGURE 2-5 Rapid infiltration basins at the Water CONSERV II facility in Orlando, Florida, which recharged 31 MGD (120,000 m³/d) of reclaimed water in 2006.
SOURCE: Alley et al. (1999).

for removal of particle-associated contaminants (e.g., pathogens, mineral particles). In addition, contaminants may be transformed by microbes as they undergo infiltration. Recharge basins are attractive to water utilities because they are relatively inexpensive to build and do not require extensive maintenance (EPA, 2004). However, compared to other means of introducing water into the subsurface (e.g., direct injection, vadose wells) recharge basins take up more space. As a result, they are often impractical in dense urban settings. Furthermore, spreading basins cannot be used in locations with shallow water tables or where local geological conditions (e.g., impermeable zones close to the land surface) limit rates of water infiltration.

In the United States, many of the pioneering efforts associated with aquifer recharge with reclaimed water have occurred in Southern California. The first major recharge project was conducted by the County Sanitation Districts of Los Angeles County and the Water Replenishment District of Southern California when they established a spreading basin in Whittier, California, in 1962. The 570-acre (220-ha) complex of spreading basins recharges a mix of reclaimed water, local stormwater runoff, and imported water to an aquifer that serves as a potable water supply for residents located as close as approximately 65 ft (20 m) downgradient of the spreading basins. On an annual basis, reclaimed water accounts for approximately 60 percent of the water recharged at this site.

Surface spreading basins also are used to recharge water from an effluent-dominated river into a potable aquifer in a community located south of Los Angeles. Since 1933, the Orange County Water District has diverted water from the nearby Santa Ana River into a series of spreading basins in the city of Anaheim. At this location, Santa Ana River water typically consists of over 90 percent wastewater effluent from the upstream communities of the Inland Empire Region during the dry season (i.e., April through October). Prior to reaching the location where the water is diverted, about half of the flow of the river passes through an engineered treatment wetland that has a hydraulic residence time of approximately 3 days (Lin et al., 2003). The remaining half of the dry season, Santa Ana River flow travels from the upstream advanced-treated wastewater effluent outfalls to the infiltration basins, in some cases with slightly less than 1-day transport time.

After percolating through the soil, the water enters an aquifer that is used as the potable supply for a well field located downgradient of the infiltration basins.

Subsurface Injection. Reclaimed water can also be directly injected into the subsurface to replenish an aquifer. Direct injection usually requires more treatment of wastewater effluent than is required for surface spreading because the injected water is pumped directly into the aquifer without the benefit of soil aquifer treatment. A high level of treatment also is needed to reduce the potential for aquifer clogging. Direct injection can occur via direct-injection wells, deep vadose zone wells that discharge water into the unsaturated zone, or aquifer storage and recovery wells, which are designed for both injection and withdrawal.

The first project in the United States that employed direct injection of reclaimed water into a potable aquifer started in Orange County, south of Los Angeles, in 1976. The Orange County Water District's Water Factory 21 facility employed a state-of-the-art treatment system for water reclamation prior to injection into a seawater intrusion barrier. Water Factory 21 injected two-thirds reclaimed water and one-third groundwater, obtained from a deep aquifer, into the barrier. The seawater barrier was a potable water reuse project because the water in the seawater intrusion barrier also flowed toward nearby potable water supply wells. For example, water supply wells located approximately 0.3 mile (500 m) from a seawater intrusion barrier in Orange County, California, exhibit chloride concentrations equal to those of the water injected into the barrier (Fujita et al., 1996), indicating that most of the water delivered by these wells originated in the injection well. Subsequent to the success of Water Factory 21, the Orange County Water District developed the new Groundwater Replenishment System, which expanded the utility's potable reuse capacity from 16 MGD (61,000 m³/d) to 70 MGD (260,000 m³/d) in 2008 (see Box 2-11).

Other projects that use a combination of advanced treatment processes similar to those practiced at Orange County's Groundwater Replenishment System have been built in Southern California and Arizona. The West Basin Water District's Recycling Plant was built near Los Angeles Airport in 1993. The project initially used deep wells to inject a mixture of equal volumes of reclaimed water and water imported from

---

BOX 2-11
Orange County Water District, California

Groundwater withdrawals make up about 70 percent of the water supply in the Orange County Water District's service area, with the remaining demand being met by imported water from the Colorado River and Northern California. Historically, imported water from the Colorado River and Northern California and water from the Santa Ana River have been the source waters for groundwater recharge in Orange County. Seawater intrusion has been a problem since the 1930s as a consequence of groundwater basin overdraft. Injection of reclaimed water from an advanced wastewater treatment facility (Water Factory 21) to form a seawater intrusion barrier in the Talbert Gap area of the groundwater basin began in 1976. The project served the dual purpose of seawater intrusion barrier and potable supply augmentation. Agency leaders acknowledged both of these purposes and did not encounter public opposition to the potable augmentation.

A recharge project called the Groundwater Replenishment (GWR) System was conceived in the 1990s to replace Water Factory 21 and provide additional water to recharge the Orange County Groundwater Basin. The GWR System consists of three major components: the Advanced Water Purification Facility (AWPF); the Talbert Gap Seawater Intrusion Barrier; and the Miller and Kraemer spreading basins. The AWPF began producing reclaimed water in January 2008 for injection at the Talbert Gap and spreading at Kraemer and Miller basins.

The source water for the 70-MGD (260,000-m³/d) advanced treatment facility is secondary effluent from the adjacent Orange County Sanitation District Plant No. 1. The AWPF provides further treatment by microfiltration, reverse osmosis, and advanced oxidation. The treated water is stabilized by decarbonation and lime addition to raise the pH and add hardness and alkalinity to make the water less corrosive and more stable.

In 2009, production of reclaimed water averaged 54 MGD (200,000 m³/d). Plans are under way to increase the capacity of the GWR System in phases, with an ultimate capacity of 130 MGD (490,000 m³/d). Half of the water produced by the advanced treatment plant is injected into the Talbert Gap Seawater Intrusion Barrier and half is pumped approximately 13 miles (21 km) to the Kraemer and Miller basins in Anaheim, which are deep spreading basins in the Orange County Forebay area. The nearest downgradient extraction well is more than 5,200 ft (1,580 m) from the percolation basins, and the retention time underground prior to extraction in excess of 6 months.

SOURCES: Crook (2007); Alan Plummer Associates (2010).

the Colorado River into the West Coast Barrier (see Figure 2-4). Projects in Scottsdale, Arizona, Los Angeles, and Denver were initiated in 1999, 2005, and 2009, respectively. The Scottsdale and Los Angeles projects employ reverse osmosis prior to groundwater injection whereas the Denver project applies reverse osmosis to the abstracted groundwater.

In light of the trend to employ reverse osmosis prior to groundwater injection, it is noteworthy that the groundwater recharge project operated by El Paso Water Utilities since 1985 employs activated carbon and ozonation as barriers against waterborne pathogens and chemical contaminants in a potable reuse project. By avoiding the use of reverse osmosis, the El Paso facility does not produce a brine waste that requires disposal. The reclaimed water produced by the advanced treatment plant is injected into the aquifer, where it spends approximately 6 years underground before abstraction. According to estimates from the operators of the system, reclaimed water accounts for approximately 1 percent of the water abstracted in the nearest downgradient wells (Ed Archuleta, El Paso Water Utilities Public Service Board, personal communication, 2010).

Given the rapid growth in population in communities that do not have access to ocean outfalls for brine disposal, projects such as the system in El Paso may become more common in the near future. For example, the 190-MGD (720,000-m^3/d) potable reuse project initiated in Aurora, Colorado, near Denver (see Table 2-3) in 2010 employs advanced treatment after groundwater recharge and extraction, without reverse osmosis. In situations where salt removal is not required, similar projects may offer distinct advantages over reverse osmosis followed by direct injection.

Riverbank Filtration

Riverbank filtration is a process that has been used to treat surface waters that have been subject to contamination from upstream sources. During riverbank filtration, aquifer sediments act as a natural filter removing contaminants as river water recharges groundwater. The hydraulic gradient driving the flow of water through the riverbank is often induced by pumping nearby water supply wells (Hiscock and Grischek, 2002; Kim and Corpcioglu, 2002). Because water follows different flow paths as it moves into ex-

traction wells, the peak concentrations of contaminants sometimes encountered in water supplied from rivers or lakes are moderated. In addition, physical and biological processes in the subsurface result in decreases in the concentrations of many contaminants as water flows toward the extraction wells (Sontheimer and Nissing, 1977; Sontheimer, 1980; Sontheimer, 1991; Kühn and Müller, 2000; Wang et al., 2002; Schmidt et al., 2004; Hoppe-Jones et al., 2010).

Riverbank filtration has been used for public and industrial water supply in Europe (Kühn and Müller, 2000; Grischek et al., 2002; Ray et al., 2002a,b) for more than a century. Riverbank filtration has been practiced to a lesser extent in the United States for more than 50 years in communities along the Ohio, Wabash, and Missouri Rivers (Weiss et al., 2002). In Europe, it provides 50 percent of potable supplies in the Slovak Republic, 45 percent in Hungary, 16 percent in Germany, and 5 percent in The Netherlands (Hiscock and Grischek, 2002). For example, Berlin obtains approximately 75 percent of its drinking water supply from riverbank filtration of effluent-dominated rivers. Düsseldorf has been using riverbank filtration of an effluent-impacted section of the Rhine River water as a potable water supply since 1870.

Site-specific factors can affect the performance of riverbank filtration systems (see Chapter 4 for additional discussion of treatment performance). As a result, riverbank filtration is mainly practiced in locations with the appropriate geological characteristics (e.g., high-permeability sediments located adjacent to a river). In addition, riverbed characteristics and operational conditions (e.g., well type, pumping rates, travel time in the subsurface) are important factors affecting water yields and water quality. Although some of these factors can be influenced by engineering design, others depend on the individual site and local hydrogeological conditions.

In the context of water reclamation, riverbank filtration offers a means of improving the quality of effluent-dominated surface waters (e.g., systems in which de facto reuse is practiced). The process also has the potential to serve as a means of attenuating contaminants in planned potable reclamation systems. However, additional research is needed to develop a better understanding of factors affecting the performance of riverbank filtration systems.

Recent Trends with Respect to Environmental Buffers

As discussed previously, environmental buffers were important features of potable water reuse projects constructed in the United States between 1960 and 2009. Over the five decades, treatment technologies have improved and their costs have decreased. In addition, the continued success of an environmental-buffer-free potable reuse project in Windhoek Namibia (see Box 2-12) has provided evidence that environmental buffers are not always necessary in potable reuse projects. As utilities have become more confident in their ability to meet potable water standards and guidelines, potable reuse projects have been proposed, designed, and in several cases built in the United States without environmental buffers.

The increasing interest of utilities in operating potable reuse projects without environmental buffers is driven by a number of factors, including water rights, lack of suitable buffers near the locations where reclaimed water is produced, potential for contamination of the reclaimed water when it is released into the environmental buffer, and costs associated with maintenance, operation, and monitoring of environmental buffers. For example, recent controversies about water rights in Lake Lanier, Georgia, could jeopardize the Gwinett County Water Authority's rights to the reclaimed water that it currently discharges to the lake. As a result, it is considering the possibility of piping the reclaimed water directly to a blending pond that is not connected to the reservoir, thereby allowing them to maintain ownership of the water. Because the blending pond would be a manmade structure that does not receive water from other sources, this potable reuse project would not include an environmental buffer.

Another example of this trend is the potable reuse project being built by the Colorado River Municipal Water District in Texas in which a series of water reclamation plants will return reclaimed water directly to its drinking water reservoir (Sloan et al., 2010). The first of these projects, which is scheduled to begin operating in 2012, will deliver 2.5 MGD (9,500 m³/d) of reclaimed water to its surface water reservoir through a transmission canal. In addition to decreasing the water district's reliance on the Colorado River, the reuse of water avoids the need to pump water up to the reservoir from water sources lower in the watershed. As a result, after including energy used by the advanced treatment plant, energy consumption for the reclamation project is approximately equal to that of other available water sources.

While the surface water reservoir employed by the Colorado River Water District or the blending pond used by the Gwinett County Water Authority have characteristics of environmental buffers, a recently built project in the community of Cloudcroft, New Mexico, in which 0.1 MGD (380 m³/d) of reclaimed water is blended with local spring water in a covered reservoir does not have many attributes normally associated with environmental buffers (see Box 2-13). This project was approved by the local community and underwent review without a requirement for an environmental buffer.

The characteristics of an environmental buffer affect the impacts on public acceptance and contaminant attenuation. For example, a wetland populated with healthy plants, birds, and fish is likely to be more acceptable to the public than a sandy-bottomed river with steeply sloped concrete flood control levees. Likewise, percolation of reclaimed water through 16 ft (5 meters) of soil followed by mixing with local groundwater and a year in the subsurface is more likely to result in contaminant attenuation than direct injection with no dilution followed by days or weeks in an aquifer consisting of fractured bedrock. Environmental buffers used in IPR projects fall along a continuum and each should be judged within the context of the entire water system. Manufactured water storage structures, such as blending ponds or artificial aquifers, employed in direct potable water reuse systems, can provide many of the same benefits as natural environmental buffers, both in terms of public perception and contaminant attenuation.

The direct connection of an advanced water reclamation plant to a water distribution plant, without an intermediate water storage structure for blending with water from other sources, would provide none of the aforementioned benefits related to public acceptance or contaminant attenuation. As a result, such structures are unlikely to be built in the near term. After the nation has more experience with potable reuse systems that employ blending structures, decisions can be made about the merits of direct "pipe-to-pipe" potable reuse systems (see also Chapter 5 discussions on quality assurance).

BOX 2-12
Windhoek, Namibia, Potable Reuse System

The Windhoek, Namibia, advanced wastewater treatment plant returns reclaimed water directly to the city's drinking water system. The average rainfall is 14.4 inches (37 cm) while the annual evaporation is 136 inches (345 cm), and this city of 250,000 people relies on three surface reservoirs for 70 percent of its water supply. First implemented in 1968 with an initial flow of 1.3 MGD (4,900 m³/d; Haarhof and Van der Merwe, 1996), the Goreangab water reclamation plant, which receives secondary effluent from the Gammans wastewater treatment plant, has been upgraded through the years to its current capacity of 5.5 MGD (21,000 m³/d). Industrial and potentially toxic wastewater is diverted from the wastewater entering the plant. There have been four distinct treatment process configurations since 1968. The current treatment train was placed in operation in 2002 and includes the following processes:

- Primary sedimentation
- Activated sludge secondary treatment with nutrient removal
- Maturation ponds (4 days)
- Powdered activated carbon, acid, polymers (used when required)
- Preozonation
- Coagulation/flocculation with ferric chloride ($FeCl_3$)
- Dissolved air flotation
- Rapid sand/anthracite filtration preceded by potassium permanganate ($KMnO_4$) and sodium hydroxide (NaOH) addition
- Ozonation preceded by hydrogen peroxide (H_2O_2) addition
- Biological and granular activated carbon
- Ultrafiltration (0.035-micrometer [μm] pore size)
- Chlorination
- Stabilization with NaOH
- Blending prior to distribution

Blending occurs at two locations. The first blending takes place at the Goreangab water treatment plant, where reclaimed water is blended with conventionally treated surface water. This mixture is then blended with treated water from other sources prior to pumping to the distribution system.

Prior to recent upgrades in 1991 the percentage of reclaimed water in the drinking water averaged 4 percent (Odendaal et al., 1998). Following the plant upgrades, reclaimed water represents up to 35 percent of the drinking water supply during normal periods, and as much as 50 percent when water supplies are limited (Lahnsteiner and Lempert, 2005; du Pisani, 2005). Extensive microbial and chemical monitoring is performed on the product water, with continuous monitoring of several constituents.

Both *in vitro* and *in vivo* toxicological testing has been conducted on product water from the Goreangab treatment plant (such as Ames test, urease enzyme activity, bacterial growth inhibition, water flea lethality, and fish biomonitoring). An epidemiological study (1976 to 1983) was also conducted, which found no relationships between cases of diarrheal diseases, jaundice, or deaths to drinking water source (Isaacson and Sayed, 1988; Odendaal et al., 1998; Law, 2003). However, a prior NRC committee concluded that because of limitations in the Windhoek epidemiological studies and its "unique environment and demographics, these results cannot be extrapolated to other populations in industrialized countries" (NRC, 1998). There was some initial public opposition to the Windhoek project, but over time, opposition has faded, and no public opposition to the project has emerged in recent years.

EXTENT OF WATER REUSE

The current extent of reuse is summarized in the following section, focusing on the United States, with additional information on other countries with large reuse initiatives. Available reuse data, however, are sparse, and most of the figures cited below should be considered estimates.

United States

Statistics on the extent of water reuse in the United States remain somewhat limited. Every 5 years, the USGS releases data on U.S. water use, and for 1995, the last year for which reclaimed water use data were included, 1,057 MGD (4 million m³/d) of wastewater was reused. This amount represented approximately

BOX 2-13
Potable Reuse in Cloudcroft, New Mexico

The village of Cloudcroft, New Mexico, is a mountain community at 8,600-ft (220-m) elevation with a permanent population of 750. As a winter resort community, population can increase during holidays and weekends to more than 2,000 with a peak demand of 0.36 MGD (1,400 m³/d). Recent drought conditions had resulted in a reduction of spring flows and groundwater tables. Because of limited local supplies and Cloudcroft's elevation, which limit use of water sources from outside the community, the village decided to reuse their local wastewater to augment their drinking water supply. In 2009, an advanced water treatment plant with a capacity of 0.10 MGD (380 m³/d) was established to treat the community's wastewater and blend it with natural spring and well water (up to 50 percent wastewater) prior to consumption.

The wastewater generated in the community is treated by a membrane bioreactor. After disinfection using chloramination, the filtered effluent is treated by reverse osmosis followed by advanced oxidation (ultraviolet radiation/hydrogen peroxide). The ultrafiltration and reverse osmosis units are located away from the membrane bioreactor at a lower elevation, allowing gravity feed to the reverse osmosis units. The plant effluent is subsequently blended with other source water from local springs and wells in a covered reservoir that provides a retention time of 40 to 60 days. The blended water is then treated by ultrafiltration followed by ultraviolet radiation and granular activated carbon prior to final disinfection. The reverse osmosis concentrate with a TDS concentration of approximately 2,000 mg/L is currently blended with membrane bioreactor filtrate and held in storage ponds for use in snow making, irrigation of the ski area, and dust control. The operations and maintenance cost for the production of this water was $2.40/ kgal ($0.63/m³) during its first year of operation.

The community provided input through public meetings, and the state regulator has approved the project.

SOURCE: Livingston (2008).

2 percent of wastewater discharged and less than 0.3 percent of total water use in 1995 (Solley et al., 1998).[5] In the 2004 EPA Guidelines for Water Reuse (EPA, 2004), total water reuse in the United States was estimated at 1,690 MGD (6.4 million m³/d), and they

---

[5] Solley et al. (1998) reported that in the United States, 155 × 10⁶ m³/d of treated water were discharged in 1995, and total water use was approximately 1.5 × 10⁹ m³/d.

estimated that water reuse was growing at a rate of 15 percent per year.

As of 2002, EPA estimated that Florida reused the largest quantities of reclaimed water, followed by California, Texas, and Arizona. At that time, these four states accounted for the majority of the nation's water reuse, although EPA reported that at least 27 states had water reclamation facilities as of 2004, with growing programs in Nevada, Colorado, Washington, Virginia, and Georgia (EPA, 2004). Three of the four states with the largest reclaimed water use are located in the arid southwest where population growth and climate variability have created recent water supply challenges. Water reuse in these states has become commonplace as a means to expand the water supply portfolio and provide an additional drought-resistant supply. Florida originally launched its water reuse program to address nutrient pollution concerns in its streams, lakes, and estuaries, but increasingly, new projects are being considered for their water supply benefits as well.

The end uses of reclaimed water are not well documented on a national scale. The WateReuse Foundation is working on a national database of reuse facilities that could help address this data gap, although as of early 2011, the database was still being refined. Some states have additional inventory data, described below, that reflect the varied uses of reclaimed water across different states.

Florida

The state of Florida conducts a comprehensive inventory of water reuse each year and reports that approximately 659 MGD (2.5 million m³/day) of wastewater was reused for beneficial purposes in 2010 (FDEP, 2011). Over half of Florida's reclaimed water is used for public access irrigation, with additional uses in agricultural irrigation, groundwater recharge, and industrial applications (Figure 2-6). In Florida, groundwater recharge consists largely of rapid infiltration basins and absorption field systems that are not specifically designated as indirect potable reuse projects. In several Florida counties, nonpotable reuse accounts for 30–60 percent of the freshwater supplied for public water supply, industry, agriculture, and power generation (FDEP, 2006; Marella, 2009).

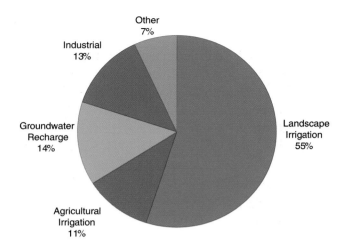

FIGURE 2-6 Water reuse in the state of Florida as of 2010, by flow volume and by application.
SOURCE: Data from FDEP (2011).

California

The California State Water Resources Control Board reported 646 MGD (2.44 million m³/day) of water reuse in California in 2009.[6] California's end uses, depicted in Figure 2-7, appear more diverse than Florida's, including recreational impoundments and geothermal energy. In general, agricultural irrigation makes up a larger percentage of water reuse in California compared with Florida, while landscape irrigation and industrial reuse represent smaller portions of the overall portfolio. Both states have comparable extents of reuse in the area of groundwater recharge (including seawater intrusion barriers in California). Nevertheless, the California data include a large percentage (20 percent) of unclassified ("other") reuse applications that may affect these comparisons.

Texas

A recent report for the Texas Water Development Board estimated 320 MGD (1.2 million m³/day) of water reuse in Texas in 2010 (Alan Plummer Associates, 2010). No additional details are provided on how this reclaimed water is used.

___
[6] See http://www.waterboards.ca.gov/water_issues/programs/grants_loans/water_recycling/munirec.shtml.

International Reuse

Crook et al. (2005) and Jiménez and Asano (2008) recently reviewed international reuse practices. According to their findings, major water reuse facilities are in place in at least 43 countries around the world, including Egypt, Spain, Syria, Israel, and Singapore. Based on the statistics given by Jiménez and Asano, approximately 13 BGD (50 million m³/d) of wastewater are reused worldwide. The authors identified 47 countries that engaged in reuse. Of these, 12 engaged in reuse of untreated municipal effluent, 7 engaged in the reuse of both treated and untreated effluent, and 34 reuse wastewater only after treatment. Of the total volume, 7.7 BGD (29 million m³/d) or 58 percent was untreated (raw) sewage used for irrigation, mostly in China and Mexico (see Figure 2-8).

Jiménez and Asano (2008) reported that 5.5 BGD (21 million m³/d) of treated municipal wastewater was reused globally in 43 countries. The United States was first among them in total volume of water reused (see Figure 2-9). Although the United States reused the largest volume of treated wastewater, per capita water

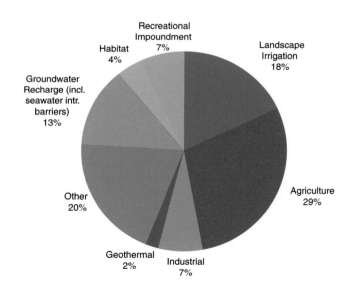

FIGURE 2-7 Water reuse in the state of California as of 2009, by flow volume and application.
SOURCE: Data from California Environmental Protection Agency http://www.waterboards.ca.gov/water_issues/programs/grants_loans/water_recycling/munirec.shtml.

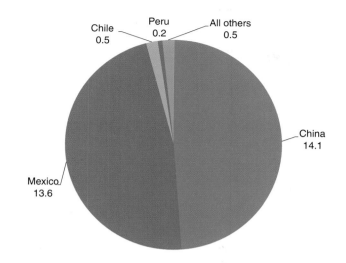

FIGURE 2-8 Countries with the most reuse of untreated wastewater in millions of cubic meters per day.
SOURCE: Data from Jiménez and Asano (2008).

Although statistics on international reuse practice provide insight into global trends, it should be recognized that local history, geography, and cultural influences have played an important role in the types of reuse practices pursued in different countries. To illustrate these differences, Israel, Australia, and Singapore are considered here—three leading practitioners of reuse where differences in climate, population density, water resources, and history have led to different outcomes with respect to water reuse. Reuse practices in other developed countries follow similar patterns. However, the acute need for water in these three countries has led them to embrace innovative water resource management approaches that are particularly relevant to the consideration of reuse in the United States.

reuse in the United States ranked 13th globally. In at least five countries—Kuwait, Israel, Qatar, Singapore, and Cyprus—water reuse represented more than 10 percent of the nation's total water extraction (Jiménez and Asano, 2008).

Israel

Since the time of its founding in 1948, Israel has relied upon agricultural water reuse as part of its water supply portfolio. Initially, wastewater from urban areas was used directly for irrigation. In recognition of potential health risks associated with this practice, Israel's

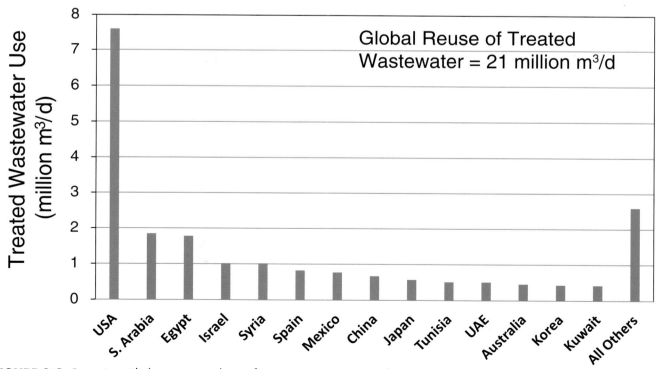

FIGURE 2-9 Countries with the greatest volume of water reuse using treated wastewater.
SOURCE: Data from Jiménez and Asano (2008).

nonpotable reuse practices were upgraded through the construction of wastewater treatment plants and groundwater recharge basins near agricultural areas. Today, approximately 75 percent of Israel's wastewater is reused, with almost all of it going for agricultural irrigation. This outcome was likely affected by several factors. First, Israel's arid climate and sparse water resources have made the public aware of the need to use water efficiently. Second, the relatively high population density and proximity of the country's cities to its farms makes it efficient to reuse municipal wastewater for agricultural irrigation. Finally, Israel's concerns about food security and uncertainty associated with its water resources have made agricultural reuse a national priority (Shaviv, 2009).

Australia

Like Israelis, Australians are highly aware of their nation's limited water resources. However, Australia's population density is much lower, and much of its agricultural activity occurs far from urban centers (e.g., most of the farming in the Murray-Darling Basin takes place hundreds of miles from coastal cities). As a result, agricultural reuse has not played a major role in the country's water reuse planning process. In contrast, nonpotable reuse projects, such as landscape irrigation and industrial reuse, are quite popular, as epitomized by Sydney's high-profile reuse project at the facility built as part of the Olympic Park for the games in 2000. Currently, approximately 10 percent of the water used in Australia's mainland capital cities is reused, mainly for landscaping and industrial applications. Until recently, potable water reuse was not considered a viable option by most water managers in Australia, but the extreme drought that lasted from 2003 to 2009 coupled with high rates of urban population growth forced several of Australia's biggest cities to reconsider (Radcliffe, 2010). At the height of the drought, Brisbane (C. Rodriguez et al., 2009), Canberra (Radcliffe, 2008), and Perth (C. Rodriguez et al., 2009) were all considering potable water reuse projects. A distinctive aspect of the planned water reuse projects in Brisbane and Canberra was blending of reclaimed water directly in drinking water reservoirs—a practice that deviated from the established soil aquifer treatment and groundwater injection projects that had been pioneered in the southwestern

United States. After the drought ended, the projects in Brisbane and Canberra were put on hold.

Singapore

The high population density, near absence of agricultural water demand, and heavy reliance on water imported from a neighboring country has led to a different outcome for water reuse in Singapore. In particular, early recognition that the country's population growth would soon outstrip its local water resources led Singapore to pursue an approach that they refer to as the "four taps": (1) local runoff, (2) imported water from Malaysia, (3) desalinated seawater, and (4) reclaimed water. As a result of its frequent rain and high population density, there is little irrigation water demand for reclaimed water. Instead, the country's water reuse program has focused on industrial and potable reuse. Given Singapore's access to seawater for cooling purposes and its growing high-tech industry, the Public Utilities Board recognized the need for high-quality reclaimed water. The resulting advanced water treatment system (see Box 2-14) delivers reclaimed water to industrial users and local reservoirs. As was the case in Brisbane and Canberra, groundwater recharge or aquifer storage and recovery were not viable options because of Singapore's local geology and geography.

CONCLUSIONS AND RECOMMENDATIONS

Water reuse is a common practice in the United States with numerous approaches available for reusing wastewater effluent to provide water for industry, agriculture, and potable supplies. However, there are considerable differences among the approaches employed for water reuse with respect to costs, public acceptance, and potential for meeting the nation's future water needs.

Water reclamation for nonpotable applications is well established, with system designs and treatment technologies that are generally accepted by communities, practitioners, and regulatory authorities. Nonpotable reuse currently accounts for a small part of the nation's total water use, but in a few communities (e.g., several Florida cities), nonpotable water reuse accounts for a substantial portion of total water use. New developments and growing communities provide op-

BOX 2-14
Singapore Public Utilities Board NEWater Project, Republic of Singapore

The Republic of Singapore has a population of about 5 million people. Although rainfall averages 98 inches (250 cm) per year, Singapore has limited natural water resources because of its small size of approximately 270 square miles (700 km²). Reclaimed water (referred to by the local utility as NEWater; see figure below) is an important element of Singapore's water supply portfolio.

Currently, there are five NEWater treatment plants in operation, all of which include nearly identical treatment processes. Feedwater to the treatment plants is activated sludge secondary effluent. The advanced water treatment processes included microscreening (0.3-mm screens), microfiltration (0.2-mm nominal pore size) or ultrafiltration, reverse osmosis, and ultraviolet disinfection. Chlorine is added before and after microfiltration to control membrane biofouling. The reclaimed water is either supplied directly to industry for nonpotable uses or discharged to surface water reservoirs, where the water is blended with captured rainwater and imported raw water. The blended water is subsequently treated in a conventional water treatment plant of coagulation, flocculation, sand filters, ozonation, and disinfection prior to distribution as potable water.

The NEWater factories all produce high-quality product water with turbidity less than 0.5 nephelometric turbidity units; TDS less than 50 mg/L; and total organic carbon less than 0.5 mg/L. The water meets all Environmental Protection Agency and World Health Organization drinking water standards and guidelines. Additional constituents monitored include many organic compounds, pesticides, herbicides, endocrine-disrupting compounds, pharmaceuticals, and unregulated compounds. None of these constituents have been found in the treated water at health-significant levels.

The NEWater facilities at the Bedok and Kranji went into service in 2003 and have since been expanded to their current capacities of 18 MGD and 17 MGD (68,000 and 64,000 m³/d), respectively. A third NEWater factory at the Seletar Water Reclamation Plant was placed in service in 2004 and has a capacity of 5 MGD (19,000 m³/d). The fourth NEWwater factory (Ulu Pandan) has a capacity of 32 MGD (121,000 m³/d) and went into operation in 2007. A fifth facility, the Changi NEWater Factory, is being commissioned in two stages: the first 15 MGD (57,000 m³/d) phase was commissioned in 2009, with an additional 35 MGD (130,000 m³/d) phase to be commissioned in 2010. Once completed, these five plants will have a combined capacity of 122 MGD (462,000 m³/d).

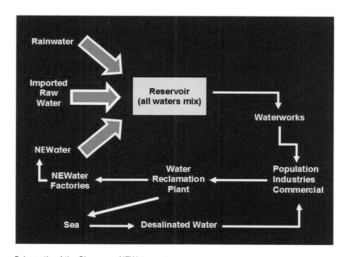

Schematic of the Singapore NEWater system.
SOURCE: Ong and Seah, 2003.

Most of the reclaimed water from the NEWater Factories is supplied directly to industries. These industries include wafer fabrication, electronics and power generation for process use, as well as commercial and institutional complexes for air-conditioning cooling purposes. Less than 10 MGD (38,000 m³/d) of NEWater currently is used for potable reuse via discharge to raw water reservoirs, accounting for slightly more than 2 percent of the total raw water supply in the reservoirs. However, the contribution of NEWater to the potable water supply is expected to increase in the coming decades.

The capital costs for all of the NEWater factories averaged about $6.03/kgal per year capacity (or $1.59/m³ per year). Annual operation and maintenance costs for the water are about $0.98/kgal ($0.26/m³) produced. The Public Utilities Board charges industries and others $2.68/kgal ($0.71/m³) for NEWater on a full cost recovery approach. This includes the capital cost, production cost, and transmission and distribution cost.

SOURCE: A. Conroy, Singapore Public Utilities Board, personal communication, 2010.

portunities to expand nonpotable water reuse because it is more cost-effective to install separate nonpotable water distribution systems at the same time the primary drinking water distribution system is installed. In existing communities nonpotable water reuse is often restricted by the high costs associated with constructing the distribution system and retrofitting existing plumbing (see also Chapter 9).

The use of reclaimed water to augment potable water supplies has significant potential for helping to meet the nation's future needs, but potable water reuse projects only account for a relatively small fraction of the volume of water currently being reused. However, potable reuse becomes more significant to the nation's current water supply portfolio if de facto or unplanned water reuse is included. **The de facto reuse of wastewater effluent as a water supply is common in many of the nation's water systems, with some drinking water treatment plants using waters from which a large fraction originated as wastewater effluent from upstream communities, especially under low-flow conditions.**

An analysis of the extent of de facto potable water reuse should be conducted to quantify the number of people currently exposed to wastewater contaminants and their likely concentrations. Despite the growing importance of de facto reuse, a systematic analysis of the extent of effluent contributions to potable water supplies has not been made in the United States for over 30 years. Available tools and data sources maintained by federal agencies would enable this to be done with better precision, and such an analysis would help water resource planners and public

health agencies understand the extent and importance of de facto water reuse. Furthermore, an analysis of de facto potable reuse may spur the additional development of contaminant prediction tools and improved site-specific monitoring programs for the betterment of public health. USGS and EPA have the necessary data and expertise to conduct this analysis on large watersheds that serve as water supplies for multiple states. For smaller watersheds or watersheds with existing monitoring networks, state and local agencies may have additional data to contribute to these analyses.

Environmental buffers can play an important role in improving water quality and ensuring public acceptance of potable water reuse projects, but the historical distinction between direct and indirect water reuse is not meaningful to the assessment of the quality of water delivered to consumers. Potable reuse projects built in the United States between 1960 and 2010 employed environmental buffers in response to concerns about public health risks and the possibility of adverse public reaction to potable water reuse. In the last few years, a potable reuse project was built and another is being built without environmental buffers, and the trend toward operating potable reuse projects without buffers is likely to continue in the future. An environmental buffer should be considered as one of several design features that can be used to ensure safe and reliable operation of potable reuse systems. As a result, they need to be designed, evaluated, and monitored like other elements of the water treatment and delivery system. See Chapters 4 and 5 for additional details on the treatment effectiveness of environmental buffers and their role in quality assurance.

3

Water Quality

The wastes discharged into municipal wastewater collection systems include a wide range of biological, inorganic, and organic constituents. Some of these constituents can be harmful to persons and/or ecosystems depending on concentration and duration of exposure (see also Chapter 6 for a discussion of risk in the context of hazards and exposure types). Some are essential nutrients at low concentrations (e.g., certain trace elements), but may become hazardous at higher concentrations. In this chapter the committee briefly describes the key water quality constituents of concern when municipal wastewater is reused or when treated municipal wastewater is discharged to a watercourse that is later used as a source of municipal water supply. Because water reuse involves multiple potential applications (see Chapter 2, Table 2-2), the constituents of concern depend upon the final use of the water. For instance, some constituents in drinking water that may affect human health may not be of concern in certain landscape irrigation or industrial applications where risk to human health from incidental consumption is negligible. Other constituents may have an adverse impact on aquatic species but no adverse impact on human health at the same concentration. It is also important to remember that the occurrence and concentration of these chemicals and microorganisms are likely to vary from one location to another, with the treatment methods applied, and according to post-reclamation storage and conveyance practice. Depending on the reuse application, these constituents may need to be addressed to differing degrees in water reuse system designs (see Chapters 4 and 5), considering that individual contami-

nants pose different hazards in one context than they do in another and their associated risks depend on the dose and paths of exposure (see Chapter 6). Although the committee provides examples below for a diversity of potential pathogens and chemical contaminants in reclaimed water, it is important to keep in mind that there are often other sources of exposure (e.g., food, distribution system failures, household products) that are not discussed here.

PATHOGENS

Wastewater contains many microorganisms but only a subportion of the organisms are potential human health hazards, notably enteric pathogens. Classes of microbes that can cause infection in humans include helminths (wormlike parasites), parasitic protozoa, bacteria, and viruses. Some microorganisms are obligate pathogens (i.e., they must cause disease to be transferred from host to host), whereas others are opportunistic pathogens, which may or may not cause disease. In the United States, the enteric protozoa *Cryptosporidium* and *Giardia*, the enteric bacteria *Salmonella*, *Shigella*, and toxigenic *Escherichia coli* O157:H7, and the enteric viruses enteroviruses and norovirus are the most frequently documented waterborne enteric pathogens (Craun et al., 2006). They cause acute gastrointestinal illness and have the potential to create large-scale epidemics. Table 3-1 lists the microbial agents that have been associated with waterborne disease outbreaks and also includes some agents in wastewater thought to pose significant risk.

TABLE 3-1 Microbial Agents of Known Hazard Via Water Exposures

Agent	Associated Illnesses
Viruses	
• Noroviruses	Gastroenteritis
• Adenoviruses	Conjunctivitis, gastroenteritis, respiratory disease, pharyngoconjunctival fever
• Coxsackieviruses	Meningitis, pharyngitis, conjunctivitis, encephalitis
• Echoviruses	Gastroenteritis, encephalitis, meningitis
• Hepatitis A virus	Hepatitis
• Astroviruses	Gastroenteritis
Bacteria	
• *E. coli* O157	Hemorrhagic diarrhea
• *Campylobacter jejuni*	Campylobacteriosis
• *Salmonella*	Salmonellosis
• *Shigella*	Shigellosis
• *Vibrio*	Gastroenteritis, wound infection
• *Legionella*	Legionellosis
Protozoa	
• *Cryptosporidium*	Cryptosporidiosis
• *Giardia*	Giardiasis
• *Microsporidia*	Microsporidiosis

NOTE: These agents are known to be present in treated wastewaters or surface water and therefore are considered to be potentially present in waters used for the production of reclaimed water.
SOURCE : Asano et al. (2007).

The occurrence and concentrations of microbial pathogens in reclaimed water depend on the health of the tributary population and the applied wastewater treatment processes (see Table 3-2). Primary and secondary treatment (see Chapter 4) attenuate microbial pathogens but do not eliminate them. For pathogenic bacteria, viruses, and protozoa that can cause acute diseases with even a single exposure, additional physiochemical treatment processes (discussed in Chapter 4) may be required to achieve acceptable levels of removal or inactivation, depending on the beneficial use.

Helminths

Often known as parasitic worms, helminths pose significant health problems in developing countries where wastewater reuse is practiced in agriculture using raw sewage or primary effluents (Shuval et al., 1986). The World Health Organization (WHO) has pointed to the need to study the transmission of intestinal parasites, particularly nematodes, in children living in areas where untreated wastewater is used for vegetable irrigation (WHO, 1989). Human exposures to helminths are mainly through ingestion of helminth eggs in food or water contaminated with untreated wastewater or sewage-derived sludge, and these exposures can cause acute gastrointestinal illness. There are over 100 different types of helminths that can be present in sewage, although the number of helminth eggs in untreated wastewater is typically much higher in developing countries than in developed countries. The concentration of helminth eggs can range from <1 to >1,000 per 0.3 gallon (1.0 L) of sewage, depending on the source of sewage (Jiménez, 2007; Ben Ayed et al., 2009). Helminth eggs can be largely removed through

TABLE 3-2 Reported Ranges of Reclaimed Water Quality for Key Water Quality Parameters After Different Degrees of Treatment

Constituent	Units	Untreated Wastewater	Range of Effluent Quality After Indicated Treatment				
			Conventional Activated Sludge (CAS)	CAS with Filtration	CAS with Biological Nutrient Removal (BNR)	CAS with BNR and Filtration	Membrane Bioreactor (MBR)
Total suspended solids (TSS)	mg/L	120-400	5-25	2-8	5-20	1-4	<2
Total organic carbon (TOC)	mg-C/L	80-260	10-40	8-30	8-20	1-5	0.5-5
Total nitrogen	mg-N/L	20-70	15-35	15-35	3-8	2-5	<10[a]
Total phosphorus	mg-P/L	4-12	4-10	4-8	1-2	≤2	<0.3[b]-5
Turbidity	NTU	—	2-15	0.5-4	2-8	0.3-2	≤1
Volatile organic compounds (VOCs)	µg/L	<100->400	10-40	10-40	10-20	10-20	10-20
Trace constituents	µg/L	10-50	4-40	5-30	5-30	5-30	0.5-20
Total coliforms	No./100 mL	10^6-10^9	10^4-10^5	10^3-10^5	10^4-10^5	10^4-10^5	<100
Protozoan cysts and oocysts	No./100 mL	10-10^4	10-10^2	0-10	0-10	0-1	0-1
Viruses	PFU/100 mL	10-10^4	10-10^3	10-10^3	10^1-10^3	10-10^3	1-10^3

NOTE: None of the treatments in the table include disinfection.
[a]With anoxic zone.
[b]With coagulant.

SOURCE: Asano et al. (2007).

secondary treatment supplemented by finishing ponds or filtration and disinfection (Blumenthal et al., 2000).

Protozoa

Protozoa are single-celled eukaryotes that are heterotrophic and generally larger in size than bacteria. Some protozoa are mobile using flagella, cilia, or pseudopods, whereas others are essentially immobile. Malaria, probably the best-known disease caused by protozoa, is caused by the genus *Plasmodium*. In U.S. water systems, *Giardia lamblia*, *Cryptosporidium parvum*, and *C. hominis* have been associated with gastrointestinal disease outbreaks through contaminated water. In 1993, an outbreak of cryptosporidiosis caused an estimated 400,000 illnesses and more than 50 deaths through contaminated drinking water in Milwaukee, Wisconsin (Mac Kenzie et al., 1994; Hoxie et al., 1997). Part of the protozoan life cycle often involves spores, cysts, or oocysts, which can be highly resistant to chlorine. *Cryptosporidium* oocysts and *Giardia* cysts of human origin are frequently detected in secondary wastewater effluent (Bitton, 2005), and these may still persist in disinfected effluent after granular media or membrane filtration (e.g., Rose et al., 1996). Thus, in potable reuse applications, additional treatment processes (see Chapter 4) are needed to reduce the risk of infection from Cryptosporidium and Giardia.

Bacteria

Bacteria are single celled prokaryotes and are ubiquitous in the environment. However, domestic wastewaters contain many pathogenic bacteria that are shed by the human population in the sewershed. Particularly important are pathogenic bacteria that cause gastroenteritis and are transmitted by fecal-oral route (enteric bacterial pathogens). From 1970 to 1990, enteric bacteria were estimated to account for 14 percent of all waterborne disease outbreaks in the United States between 1971 and 1990 (Craun, 1991) and 32 percent between 1991 to 2002 (Craun et al., 2006). Based on hospitalization records, the most severe bacterial infections result from *E. coli* (14 percent), *Shigella* (5.4 percent), and *Salmonella* (4.1 percent) (Gerba et al., 1994).

Because of the public health significance of bacterial pathogens, monitoring systems and water quality standards have been established based on fecal coliforms (a classification that includes *E. coli*) and *enterococcus* in the United States and in many nations around the world (NRC, 2004). It is important to note that most *E. coli* and *enterococcus* are not pathogenic. Rather they are part of the normal microflora in the human digestive tract and are necessary for proper digestion and nutrient uptake. *E. coli* and *enterococcus* are employed as indicators of the presence of human waste (also called fecal indicator bacteria) in water quality monitoring and protection because they are present in high concentrations in human feces and sewage and they are more persistent than most bacterial pathogens. They are, therefore, used to indicate inadequate treatment of sewage to remove bacterial pathogens (NRC, 2004). Fecal indicator bacteria in undisinfected secondary effluent range from 10^2 to 10^5/100 mL depending on the quality of the influent water (Bitton, 2005). However, the concentration of fecal indicator bacteria (i.e., total coliform, fecal coliform, *enterococcus,* and *E. coli*) in filtered, disinfected secondary effluent can be brought below the nominal detection limit of 2.2 organisms/100 mL and with advanced treatment, they can be brought even lower.

Viruses

Viruses are extremely small infectious agents that require a host cell to replicate. They are of special interest in potable reuse applications because of their small size, resistance to disinfection, and their low infectious dose. There are many different viruses, and they infect nearly all types of organisms, including animals, plants, and, even bacteria. Aquatic viruses can occur at concentrations of 10^8 to 10^9 per 100 mL of water in the ocean (Suttle, 2007) and 10^9 to 10^{10} per 100 mL in sewage (Wu and Liu, 2009); however, most of these are bacteriophages—viruses that infect bacteria. The viruses of concern in water reuse or in the discharge of treated wastewater to drinking water sources are human enteroviruses (e.g., poliovirus, hepatitis A), noroviruses (i.e., Norwalk virus), rotaviruses, and adenoviruses. Human viruses are usually present in undisinfected secondary effluent and may still persist in effluents after some advanced treatment (e.g., Blatchley et al., 2007; Simmons and Xagoraraki, 2011). Fecal indicator bacteria that are currently used for water quality monitoring are not an adequate indication of the presence or absence of viruses because bacteria are more efficiently removed or inactivated by some wastewater treatment processes than are enteric viruses (Berg, 1973; Harwood et al., 2005). Thus, viruses need to be carefully addressed whenever treated municipal wastewater is discharged or reused in a context where there may be human contact, particularly when it makes up all or part of a drinking water supply.

Prions

A prion is an infectious agent that is primarily a protein. The prion causes a morphological change to native proteins, which can, in turn, lead to disease symptoms. The best-known example of prion-based disease is bovine spongiform encephalopathy ("mad cow disease"). In animals, prions can cause a variety of diseases including scrapie and chronic wasting disease (CWD); however, the spectrum of cross transmission of different prion agents is not clear. It has been demonstrated that CWD can be transmitted to animals by direct oral ingestion of prion-containing animal tissue (Mathiason et al., 2009). It has not been demonstrated that prions can be transmitted by the ingestion of drinking water, and their occurrence in water is poorly understood.

Currently, sparse data exist on the occurrence of prions outside of animal flesh or on the fate of prions in water or wastewater treatment. Prions are thought to substantially partition into the sludge during biological wastewater treatment, although according to a pilot study reported by Hinckley et al. (2008), some remain in effluent. Nichols et al. (2009) developed an analytical technique for measuring prions in water and environmental samples. Using this assay they reported detection of prions in one of two surface water samples in an area known to be endemic for CWD. They also reported detection of prions from water drawn from the flocculation stage of a water treatment plant using this source, but none in the water in subsequent stages of treatment.

INORGANIC CHEMICALS

Wastewater contains a variety of inorganic constituents including metals, oxyhalides, nutrients, and salts. Generally, aggregate measures of inorganic con-stituents in water are total dissolved solids (TDS) and conductivity, although both TDS and conductivity measurements may include contributions from some organic constituents. Because human and industrial activities consistently increase the TDS in water, the reuse of water will increase the TDS in the water supply.

Metals and Metalloids

Metals and metalloids, such as lead, mercury, chromium, arsenic, and boron, can result in adverse effects to human health when consumed in excessive amounts. However, regulatory statutes and industrial pretreatment regulations promulgated through the Clean Water Act specifically target toxic metals and, as a result, most municipal effluents have concentrations of toxic metals below public health guidelines and standards. Therefore, toxic metals in contemporary treated domestic wastewaters in the United States do not generally exceed human health exposure.

Boron (a metalloid) occurs in domestic wastewater, most likely resulting from its use in household products such as detergents (WHO, 2009). However, boron typically is not an issue for water reuse systems because concentrations are generally less than 0.5 mg/L (Asano et al., 2007), although in certain unique geologies or coastal communities boron can be elevated. Boron is of particular interest because no removal occurs during conventional biological treatment, and even advanced water treatment processes (i.e., reverse osmosis) are not highly effective at ambient pH. Although boron is not regulated in drinking water in the United States, the U.S. Environmental Protection Agency (EPA) published a health advisory level of 7 mg/L for adults and a level of 3 mg/L for 10-kg children (EPA, 2009b). Similarly, WHO has established a human health guideline for boron of 2.4 mg/L (WHO, 2009). Thus, typical boron levels in domestic wastewaters are well below drinking water guidelines.

There are many ornamental plants, however, that are more sensitive to boron (Tanji et al., 2008). Although boron is essential for plant growth and development, it can be toxic to plants at concentrations above 0.5 to 1 mg/L (Brown et al., 2002). In some settings, boron may place limits on the types of plants that can be successfully irrigated with reclaimed water.

Salts

The reuse of water generally increases the concentration of dissolved salts because of significant contributions of various salts through municipal and industrial water uses. In general, the levels of salts as measured as TDS do not exceed thresholds of concern to human health; however, excess salt concentrations can result in aesthetic concerns (i.e., unpalatable water) as well as agricultural and infrastructure damage. Certain salts in elevated concentrations can lead to scaling and corrosion issues. Calcium and magnesium concentrations are primarily responsible for hardness, and excess levels can cause damage to household appliances and industrial equipment (Hudson and Gilcreas, 1976). In service areas with elevated hardness, households commonly employ ion exchange–based water softeners as a local remedy for "hard water," but these units significantly increase the total salinity of the wastewater, particularly chloride. High levels of chloride are of concern because these ions exacerbate the corrosion of metals and reinforced concrete (Crittenden et al., 2005; Basista and Weglewski, 2009). The U.S. Bureau of Reclamation estimated in 2004 that excess salinity in the Colorado River caused more than $300 million per year in economic damages in the United States (U.S. Bureau of Reclamation, 2005).

Excess salinity can also be detrimental to plant growth (Tanji et al., 2008; Goodman et al., 2010). High sodium and chloride concentrations in reclaimed water used for irrigation can cause leaf burn, and high sodium concentrations can also reduce the permeability of clay-bearing soils and adversely affect the soil structure. The suitability of a water source for irrigation can be assessed by the electrical conductivity and the sodium adsorption ratio (SAR), a calculated ratio of sodium to calcium and magnesium ions;[1] the higher the electrical conductivity and the SAR, the less suitable the water is for use in irrigation. Therefore, careful control of salts and salt compositions is critical to water reuse, with specific limits dictated by end-use applications (i.e., irrigation vs. potable).

Salinity control is quite challenging because treatment options are limited and costly and because significant residuals are produced. Virtually all processes employed for salinity reduction result in a concentrated liquid waste (brine), which must subsequently be disposed (see also Chapter 4).

Oxyhalides

Oxyhalides are anionic salts consisting of a halogen covalently bonded to one or more oxygen atoms. In water reuse, the primary oxyhalides of concern are bromate, chlorite, chlorate, and perchlorate. Bromate is of primary concern when water containing bromide is ozonated, because its maximum contaminant level (MCL) is 10 μg/L and EPA has been made it clear it will seek even lower levels when feasible (EPA, 2006b). Sodium hypochlorite, commonly known as bleach, can contain elevated levels of bromate, chlorate, and perchlorate, depending upon the manufacturing and storage conditions (Asami et al., 2009).

Neither chlorate nor perchlorate is currently regulated under EPA's primary drinking water standards, although both are included on EPA's Contaminant Candidate List 3. Additionally, the state of California has established a notification level of 800 μg/L for chlorate and an enforceable MCL of 6 μg/L for perchlorate. Excess exposure to chlorate and perchlorate can result in inhibition of iodide uptake, resulting in decreased production of thyroid hormones (Snyder et al., 2006b). Chlorate is generally associated with the decomposition of bleach, where bleach age and handling procedures greatly influence the degree of chlorate formation (Gordon et al., 1997).

Perchlorate as a water contaminant is generally associated with anthropogenic activities, including solid propellants for missiles and spacecraft, flares, and fireworks (Urbansky, 2000). More recent data have demonstrated that perchlorate also is found in bleach, with the concentration dependent primarily upon bleach storage conditions and age (Snyder et al., 2009). Although, there is no federal regulation for perchlorate in drinking water, several states have promulgated enforceable regulations, with Massachusetts having the most stringent standard at 2 μg/L (Pisarenko et al., 2010). Perchlorate has been demonstrated to accumulate in certain plants (Sanchez et al., 2005);

---

[1] $SAR = [Na^+]/\{([Ca^{2+}] + [Mg^{2+}])/2\}^{1/2}$, where the concentrations are provided in milliequivalents per liter.

therefore, irrigation of food crops with reclaimed water containing elevated levels of perchlorate could result in elevated levels of perchlorate in certain food products. However, perchlorate also is naturally occurring as the result of formation in the atmosphere and subsequent deposition with rainfall (Dasgupta et al., 2005), thus complicating investigations of perchlorate bioaccumulation from natural versus artificial irrigation. Water reuse practitioners employing ozonation should be aware of the potential for bromate formation, and those using bleach should be cautious purchasing and storing bleach, to avoid excess chlorate and perchlorate formation. As in drinking water treatment, with the exception of perchlorate, the oxyhalide problem is not so much a problem of source water quality but one that requires proper design and operation of treatment facilities to minimize their formation during treatment. Among the processes that are employed in conventional drinking water treatment and in advanced wastewater treatment, oxidation and disinfection processes are those that have the greatest potential for creating oxyhalides. Disinfection is especially important in potable reuse projects; therefore, the formation of oxyhalides will be a key consideration in process train selection and design.

Nutrients

Human waste products are rich in nitrogen and phosphorus, and the human body metabolizes and excretes both phosphorus and nitrogen in various forms. The primary forms of nitrogen in wastewater effluent are ammonia, nitrate, nitrite, and organic nitrogen. Phosphorus also occurs in wastewater mainly in inorganic forms. These nutrients can pose environmental concerns but also carry potential benefits to nonpotable water reuse applications that involve irrigation. Elevated nitrate in drinking water can also present public health issues, especially in infants. To protect human health, EPA established an MCL in drinking water of 10 mg (as N)/L for nitrate and 1 mg (as N)/L for nitrite.[2]

Therefore, the need for removal of nutrients during treatment of wastewater for subsequent reuse depends

largely on the intended use of the produced water. In water reuse for irrigation, the presence of nitrogen and phosphorus are generally beneficial and promote growth of plants or crops. However, ammonia, particularly in its un-ionized form (i.e., as NH_3), is highly toxic to fish; therefore, wastewater discharges to surface waters generally are regulated to prevent excess ammonia release. Ammonia can reach levels of 30 mg/L in secondary treated effluents; however, ammonia can be oxidized to nitrite and further to nitrate by aerobic autotrophic bacteria during wastewater treatment. Although the nitrification process leads primarily to nitrate, water reuse facilities often also denitrify to reduce nitrate levels, converting nitrate to nitrite and ultimately to nitrogen gas. When nitrogen is not removed, it is usually present at levels that are above the EPA MCL for nitrate (as N). This can be a concern because in the natural environment, all forms of nitrogen in effluent are generally transformed to nitrate.

Although reclaimed water is frequently desirable for irrigation, excess irrigation can lead to nutrient contamination of underlying aquifers and of surface waters through runoff. An additional concern for nutrients in reclaimed water stored or reused in ponds, lakes, or streams arises from eutrophication wherein excess nitrogen and phosphorus stimulate the rapid growth of algae, which can cause problems including a depletion of oxygen concentrations in water, alteration of the trophic state of the system, impairment of the operation of drinking water treatment plants, and production of compounds that affect taste and cause odors in drinking water. The processes for management of nitrogen in wastewater treatment are now well-understood (Tchobanoglous, 2003). As a consequence the challenge is matching the appropriate treatment with the intended use and assessing the affordability of the project.

Engineered Nanomaterials

Nanomaterials are generally considered to be materials with at least one dimension from 1 to 100 nm (Jiménez et al., 2011). Nanomaterials exhibit this geometry in one dimension (i.e., nanofilms), two dimensions (i.e., nanotubes, nanowires), or three dimensions (i.e., nanoparticles). Nanoscale particles are not new to the water and wastewater field. Many natural subcolloidal particles in this range, including viruses

[2] See http://water.epa.gov/drink/contaminants/index.cfm. Additionally, WHO (2011) set a guideline value of 11 mg/L nitrate as N (or 50 mg/L as nitrate) and 1 mg/L nitrate as N (or 3 mg/L as nitrite).

and natural organic matter (Baalousha and Lead, 2007; Song et al., 2010), have been dealt with for decades in water and wastewater treatment. More recent examples of natural nanoscale particles include oxidation products of manganese, iron, and perhaps lead (Lytle and Snoeyink, 2004; Lytle and Schock, 2005). However, the purposeful manufacturing of nanoscale materials (called engineered nanomaterials) for consumer products is rapidly increasing.[3] Because nanoscale particles have an extraordinary surface-to-volume ratio, they are of interest in many applications where surface chemistry or catalysis is important (Weisner and Bottero, 2007). Potential applications of nanotechnology in the environmental industry itself are also evolving (Savage and Diallo, 2005; Chong et al., 2010; Pendergast and Hoek, 2011). As a result, many new questions have emerged about the fate of engineered nanomaterials when released to the environment.

Engineered nanomaterials can be organic, inorganic, or a combination of organic and inorganic components. Because of the complexity and diversity of engineered nanomaterial structure and composition, the behavior and toxicity of particles released to the environment will vary greatly. A recent review discusses the potential implications of engineered nanomaterials in the environment (Scown et al., 2010). However, specific information is limited regarding the occurrence and fate of engineered nanomaterials in municipal wastewater, their response to treatment, and their public health and environmental significance.

Some research has been conducted on the fate of engineered nanoparticles in wastewater treatment. Kaegi et al. (2011) studied the fate of silver nanoparticles added to the inflow of a pilot-scale conventional wastewater treatment plant. Most of the silver nanoparticles became associated with sludge and biosolids and were not detected in the pilot plant effluent. Another study investigated the removal of titanium nanoparticles at wastewater treatment plants. Kiser et al. (2009) found that the majority of titanium in raw sewage was associated with particles >0.7 μm, which were generally well removed through a conventional process train. However, titanium associated with particles <0.7 μm (near the nanoscale) were found in the treated wastewater effluents.

Ongoing research is exploring possible health effects from engineered nanoparticles (and associate mechanisms of effect) via various exposure pathways (NRC, 2009a, 2011b). So far, the trace levels of engineered nanoparticles in wastewater have not been linked to adverse human health impacts (O'Brien and Cummins, 2010). At present, most engineered nanoparticles in municipal wastewater originate from household and personal care products, and for these, direct exposure in the household itself is likely far greater than from potential ingestion of wastewater-influenced drinking water. Because the use of engineered nanoparticles in consumer products is expected to continue to rise, continued exposure and risk assessments will be important for assessing impacts on the environment and public health.

ORGANIC CHEMICALS

Wastewater is generally rich in organic matter, which is measured as TOC, dissolved organic carbon (DOC; that portion of the TOC that passes a 0.45-mm pore-size filter), and particulate organic carbon (POC; that portion of the TOC that is retained on the filter). Of the DOC present in highly treated reclaimed water, the vast majority is generally natural organic matter and soluble microbial products, with small concentrations of a variety of individual organic chemicals (Table 3-3; Namkung and Rittman, 1986; Shon et al., 2006).

Trace organic chemicals originate from industrial and domestic products and activities (e.g., pesticides, personal care products, preservatives, surfactants, flame retardants, perfluorochemicals), are excreted by humans (e.g., pharmaceutical residues, steroidal hormones), or are chemicals formed during wastewater and drinking water treatment processes. The vast majority of these trace organic chemicals occur at microgram per liter and lower levels. This complex mixture of low concentrations of contaminants has long been recognized; Ram (1986) reported that 2,221 organic chemicals had been identified in nanogram per liter to microgram per liter concentrations in water around the world, including 765 in finished drinking water. Modern analytical tools are extremely sensitive and often capable of detecting nanogram per liter or lower concentrations of organic contaminants in water. In this report, these compounds are termed trace organic contaminants,

---

[3] http://www.nanotechproject.org/inventories/consumer/analysis_draft.

TABLE 3-3 Categories of Trace Organic Contaminants (Natural and Synthetic) Potentially Detectable in Reclaimed Waters

Category	Examples
Industrial Chemicals	1,4-Dioxane, perflurooctanoic acid, methyl tertiary butyl ether, tetrachloroethane
Pesticides	Atrazine, lindane, diuron, fipronil
Natural chemicals	Hormones (17β-estradiol), phytoestrogens, geosmin, 2-methylisoborneol
Pharmaceuticals and metabolites	Antibacterials (sulfamethoxazole), analgesics (acetominophen, ibuprofen), beta-blockers (atenolol), antiepileptics (phenytoin, carbamazepine), antibiotics (azithromycin), oral contraceptives (ethinyl estradiol)
Personal care products	Triclosan, sunscreen ingredients, fragrances, pigments
Household chemicals and food additives	Sucralose, bisphenol A (BPA), dibutyl phthalate, alkylphenol polyethoxylates, flame retardants (perfluorooctanoic acid, perfluorooctane sulfonate)
Transformation products	N-Nitrosodimethylamine (NDMA), bromoform, chloroform, trihalomethanes

but they are also commonly called micropollutants or contaminants of emerging concern (CECs). EPA has defined CECs as "pollutants not currently included in routine monitoring programs" that "may be candidates for future regulation depending on their (eco)toxicity, potential health effects, public perception, and frequency of occurrence in environmental media" (EPA, 2008a). Trace organic contaminants and CECs are not always newly discovered waterborne contaminants. They also include constituents that have been present in the environment for long periods of time, but for which analytical or health data have only recently become available.

With modern analytical technology, nearly any chemical will likely be detectable at some concentration in wastewater, reclaimed water, and drinking water. The challenge is not so much one of detection, but rather determination of human and environmental health relevance. The following section provides information on representative classes of trace organic chemicals present in reclaimed water, although the committee acknowledges that there may be many other classes and substances present.

Industrial Chemicals

Many chemicals originating from industrial activities that have been detected in wastewater need to be considered when that wastewater becomes part of a domestic water supply. These include solvents, detergents, petroleum mixtures, plasticizers, flame retardants, and a host of other products or product ingredients. A few of these chemicals are not completely removed by conventional water and wastewater treatment processes. For example, an industrial chemical that has caused concern in water reuse programs in California is 1,4-dioxane, a common industrial solvent considered a probable carcinogen, which has been shown to break through reverse osmosis membranes.

In 1986, EPA estimated that as much as one-third of all priority pollutants entering U.S. waters from wastewater discharges were the result of industrial discharges into public sewers (EPA, 1986). Additionally, pulsed releases from certain industries have been known to disrupt the biological processes at wastewater treatment plants, resulting in reduced treatment efficiency (Kelly et al., 2004; Kim et al., 2009; You et al., 2009). For these reasons, under the authority of the Clean Water Act, EPA established the industrial pretreatment program, which requires wastewater treatment plants processing 5 million gallons per day (19,000 m³/d) or greater to establish pretreatment programs (see also Box 10-1). The pretreatment program also applies to smaller systems with known industrial input. This program was specifically designed to address priority pollutants, which are defined under the Clean Water Act in section 307(a). Although the pretreatment program has been largely successful at reducing the loading of contaminants into municipal wastewater treatment plants, a much smaller, but perhaps significant, input of these chemicals also enters the sewer system from household use, leaking sewage conveyance pipes, and illegal connections/dumping (mostly from the former).

Pesticides

Despite the fact that pesticides are generally used outdoors and would not be expected to be discharged directly to the sewer, some pesticides have been detected in wastewater effluents. The sources are not fully characterized, but some loading could be expected through residues in food products, head lice treatments, veterinary/pet care applications, manufacturing or handling facilities, and infiltration of landscape runoff into sewer conveyance lines. The herbicide atrazine, which is used primarily on corn and soybean crops, recently has been shown to be a contaminant in nearly all U.S. drinking water, appearing in regions far removed from agricultural activities (Benotti et al., 2009). Subsequent research also has demonstrated that atrazine also occurs in most wastewater treatment effluents (Snyder et al., 2010a), yet the levels detected are generally in the nanogram-per-liter range, far lower than the EPA MCL of 3 µg/L. Considering that wastewater effluents are generally low in pesticide residues and that reclaimed water employed in potable reuse projects is regularly surveyed for all pesticides regulated in drinking water, it is unlikely that these compounds will pose a unique risk to water reuse.

Pharmaceuticals and Personal Care Products

Recently, a great deal of attention has been given to the occurrence of pharmaceuticals in wastewater effluents. Although pharmaceuticals were detected in U.S. waters as early as the 1970s (Garrison et al., 1975, Hignite and Azarnoff, 1977), much of the recent interest was evoked when Kolpin et al. (2002) in a nationwide stream sampling study documented the occurrence of 82 trace organic chemicals of wastewater origin. Commonly detected chemicals included triclosan (an antimicrobial compound), 4-nonylphenol (a metabolite of a chemical found in detergents, see Box 3-1), and synthetic estrogen from birth control, which has been implicated as a causative agent in fish feminization (Purdom et al., 1994). Laboratory studies have confirmed that ethinyl estradiol (EE2) is capable of affecting fish physiology at subnanogram-per-liter concentrations, with a predicted no-effect concentration of 0.35 ng/L (Caldwell et al., 2008). It is now quite clear that a wide range of pharmaceuticals can and will be detected in reclaimed water samples (see Table 3-3 for examples).

Personal care products (e.g., shampoo, lotions, perfumes) represent the source of another class of chemicals that have been widely detected in wastewater treatment plant effluents. It is logical that a substance used as an ingredient of a personal care product will enter the sewer system. For instance, several studies have demonstrated that certain synthetic musks used as fragrances in personal care products not only are incompletely removed by conventional wastewater treatment (Heberer, 2002) but also bioaccumulate in fish residing in effluent-dominated streams (Ramirez et al., 2009). There are many other examples of personal care products, which have been detected in treated wastewater. Many of these key ingredients may also be classified as household or industrial chemicals as well.

Household Chemicals and Food Additives

Within the typical household, many chemicals are used for cleaning, disinfecting, painting, preparation of meals, and other applications. Many of these chemicals find their way into the wastewater collection system, and some are detectable in reclaimed water as well. An interesting illustration is the artificial sweetener sucralose (1,4,6-trichlorogalactosucrose), which is widely used in the United States. This chlorinated sucrose molecule is predictably difficult to remove through biological treatment and is largely resistant to oxidation during water treatment as well. Therefore, concentrations in wastewater are generally in the microgram-per-liter range, and sucralose has been detected at similar concentrations in potable water (Buerge et al., 2009; Mawhinney et al., 2011).

Of the household chemicals of interest, those chemicals with the potential to disrupt the function of the endogenous endocrine system have been of particular interest. One particular class of surfactants, aklylphenol polyethoxylates (APEOs), has become of concern because of the estrogenic potency of some of its degradation products (see Box 3-1). Another compound of increasing interest is bisphenol A (BPA), which is used in a variety of consumer products and has been shown to be estrogenic (Durando et al., 2007). BPA has been detected in drinking water, but the concentrations are extremely low (Benotti et al., 2009), in part because of BPA's rapid oxidation by chlorine and ozone disinfectants commonly used in water treatment (Lenz et al., 2004). In terms of human exposure, the contribution of BPA from drinking water is minute compared with exposure from food packaging and storage materials (Stanford et al., 2010). Household products and pharmaceuticals often contain inert substances at much higher concentrations than the active product. In some cases, these inert substances may also warrant further investigation as to potential impacts to water treatment systems and environmental health.

BOX 3-1
Alkylphenol Polyethoxylates

Alkylphenol polyethoxylates (APEOs) are a family of surfactants that were once widely used in domestic and industrial cleaning products. This family of relatively benign chemicals serves as an example of how transformation reactions in engineered and natural systems can produce compounds that pose potential risks to aquatic organisms or human health.

The most common members of this family of compounds contain either eight or nine carbon atoms in their alkyl functional group (Montgomery-Brown et al., 2003; Loyo-Rosales et al., 2009) and are referred to as octylphenol polyethoxylates (OPEO) and nonylphenol polyethoxylates (NPEO), respectively (see figure below). Most OPEOs and NPEOs in commercial products consist of a mixture of compounds with between 1 and 20 ethoxylate groups. The surfactants with more than two carbons in their ethoxylate chain exhibit relatively low toxicity to aquatic organisms in standard toxicity tests (Staples et al., 2004; Loyo-Rosales et al., 2009). However, the compounds undergo biotransformation in wastewater treatment plants that employ anaerobic treatment processes (e.g., nitrate removal by denitrification) and in aquifer recharge systems in which anoxic (anaerobic) conditions occur. Anaerobic biotransformation of OPEO and NPEO occurs through sequential cleavage of the ethoxylate carbons, ultimately leading to formation of octylphenol or nonylphenol (Ahel et al., 1994a,b). Nonylphenol typically occurs in wastewater effluent at concentrations about 10 times higher than those of octylphenol (Loyo-Rosales et al., 2009). Nonylphenol and the transformation products with only one or two carbons in the polyethoxylate chain are substantially more toxic to aquatic life than the corresponding OPEO and NPEO surfactants (Staples et al., 2004). Octylphenol, nonylphenol, and the short polyethoxylate chains have been implicated in the feminization of fish observed in effluent-dominated streams (Johnson et al., 2005), although steroid hormones (e.g., 17β-estradiol) typically account for about 10 times more estrogenic activity than octylphenol or nonylphenol.

In recognition of the risks to aquatic life associated with APEOs and their transformation products, their use was restricted in the European Union in the 1990s. In 2005, EPA set a water quality criterion for freshwater aquatic life of 6.6 µg/L for chronic exposure to nonylphenol (EPA, 2005a) that is approximately equal to or slightly higher than concentrations typically detected in wastewater effluent in the United States (Montgomery-Brown et al., 2003; Loyo-Rosales et al., 2009). As a result, many manufacturers have replaced APEOs in consumer products or have reduced their concentrations. The compounds are still used for certain industrial applications and for specialty cleaning products.

$$CH_3-(CH_2)_x-\hspace{-1em}\bigcirc\hspace{-1em}-O\text{-}(CH_2\text{-}CH_2\text{-}O)_y\text{-}CH_2CH_2OH$$

General structure of alkylphenol polyethoxylate surfactants. For the alkyl group, x = 7 for octylphenol and 8 for nonylphenol. For the ethoxylate group, y = 0 to 19.

Naturally Occurring Chemicals

Estrogen hormones (e.g., 17β-estradiol) are endogenous[4] compounds that are excreted in relatively large concentrations by animals. In studies of wastewater effluent, the measured concentrations of endogenous estrogen hormones in most cases far exceeded those of the synthetic steroid hormones (Snyder et al., 1999; Huang and Sedlak, 2001). Huang and Sedlak (2001) reported that reverse osmosis treatment (see Chapter 4) removed more than 95 percent of estrogen hormones. Additionally, free chlorine or ozone disinfection will effectively attenuate estrogen hormone concentrations in water (Westerhoff et al., 2005).

Naturally occurring compounds that affect taste and odor represent another important class of natural chemicals that may pose challenges in water reuse. Of these, the best characterized are geosmin and 2-methylisoborneol (MIB), which are generally found in lakes and reservoirs (Medsker et al., 1968, 1969). However, geosmin is also naturally occurring in certain vegetables, such as red beets (Lu et al., 2003). Although geosmin and MIB are not considered toxic at the concentrations found in water, the olfactory displeasure can create great public resistance to water. Compounds that affect taste and odor can be present through naturally occurring compounds or through anthropogenic substances. However, these two odoriferous compounds that cause great public resistance to water can and should be considered in reuse planning for both potable and nonpotable applications in urban environments (Agus et al., 2011).

---

[4] Synthesized within an organism.

Transformation Products

Wastewater effluents are generally rich in organic constituents, and during most wastewater treatment processes, the majority of organic chemicals are not completely removed or mineralized. Although some treatment processes separate contaminants for subsequent disposal (i.e., sludge, reverse osmosis concentrate, spent activated carbon), both biological and oxidative processes commonly employed in water and wastewater treatment result in the formation of transformation products. When they result from disinfection processes, these products are generally referred to as disinfection byproducts; however, some oxidative processes (e.g., ozonation, ultraviolet [UV] irradiation–advanced oxidation processes [UV-AOP]) are used specifically for contaminant attenuation and not disinfection. Therefore, the term transformation product is more applicable to the range of water reclamation processes.

Through most oxidation processes, the total concentration of DOC remains relatively unchanged (Wert et al., 2007), although the attenuation of many specific trace organic chemicals is observed (Snyder et al., 2006c). This empirical observation dictates that the vast majority of chemicals attenuated during oxidative processes are not truly removed, but rather transformed into oxidation products. Most biological and oxidative transformation products have not been characterized. For instance, in drinking water it has been estimated that the majority of total organic halides (TOX) formed during disinfection with chlorine have not been identified (Krasner et al., 2006; Hua and Reckhow, 2008).

One example is triclosan, an antimicrobial compound used frequently in soap and other personal care products and thus commonly detected in wastewater (Singer et al., 2002). Triclosan is known to react with chlorine to form various disinfection byproducts, including chloroform (Rule et al., 2005; Greyshock and Vikesland, 2006). Studies have also demonstrated that when triclosan is exposed to UV irradiation, it can form dioxin-like compounds that may be toxicologically significant (European Commission, 2009) but are easily biodegraded.

It is also well known that certain compounds, which may be innocuous in their original form, can transform into toxic substances through water or wastewater treatment processes. The disinfection byproducts of chlorine first identified in the 1970s are a good example (Trussell and Umphres, 1978). N-Nitrosodimethylamine (NDMA; see Box 3-2) is a more contemporary example. NDMA can be an especially challenging contaminant for water reuse applications because chloramination, a common method of wastewater disinfection, has been linked to NDMA formation and because NDMA is not well rejected by reverse osmosis membranes (Mitch et al., 2003) and must be removed by subsequent photolysis. There is some evidence that polymers used in the management of biological wastewater treatment may serve as important NDMA precursors (Kohut and Andrews, 2003; Neisess et al., 2003). Continued research examining how NDMA is formed, how it can be removed, what its precursors are, and how they can be better managed in processes upstream of disinfection is needed.

Municipal wastewater is often elevated in nitrogen, iodine, and bromine constituents as compared with ambient waters (Venkatesan et al., 2011), which may lead to increased levels of nitrogenous, iodinated, and brominated disinfection products, respectively, when chlorination is applied (Joo and Mitch, 2007; Krasner et al., 2009), but this has not yet been documented. Iodinated and brominated disinfection products are among the most genotoxic of those disinfection byproducts currently identified in water (Plewa et al., 2004; Richardson et al., 2008). Recently, medium-pressure UV-AOP has been shown to form genotoxic organic transformation products when applied to waters containing nitrate, although subsequent treatment with granulated activated carbon was able to remove the formed genotoxic products to levels below detection (Heringa et al., 2011).

As water conservation efforts grow in many urban regions, concentrations of salt and organics will likely increase in wastewater. Thus, a better understanding of disinfection byproduct precursors, ways to minimize the disinfection byproduct formation, and ways to remove them is important for enhancing the safety of water reuse scenarios, including de facto reuse. Transformation products in reclaimed water will also be widely variable in concentration and structure because of the highly complex mixtures and different source water characteristics. Water reuse projects would therefore benefit from improved methods for understanding the toxicity of complex mixtures (see Chapter 11).

BOX 3-2
***N*-Nitrosodimethylamine**

N-Nitrosodimethylamine (NDMA) has been considered a carcinogen for some time (Magee et al., 1976), and EPA has calculated the one in one million cancer risk from drinking water to occur at approximately 0.7 ng/L. Along with other members of the nitrosamine family, NDMA received attention in the 1970s in connection with processed foods and beverages, but it was not found in drinking water or domestic wastewater until the turn of the century when analytical methods improved to the point where NDMA could be identified at submicrogram-per-liter levels (Taguchi et al., 1994). Subsequently, NDMA was found in groundwater downgradient of rocket engine testing facilities, in water leaving ion exchange facilities, and in wells influenced by reuse projects (Najm and Trussell, 2001). Recently, as part of EPA's unregulated contaminant monitoring rule (UCMR2), NDMA was detected in 25 percent of the drinking water distribution systems sampled, at levels between 2 and 600 ng/L. For the most part, these drinking water systems reported that their source water was influenced by wastewater and used chloramines for disinfection (Blute et al., 2010).

NDMA often appears both in raw and treated wastewaters in the United States and Europe (Mitch et al., 2005, Krauss et al., 2009). A 2005 survey of 10 wastewater plants found NDMA in the influent up to 140 ng/L; two plants were 20 ng/L or below, but most were between 20 and 70 ng/L. Effluent samples, however, ranged as high as 960 ng/L (Valentine et al., 2005). Others have reported levels as high as 1,820 ng/L (Gan et al., 2006).

Control of NDMA in treated reclaimed water involves three components: (1) control of the sources of NDMA and its precursors in treatment plant influents, (2) management of the conventional wastewater treatment process, and (3) application of advanced treatment to remove what remains. Both Orange County and Los Angeles have had some success in identifying sources of NDMA and its precursors and have improved the quality of the influent (Valentine et al., 2005). However, it is unclear how much of the NDMA may be the result of domestic sources (e.g., pharmaceuticals, personal care products) that are more difficult to control (Sedlak et al., 2005; Krauss et al., 2009; Shen and Andrews, 2011). Wastewater disinfection practice, particularly chloramination (Pehlivanogllu-Mantas et al., 2006) appears to be an important target. Research by wastewater authorities has demonstrated several factors important to NDMA formation during wastewater chlorination and a number of strategies that may be employed to reduce it (Neisess et al., 2003; Huitric et al., 2005, 2007; Tang et al., 2006; Farée et al., 2011). Although these strategies show promise, NDMA remains an issue in wastewaters disinfected with chloramines, where levels above 100 ng/L are common (Najm and Trussell, 2001; Valentine et al., 2005; Huitric et al., 2007). As a result, facilities designed to produce reclaimed water for direct injection into groundwater include treatment processes designed to remove it (e.g., UV-AOP).

CONCLUSIONS

The very nature of wastewater suggests that nearly any substance used or excreted by humans has the potential to be present at some concentration in the treated product. Modern analytical technology allows detection of chemical and biological contaminants at levels that may be far below human and environmental health relevance. Therefore, **if wastewater becomes part of a reuse scheme (including de facto reuse), the impacts of wastewater constituents on intended applications should be considered in the design of the treatment systems**. Some constituents, such as salinity, sodium, and boron, have the potential to affect agricultural and landscape irrigation practices if they are present at concentrations or ratios that exceed specific thresholds. Some constituents, such as microbial pathogens and trace organic chemicals, have the potential to affect human health, depending on their concentration and the routes and duration of exposure (see Chapter 6). Additionally, not only are the constituents themselves important to consider but also the substances into which they may transformed during treatment. Pathogenic microorganisms are a particular focus of water reuse treatment processes because of their acute human health effects, and viruses necessitate special attention based on their low infectious dose, small size, and resistance to disinfection. Chapter 4 discusses the treatment processes often used to attenuate concentrations of chemical and biological contaminants of suspected health risk to humans.

4

Wastewater Reclamation Technology

Treatment processes in wastewater reclamation are employed either singly or in combination to achieve reclaimed water quality goals. Considering the key unit processes and operations commonly used in water reclamation (see Figure 4-1), an almost endless number of treatment process flow diagrams can be developed to meet the water quality requirements of a certain reuse application.

Many factors may affect the choice of water reclamation technology. Key factors include the type of water reuse application, reclaimed water quality objectives, the wastewater characteristics of the source water, compatibility with existing conditions, process flexibility, operating and maintenance requirements, energy and chemical requirements, personnel and staffing requirements, residual disposal options, and environmental constraints (Asano et al., 2007). Decisions on treatment design are also influenced by water rights, economics, institutional issues, and public confidence. The relative importance of some of these factors is likely going to change in the future. With the current desire to limit greenhouse gas emissions and introduction of carbon taxes, energy-intense processes likely will be viewed much less favorable than today. This chapter focuses on treatment processes—characterized as preliminary, primary, secondary, and advanced and including both natural and engineered processes—that can be used to meet water quality objectives of a reuse project and their treatment effectiveness. The efficiency in removing certain constituent classes, energy requirements, residual generation, and costs of these treatment processes are qualitatively summarized in Table 4-1. Economic, so-cial, and institutional considerations that also influence the choice of reclamation technologies are addressed in Chapters 9 and 10.

PRELIMINARY, PRIMARY, AND SECONDARY TREATMENT

Wastewater treatment in the United States typically includes preliminary treatment steps in addition to primary and secondary treatment. Preliminary steps include measuring the flow coming into the plant, screening out large solid materials, and grit removal to protect equipment against unnecessary wear. Primary treatment targets settleable matter and scum that floats to the surface. As shown in Table 2-1, only 1.3 percent of wastewater treatment plant effluents in the United States are discharged after receiving less than secondary treatment because of site-specific waivers (EPA, 2008b).

Secondary treatment processes are employed to remove total suspended solids, dissolved organic matter (measured as biochemical oxygen demand), and, with increasing frequency, nutrients. Secondary treatment processes usually consist of aerated activated sludge basins with return activated sludge or fixed-media filters with recycle flow (e.g., trickling filters; rotating biocontactors), followed by final solids separation via settling or membrane filtration and disinfection (Figure 4-1) (Tchobanoglous et al., 2002).

Advances over the past 20 years in membrane bioreactor (MBR) technologies have resulted in an alternative to conventional activated sludge processes

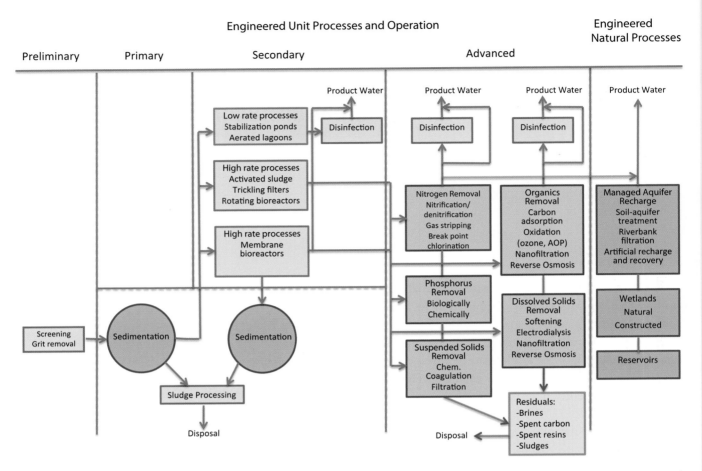

FIGURE 4-1 Treatment processes commonly used in water reclamation. Note that some or all of the numerous steps represented under advanced processes may be employed, depending on the end-product water quality desired and whether engineered natural processes are also used. All possible combinations are not displayed here.

that does not require primary treatment and secondary sedimentation (LeClech et al., 2006). Instead, raw wastewater can be directly applied to a bioreactor with submerged microfiltration or ultrafiltration membranes. These applications may only employ a fine screen as a preliminary treatment step. MBR processes combine the advantage of complete solids removal, a significant disinfection capability, high-rate and high-efficiency organic and nutrient removal, and a small footprint (Stephenson et al., 2000). In the past 10 years, reductions in the cost of membrane modules, extended life expectancy of the membranes, and advances in process design and operation have resulted in many domestic and industrial applications using MBRs. Its integrated design, which can be scaled down more easily than conventional secondary treatment processes, can facilitate decentralized water reclamation. How-

ever, membrane fouling and its consequences regarding plant maintenance and operating costs limit the widespread application of MBRs (LeClech et al., 2006; van Nieuwenhuijzen et al., 2008). Challenges that require research relate to maintaining productivity (or flux, i.e., the amount of water produced per membrane area) and minimizing the effects of membrane fouling. Other MBR research needs include the effluent quality that can be achieved and improvements in oxygen transfer and membrane aeration to lower operational costs of MBRs (van Nieuwenhuijzen et al., 2008).

In the United States, 45 percent of wastewater treatment plant effluent as of 2004 received only primary and secondary treatment (see Table 2-1). EPA (2008b) reported that 49 percent of all wastewater treatment plant effluent received "greater than secondary" treatment. This could include MBR treatment or

TABLE 4-1 Treatment Processes and Efficiencies to Remove Constituents of Concern during Water Reclamation

Process	Pathogens			Nitrate	TDS	Boron	Bromate and Chorate	Metals	DBPs	Trace Organics		Energy Requirements	Residual Generation[a]	Cost
	Protozoa	Bacteria	Viruses							Nonpolar	Polar			
Engineered Systems: Physical														
Filtration	Moderate	Moderate	Low	None	None	None	None	Low	None	None	None	Low	Low	Low
PAC/GAC	Low	Low	Low	None	None	None	Low	Low	Moderate	High	Low	Low	Low	Moderate
MF/UF	High	Moderate	Low	None	None	None	Low	Low	Low	Low	None	Moderate	Low	Moderate
NF/RO	High	High	High	High	High	Moderate	High	High	Moderate	High	High	High	High	High
Engineered Systems: Chemical														
Chloramine	Low	Moderate	Low	None	None	None	None	None	None	None	None	Low	None	Low
Chlorine	Moderate	High	High	None	None	None	None	None	None	Low to moderate	Low to moderate	Low	None	Low
Ozone	Moderate	High	High	None	None	None	None	None	Low	High	High	High	None	High
UV	High	High	Moderate	None	None	None	None	None	None	None	None	Moderate	None	Low
UV/H$_2$O$_2$	High	High	High	None	None	None	None	None	Low	High	High	High	None	High
Engineered Systems: Biological														
BAC	Low	Low	Low	None to low	None	None	Low	Low	Low to moderate	Moderate	Moderate	Low	None to low	Low
Natural Systems														
SAT	High	High	Moderate	High	None	None	Low to moderate	High	High	High	Moderate to High	Low	None	Low
Riverbank Filtration	High	High	Moderate	High	None	None	Low to moderate	High	High	High	Moderate to High	Low	None	Low
Direct inj.	Moderate	Low	Low	Low	None	None	Low to moderate	High	Moderate	Low	None	Moderate	None	Low to moderate[b]
ASR	Moderate	Moderate	Moderate	Moderate	None	None	Low	High	Moderate	Moderate	Low to moderate	Low	None	Low
Wetlands	Low to moderate	Low to moderate	Low	Moderate	None	None	Low	Moderate to high	Low	Low to moderate	Low	Low	None	Low
Reservoirs	Low to moderate	Low to moderate	Low	Low to moderate	None	None	Low	Moderate to high	Low	Low	Low	Low	None	Low

NOTE: The qualitative values in the table represent the consensus best professional judgment of the committee. [a]Low represents little generation of residuals, high represents significant amounts of residual generation; [b]High when required pretreatment is considered.

any combination of the treatment processes described in the following sections.

DISINFECTION

Disinfection processes are those that are deliberately designed for the reduction of pathogens. Pathogens generally targeted for reduction are bacteria (e.g., *Salmonella*, *Shigella*), viruses (e.g., norovirus, adenovirus), and protozoa (e.g., *Giardia*, *Cryptosporidium*) (see also Chapter 3).

Common agents used for disinfection in wastewater reclamation are chlorine (applied as gaseous chlorine or liquid hypochlorite) and ultraviolet (UV) irradiation. Only chlorine is purchased as a chemical in commerce. Chlorine dioxide, ozone, and UV are generated on-site. In drinking water applications, chlorine and hypochlorite remain the most common disinfectants, although they are decreasing in prevalence (Table 4-2). Chloramines are formed from either chlorine or hypochlorite if appropriate amounts of ammonia are present (as in wastewater) or if ammonia is deliberately added. Although chlorine or hypochlorites are still the most prevalent disinfection processes used in wastewater applications, UV is much more common and chlorine dioxide and ozone are less common than in drinking water applications (Asano et al., 2007). Membrane processes can also remove many pathogens, although they are not considered reliable stand-alone

methods for disinfection, as discussed later in this chapter.

The effectiveness of each of the disinfectants against pathogens is a function of the amount of disinfectant added, the contact time provided, and water quality variables that may compete for the disinfectant or modulate its effectiveness. Once decay (or in the case of UV, absorbance of energy) is taken into account, a first approximation to effectiveness is the product of residual concentration (C) (or in the case of UV delivered power intensity [I]) and contact time (t). There is a relationship between the $C \cdot t$ "product" (actually integrated over the contact time of a disinfection reactor, taking into account hydraulic imperfections) and the degree of microbial inactivation. This concept is schematically illustrated in Figure 4-2.

The relationships between $C \cdot t$ and microbial inactivation may be affected by water quality (e.g., temperature, turbidity, pH). For chlorine in particular, there is a strong effect of pH, with disinfection being more effective below pH 7.6 (when hypochlorous acid [HOCl] predominates) than above pH 7.6 (where hypochlorite [OCl⁻] predominates) (Fair et al., 1948). The impact of turbidity on disinfection has been known for a long time and is particularly problematic in disinfection of wastewater effluents (Hejkal et al., 1979). However, in drinking water, when the turbidity is <1 turbidity unit (TU), the effect of turbidity on disinfection is minimal (LeChevallier et al., 1981). This has also been confirmed on experiments with actual waters, demonstrating that 0.45-µm filtration had minimal effect on disinfection of water (by chlorine or chlorine dioxide) in waters with initial turbidity <2 TU (Barbeau et al., 2005). Other water quality factors, the nature of which remains unknown, may modulate disinfection effectiveness for both chlorine (Haas et al., 1996) and ozone (Finch et al., 2001). It should also be noted that disinfection and the competing decay and demand processes are nonlinear. Therefore, a more detailed consideration of these nonlinearities as coupled to hydraulics is needed for a full engineering design (Bellamy et al., 1998; Bartrand et al., 2009).

In general, in most disinfection approaches except UV, bacteria are more easily disinfected (lower required $C \cdot t$) than viruses, which are in turn more easily disinfected than protozoa. With UV, protozoa, are somewhat more sensitive than viruses (particularly

TABLE 4-2 Drinking Water Disinfection Practices According to 1998 and 2007 AWWA Surveys

Disinfectant	Percent of Drinking Water Utilities Using	
	1998	2007
Chlorine gas	70	61
Chloramines	11	30
Sodium hypochlorite	22	31
Onsite generation of hypochlorite	2	8
Calcium hypochlorite	4	8
Chlorine dioxide	4	8
Ozone	2	9
UV	0	2

NOTE: Percentages sum to more than 100 because some utilities use multiple disinfectants.
SOURCE: AWWA Disinfection Systems Committee (2008); AWWA Water Quality Division Disinfection Systems Committee (2000).

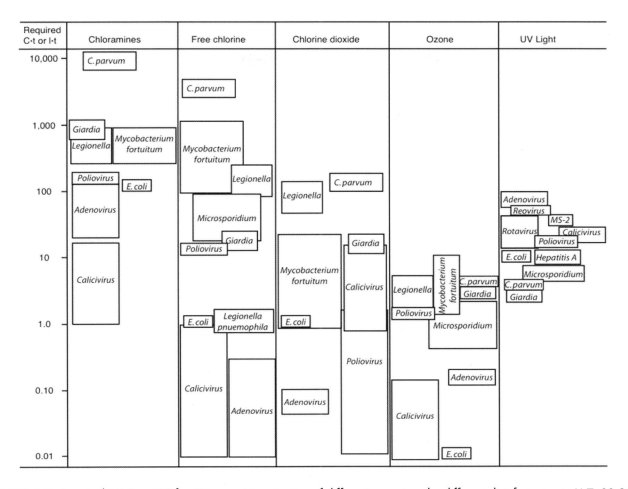

FIGURE 4-2 Required "C·t" or "I·t" for 99 percent inactivation of different organisms by different disinfectants at pH 7, 20-25 °C.
SOURCE: Crittenden et al. (2005).

adenovirus, the most UV-resistant class of viruses) (Jacangelo et al., 2002).

Chemical disinfectants (i.e., chlorine, ozone, chlorine dioxide) are known to produce characteristic disinfection byproducts (Minear and Amy, 1996; see also Chapter 3). The spectrum of these will not be reviewed in this report, but in general, chlorine and ozone can react with organic materials to produce stable disinfection byproducts (which may or may not be halogenated). For chlorine, these include trihalomethanes, trihaloacetic acids, haloaldehydes, and haloamines. Ozone can react with bromide that may be present to produce bromine and, in turn, brominated byproducts, including bromate. Chlorine dioxide can produce chlorite and chlorate, and depending on the mode of production of chlorine dioxide, chlorine may also be present, which

can produce disinfection byproducts analogous to those produced by chlorination (Tibbets, 1995; Richardson et al., 1994; van Nieuwenhuijsen et al., 2000; Hua and Reckhow, 2007).

ADVANCED ENGINEERED TREATMENT

Advanced engineered unit processes and operations can be grouped into engineering systems targeting the removal of nutrients and organic constituents, reduction of total dissolved solids (TDS) or salinity, and provision of additional treatment barriers to pathogens (Figure 4-1). Nutrients can be reduced by biological nitrification/denitrification processes, gas stripping, breakpoint chlorination, and chemical precipitation. Organic constituents can be further removed by various advanced processes, including activated carbon, chemical oxidation (ozone, advanced oxidation processes [AOPs]), nanofiltration (NF), and reverse osmosis (RO). Dissolved solids are retained during softening, electrodialysis, NF, and RO. Various processes can be combined to produce the desired effluent water quality depending on the reuse requirements, source water quality, waste disposal considerations, treatment cost, and energy needs.

Nutrient Removal

Nutrient removal is often required in reuse applications where streamflow augmentation or groundwater recharge is practiced to prevent eutrophication or nitrate contamination of shallow groundwater. Nutrient removal can be either an integral part of the secondary biological treatment system or an add-on process to an existing conventional treatment scheme.

All of the biological processes for nitrogen removal include an aerobic zone in which biological nitrification occurs. An anoxic zone and proper retention time is then provided to allow biological denitrification (conversion to nitrogen gas) to reduce the concentrations of nitrate to less than 8 mg N/L as illustrated in Table 3-2 (Tchobanoglous et al., 2002). Gas stripping for removal of ammonia or breakpoint chlorination as the primary means for nitrogen removal is not commonly employed in wastewater reclamation applications in the United States.

To accomplish biological phosphorus removal via phosphorus-storing bacteria, a sequence of an anaerobic zone followed by an aerobic zone is required (for more detailed information see Tchobanoglous et al., 2002). Phosphorus removal can also be achieved by chemical precipitation by adding metal salts (e.g., Ca(II), Al(III), Fe(III)) with a subsequent filtration following the activated sludge system. Although chemical precipitation for phosphorus removal is practiced in many water reclamation facilities, biological phosphorus removal requires no chemical input. Biological phosphorus removal, however, requires a dedicated anaerobic zone and modifications to the activated sludge process, which usually is more costly during a plant retrofit than an upgrade to chemical precipitation. A biological phosphorus removal process is also more challenging to control and maintain because it depends upon a more consistent feedwater quality and steady operational conditions. Biological and chemical phosphorus removal can result in effluent concentrations of less than 0.5 mg P/L (see Table 3-2).

Suspended Solids Removal

Filtration is a key unit operation in water reclamation, providing a separation of suspended and colloidal particles, including microorganisms, from water. The three main purposes of filtration are to (1) allow a more effective disinfection; (2) provide pretreatment for subsequent advanced treatment steps, such as carbon adsorption, membrane filtration, or chemical oxidation; and (3) remove chemically precipitated phosphorus (Asano et al., 2007). Filtration operations most commonly used in water reclamation are depth, surface, and membrane filtration.

Depth filtration is the most common method used for the filtration of wastewater effluents in water reclamation. In addition to providing supplemental removal of suspended solids including any sorbed contaminants, depth filtration is especially important as a conditioning step for effective disinfection. At larger reuse facilities (>1,000 m^3/d or >4 MGD), mono- and dual-media filters are most commonly used for wastewater filtration with gravity or pressure as the driving force. Both mono- and dual-media filters using sand and anthracite have typical filtration rates between 2,900 and 8,600 gal/ft^2 per day (4,900–14,600 L/m^2 per hour) while achieving effluent turbidities between 0.3 and 4 nephelometric turbidity units (NTU). Because large plants with many filters usually do not practice wasting of the initial filtrate after backwash (filter-to-waste), effluent qualities with elevated initial turbidity are commonly observed, and as a consequence, the overall effluent quality can be less consistent in granular media filtration plants compared with reclaimed water provided by a membrane filtration plant.

As an alternative to depth filtration, surface filtration can be used as pretreatment for membrane filtration or UV disinfection. In surface filtration, particulate matter is removed by mechanical sieving by passing water through thin filter material that is composed of cloth fabrics of different weaves, woven metal fabrics, and a variety of synthetic materials with openings between 10 to 30 m or larger. Surface filters can be operated at higher filtration rates (3,600–30,000 gal/ft^2 per day; 6,100–51,000 L/m^2 per hour) while achieving lower effluent turbidities than conventional sand filters.

Membrane filters, such as microfiltration (MF) and ultrafiltration (UF), are also surface filtration devices, but they exhibit pore sizes in the range from 0.08 to 2 m for MF and 0.005 to 0.2 m for UF. In addition to removing suspended matter, MF and UF can remove large organic molecules, large colloidal particles, and many microorganisms. The advantages of membrane filtration as compared with conventional filtration are the smaller space requirements, reduced labor requirements, ease of process automation, more effective pathogen removal (in particular with respect to protozoa and bacteria), and potentially reduced chemical demand. An additional advantage is the generation of a consistent effluent quality with respect to suspended matter and pathogens. This treatment usually results in effluent turbidities well below 1 NTU (Asano et al., 2007). The drawbacks of this technology are potentially higher capital costs, the limited life span of membranes requiring replacement, the complexity of the operation, and the potential for irreversible membrane fouling that reduces productivity. Unlike robust conventional media filters, membrane systems require a higher degree of maintenance and strategies directed to achieve optimal performance. More detail about MF and UF membranes and their operation in reuse applications is provided in the following sections as well as in Asano et al. (2007).

Removal of Organic Matter and Trace Organic Chemicals

The following sections describe processes that are designed to remove organic matter and trace organic chemicals from reclaimed water. These processes include membrane filtration (MF, UF, NF, and RO), adsorption onto activated carbon, biological filtration, and chemical oxidation (chlorine, chloramines, ozone, and UV irradiation).

Microfiltration and Ultrafiltration

MF and UF membrane processes can be configured using pressurized or submerged membrane modules. In the pressurized configuration, a pump is used to pres-

surize the feedwater and circulate it through the membrane. Pressurized MF and UF units can be operated in two hydraulic flow regimes, either in cross-flow or dead-in filtration mode. In a submerged system, membrane elements are immersed in the feedwater tank and permeate is withdrawn through the membrane by applying a vacuum. The key operational parameter that determines the efficiency of MF and UF membranes and operating costs is flux, which is the rate of water flow volume per membrane area. Factors affecting the flux rate include the applied pressure, fouling potential, and reclaimed water characteristics (Zhang et al., 2006). Flux can be maintained by appropriate cross-flow velocities, backflushing, air scouring, and chemical cleaning of membranes. Typically, MF and UF processes operate at flux rates ranging from 28 to 110 gal/ft^2 per day (48 to 190 L/m^2 per hour) (Asano et al., 2007).

MF and UF membranes are effective in removing microorganisms (Figure 4-3). It is generally believed that MF can remove 90 to 99.999 percent (1 to 5 logs) of bacteria and protozoa, and 0 to 99 percent (0 to 2 logs) of viruses (EPA, 2001; Crittenden et al., 2005). However, filtration efficiencies vary with the type of membrane and the physical and chemical characteristics of the wastewater, resulting in a wide range of removal efficiencies for pathogens (NRC, 1998). MF and certain UF membranes should not be relied upon

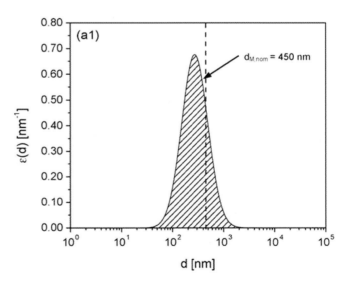

FIGURE 4-3 Pore size distribution of a microfiltration membrane.
SOURCE: Pera-Titus and Llorens (2007).

for complete removal of viruses for several reasons (Asano et al., 2007). First, whereas the terms micro- and ultrafiltration nominally refer to pore sizes that have cutoff characteristics as shown in Figure 4-3, the actual pore sizes in today's commercial membranes often vary over a wide range. Second, experience has shown that today's membrane systems sometimes experience problems with integrity during use for a variety of reasons. Although membrane integrity tests have been developed and these tests are widely used, they are not suitable for detecting imperfections small enough to allow viruses to pass.

Nevertheless, it is generally believed that the new generation of filtration systems has significantly improved performance for microbial removal. For example, Orange County Water District (OCWD) compared the MF filtration result of their current groundwater replenishment system (GWRS) operation initiated in 2008 (see Table 2-3) with data collected during Interim Water Factor 21 (IWF21), the precursor to the GWRS project, started in 2004. Although the influent water quality was similar for both projects, IWF21 MF filtrate showed breakthrough of total coliform in 58 percent of the samples and *Giardia* cysts in 23 percent of samples, whereas both were absent in the GWRS MF filtrates (OCWD, 2009). However, MF did not eliminate viruses. Coliphages were present in GWRS after MF treatment. The geometric mean of male-specific coliphage was 134 plaque-forming units (pfu)/100 mL in MF-treated water (OCWD, 2009). Combining MF with chlorination is likely to improve the rate of virus removal. The OCWD reports significant reduction of coliphage in the MF feed in the presence the chloramine residual. Male-specific coliphage dropped from a geometric mean of 1,800 pfu/100 mL in the previous year to 28 pfu/100 mL in the MF feed and they were absent in the MF filtrate (OCWD, 2010).

MF and UF membranes sometimes in combination with coagulation can also physically retain large dissolved organic molecules and colloidal particles. Effluent organic matter and hydrophobic trace organic chemicals can also adsorb to virgin MF and UF membranes, but this initial adsorption capacity is quickly exhausted. Thus, adsorption of trace organic chemicals is not an effective mechanism in steady-state operation of low-pressure membrane filters.

Nanofiltration or Reverse Osmosis

For reuse projects that require removal of dissolved solids and trace organic chemicals and where a consistent water quality is desired, the use of integrated membrane systems incorporating MF or UF followed by NF or RO may be required. RO and NF are pressure-driven membrane processes that separate dissolved constituents from a feedstream into a concentrate and permeate stream (Figure 4-4). Treating reclaimed water with RO and NF membranes usually results in product water recoveries of 70 to 85 percent. Thus, the use of NF or RO results in a net loss of water resources through disposal of the brine concentrate. RO applications in water reuse have been favored in coastal settings where the RO concentrate can be conveniently discharged to the ocean, but inland applications using RO are restricted because of limited options for brine disposal (see NRC [2008b] for an in-depth discussion of alternatives for concentrate disposal and associated issues). Thus, existing inland water reuse installations employing RO membranes are limited in capacity and commonly discharge brine to the sewer or a receiving stream provided that there is enough dilution capacity.

Most commonly used RO and NF membranes provide apparent molecular weight cutoffs of less than 150 and 300 Daltons, respectively, and are therefore highly efficient in the removal of organic matter and selective for trace organic chemicals. Some of the organic constituents that are only partially removed by NF and RO membranes while still achieving total organic carbon (TOC) concentrations of less than 0.5 mg/L are low-molecular-weight organic acids and neutrals (e.g., N-nitrosodimethylamine [NDMA], 1,4-dioxane) as well as certain disinfection byproducts (e.g., chloroform) (Bellona et al., 2008). Recent advances in membrane development have resulted in low-pressure RO membranes and NF membranes that can be operated at significantly lower feed pressure while providing approximately the same product water quality. However, certain monovalent ions (e.g., Cl^-, Na^+, NO_3^-) are only partially rejected by NF, and NF membranes result in product water with higher TDS than RO (Bellona et al., 2008).

Today, most integrated membrane systems applied in reuse employ RO rather than NF. However, certain low-pressure NF membranes offer opportunities for

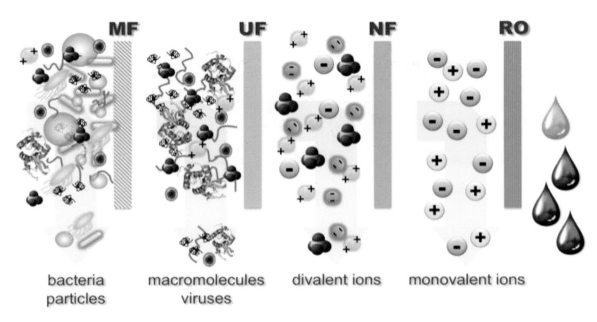

FIGURE 4-4 Substances and contaminants nominally removed by pressure-driven membrane processes. SOURCE: Cath (2010).

wider applications in water reclamation projects because they have lower energy requirements and can achieve selective rejection of salts and organic constituents that results in less concentrated brine streams. For wastewater applications, RO and ultra-low-pressure RO membrane facilities typically operate at feed pressures between 1,000 and 2,100 kPa (approximately 150–300 psi) in order to produce between 8.5 and 12.5 gal/ft^2 per day (13.5 and 20 L/m^2 per hour) of permeate (Lopez-Ramirez et al., 2006). NF membranes, while achieving a similar product water quality with respect to TOC and trace organic chemicals, can be operated at 2 to 4 times lower feed pressures, resulting in significantly greater energy savings than conventional RO membranes (Bellona and Drewes, 2007).

RO and NF membranes, in theory, should remove all pathogens from the feedwater because they are designed to remove relatively small molecules. However, some earlier testing results have shown that the removal of virus surrogates (coliphage) seeded in front of RO is sometimes incomplete. For example, studies conducted by the City of San Diego noted coliphage breakthrough in the permeate of the RO system at concentrations up to 10^3 pfu/100 mL (Adham et al., 1998). Early tests showed inadequate removal of protozoa and bacteria

as well. Leaks around the seals and connectors were suspected as the cause of reduced microbial removal efficiency, but once faulty connectors and an obviously flawed membrane element were identified and replaced, rejection of bacteria and protozoa seemed absolute, but the removal of the surrogate coliphage MS2 remained slightly above 2.5 logs (99.7 percent). Expansion of both bench- and pilot-scale testing to include a variety of manufacturers revealed that the quality of brackish water RO membranes ranged widely, with one manufacturer consistently demonstrating complete rejection in both types of tests. Though systematic tests are not available, newer RO systems may have significantly improved performance for microbial removal. Recent tests have shown promising results (Lozier et al., 2006) and data collected in 2008 at OCWD's GWRS revealed the absence of native coliphage in 1-L samples of RO effluent,[1] which indicated an improvement from the earlier pilot study (27 percent RO breakthrough rates) using an older generation of membranes (OCWD, 2010).

---

[1] Randomly selected RO permeate samples taken from each of 15 RO trains, each sampled three times (M. Wehner, OCWD, personal communication, 2011)

Activated Carbon

In water reclamation, adsorption processes are sometimes used to remove dissolved constituents by accumulation on a solid phase. Activated carbon is a common adsorbent, which is employed as powdered activated carbon (PAC) with a grain diameter of less than 0.074 mm or granular activated carbon (GAC), which has a particle diameter greater than 0.1 mm. During water reclamation, PAC can be added directly to the activated sludge process or solids contact processes, upstream of a tertiary filtration step. GAC is used in pressure and gravity filtration. Activated carbon is efficient for the removal of many regulated synthetic organic compounds as well as unregulated trace organic chemicals exhibiting properties of high and moderate hydrophobicity (e.g., steroid hormones, triclosan, bisphenol A) (Snyder et al., 2006a). Although PAC needs to be disposed of after its adsorption capacity is reached, GAC can be regenerated either on- or offsite, providing this practice is more cost-effective than disposing it via landfills. Onsite GAC regeneration is only cost-effective for large installations and is currently not practiced by any water reclamation facility in the United States. GAC adsorbents are characterized by short empty-bed contact times (i.e., 5-30 min) and preferably a large throughput volume (i.e., bed volumes of 2,000 to 20,000 m^3/m^3) (Asano et al., 2007).

Biological Filtration

As mentioned previously in this chapter, the use of strong oxidants, such as ozone or ozone/peroxide and UV/peroxide, results in the formation of various biodegradable byproducts (Wert et al., 2007). For instance, simple aldehydes, ketones, and carboxylic acids are produced as ozone oxidizes organic matter in water. The aggregate measurements commonly employed to assess the biodegradability of transformation products is assimilable organic carbon (AOC) (Hammes and Egli, 2005) and biodegradable dissolved organic carbon (BDOC) (Servais et al., 1987). This readily biodegradable carbon has been implicated in the acceleration and promotion of biofilm growth in distribution systems. Thus, drinking water treatment facilities usually employ biofiltration after ozonation

to reduce BDOC with the aid of indigenous bacteria present in the feedwater. Additionally, the use of biofiltration after ozone also has been shown to reduce the formation of some byproducts formed during secondary disinfection with chlorine (Wert et al., 2007). Some studies have also demonstrated that the byproducts from ozonation of trace organic chemicals, such as steroid hormones and pharmaceuticals, also are largely biodegradable (Stalter et al., 2010); therefore, there is growing support for the use of biofiltration after ozone or AOP. Although biofiltration alone may provide some direct benefit in terms of removing trace organic chemicals, it has generally been shown to be only marginally effective without a prior oxidation step (Juhna and Melin, 2006).

Biological filtration can be accomplished using traditional media (i.e., sand/anthracite) or using activated carbon (biologically activated carbon [BAC]). Although some studies have suggested that activated carbon is superior for supporting biological growth, mainly because of superior adherence of the biofilm to the GAC, there are some conflicting reports that show approximately equal performance using anthracite (Wert et al., 2008). Some studies have demonstrated that BAC is capable of adsorption as well as biological degradation; however, the adsorptive capacity of the BAC will eventually be reduced as the micropores in the carbon structure become blocked and the adsorptive capacity subsequently becomes exhausted. At this point, fresh GAC will be required to restore the adsorptive capacity, but effective biological activity as measured by reduction of AOC or BDOC will take time to establish. The amount of time needed to develop a biologically active filter will depend on water quality, water temperature, and operational parameters. An important factor in establishing and maintaining an active biofilm is the backwash frequency with chlorinated water.

One major disadvantage of using biological filtration is the detachment of biofilm and likely detection of bacteria in filtered water. Although these bacteria are not harmful, the detection of heterotrophic bacteria could in some cases lead to regulatory violations. In those cases, biofiltration would generally be followed by a disinfection step, such as chlorination or UV irradiation.

Chemical Oxidation

Chemical oxidation is commonly employed in water treatment to achieve disinfection, as described previously in this chapter; however, oxidants are also used to remove tastes, odors, and color and to improve the removal of metals (Singer and Reckhow, 2010). Oxidants used for water treatment include chlorine, chloramine, ozone, permanganate, chlorine dioxide, and ferrate. Advanced oxidation relies upon formation of powerful radical species, primarily hydroxyl radicals (OH·) and is rapidly gaining in use for the oxidation of more resistant chemicals, such as many trace organic chemicals and industrial solvents (Esplugas et al., 2007). The most commonly employed advanced oxidation techniques in water reclamation use hydrogen peroxide coupled with UV light or ozone gas. The UV light itself is not strictly an oxidant but it does selectively transform a small group of compounds sensitive to direct photolysis (e.g., NDMA, iohexol, triclosan, acetaminophen, diclofenac, sulfamethoxazole) (Pereira et al., 2007; Snyder et al., 2007; Yuan et al., 2009; and Sanches et al., 2010).

Very few oxidative technologies are employed at operational conditions capable of mineralizing organic materials in water. Even the most promising advanced oxidation techniques using ozone and UV irradiation combined with peroxide will result in only a minor (if any) measurable reduction of dissolved organic carbon (DOC). Regardless of the oxidation technique deployed and superior performance of trace organic chemical removal, some transformation products will result that are often uncharacterized (see Chapter 3 for additional discussions of transformation byproducts). The most commonly used oxidation methods for the removal of trace organic contaminants are described below.

Chlorine. Chlorine, defined here as the combination of chlorine gas, HOCl, and OCl⁻, reacts selectively with electron-rich bonds of organic chemicals (e.g., double bonds in aromatic hydrocarbons) (Minear and Amy, 1996). Recently, several reports have shown that many trace organic chemicals containing reactive functional groups can be oxidized by free chlorine (Adams et al., 2002; Deborde et al., 2004; Lee et al., 2004, Pinkston and Sedlak, 2004; Westerhoff et al., 2005), while ketone steroids (e.g., testosterone and progesterone) are

not as effectively oxidized (Westerhoff et al., 2005). However, the ability of chlorine to effectively oxidize trace organic chemicals, including steroid hormones, is a function of contact time and dose. More importantly, chlorine is not expected to mineralize trace organic chemicals, but rather to transform them into new products (Vanderford et al., 2008), which may in fact be more toxic than the original molecule.

Chloramines. Chloramines are not nearly as effective as oxidants and thus play a much smaller role in trace organic chemical oxidation. Snyder (2007) demonstrated that a dose of 3 mg/L chloramines and a contact time of 24 hours was able to effectively oxidize phenolic steroid hormones (e.g., estrone, estradiol, estriol, ethinyl estradiol) as well as triclosan and acetaminophen; however, the vast majority of trace organic chemicals studied were not significant oxidized. Therefore, although chloramines play an important role in reduction of membrane fouling and disinfection, only minimal expected benefit in oxidation of trace organic chemicals will result. Moreover, careful evaluation of nitrosamine formation should be undertaken when using chloramines, considering the carcinogenic potency of these byproducts (see Choi et al., 2002; Mitch et al., 2003; Haas, 2010).

Ozone. Ozone (O_3) is a powerful oxidant and disinfectant that decays rapidly and leaves no appreciable residual in reclaimed water during storage and distribution. Ozone-enriched oxygen is generally added to water through diffusers producing fine bubbles, and once dissolved in water, ozone quickly undergoes a cascade of reactions, including decomposition into hydroxyl radicals (OH·), hydroperoxyl radical (HO_2), and superoxide ion (O_2^-). These radicals along with molecular ozone will rapidly react with organic matter, carbonate, bicarbonate, reduced metals, and other constituents in water. The reactions mediated by the hydroxyl radical are relatively nonselective, whereas molecular ozone is more selective (Elovitz et al., 2000).

Because of ozone's ability to oxidize organic chemicals, it has been widely applied in water treatment for taste and odor control, color removal, and to reduce concentrations of trace organic chemicals. At dosages commonly employed for disinfection, the vast majority of contaminants can be effectively converted into transformation products (Snyder et al., 2006c). Although several studies have shown that ozone effectively reduces estrogenic potency in reclaimed water (Snyder et al., 2006c), recent publications have suggested that biologically active filters be included after the ozone process in order to remove biodegradable byproducts formed during ozonation (Stalter et al., 2010). For potable reuse applications, ozonation could also be applied after soil aquifer treatment (SAT), which combines the benefits of a more selective oxidation of remaining chemicals persistent to biodegradation and a lower ozone demand due to reduced DOC concentrations in the recovered water.

It is well known that in the presence of bromide, ozone can form bromate, a toxic byproduct. There are steps that can be employed to mitigate the formation of bromate, such as the use of chlorine and ammonia before ozone addition (von Gunten, 2003). Some reports have shown that ozone applied before chloramination also results in the oxidation of nitrosamine precursors (Lee et al., 2007). However, ozone also has been shown to form some nitrosamines directly (von Gunten et al., 2010).

Ozone can play an important role in water reclamation, but the process is more energy intensive and operationally complex than chlorination. In cases where trace organic chemical removal (e.g., pharmaceuticals, steroid hormones) is important, ozone is a viable option and does not result in a residuals stream like NF or RO membrane processes or in spent media as with activated carbon. However, ozone does not provide a complete barrier to trace organic chemicals, and there are certain chemicals that are not amendable to oxidation (e.g., chlorinated flame retardants; artificial sweeteners) (Snyder et al., 2006c).

UV irradiation. UV light at doses commonly employed for disinfection (40–80 mJ/cm^2) is largely ineffective for trace organic chemical removal. In a recent study that investigated the removal of trace organic chemicals from water, none of the target compounds investigated were well removed (>80 percent oxidized) using UV at disinfection doses (Snyder, 2007). However, when UV doses are significantly increased (generally by 10-fold) and high doses of hydrogen peroxide (5 mg/L and higher) are added, most trace organic chemicals were effectively oxidized (Snyder et al., 2006c). Activated carbon is sometimes employed to catalytically remove hydrogen peroxide, and other chemicals can be used

to remove excess peroxide from the UV-AOP effluent. Although UV-AOP does form transformation products (i.e., it does not result in mineralization of organic compounds), it does not form bromate. Additionally, UV alone at elevated dosages or in combination with hydrogen peroxide (UV-AOP) effectively removes NDMA.

UV-AOP efficacy, however, is quite susceptible to water quality and requires proper pretreatment. In many potable reuse applications, UV-AOP is applied after RO treatment to negate the detrimental impacts of water quality, such as suspended and particulate matter and DOC. UV-AOP applications generally will require extensive pretreatment to increase UV transmittance; however, recent studies have demonstrated that UV-AOP can be also effective in advanced-treated effluents (Rosario-Ortiz et al., 2010).

Removal of Dissolved Solids

Domestic and commercial uses of public water supplies result in an increase in the mineral content of municipal wastewater. This increase can be problematic where drinking water supplies are already elevated in TDS and regional water reuse is already occurring, resulting in partially closed water and salt cycles. Hard water can also be a problem because it results in the proliferation of self-regenerating water softeners, which discharge their regenerant into the wastewater collection system. To mitigate salinity problems associated with local water reuse activities, especially in inland applications, partial desalination of reclaimed water especially for potable reuse projects may be required.

In addition to pressure-driven membrane-based separation processes, such as NF and RO, as discussed above, current-driven membrane processes, such as electrodialysis (ED) or electrodialysis reversal (EDR), can be used to separate salts. Nevertheless, ED and EDR are not commonly employed in water reclamation and currently only one facility in Southern California is using EDR to remove TDS at a demonstration-scale facility. Precipitative softening can also be used for partial demineralization (mainly to remove hardness) and is currently employed for this purpose in the City of Aurora's Prairie Waters Project, Colorado (see also Box 4-1).

The Prairie Waters Project, established by the City of Aurora, Colorado, in 2010, is a potable water reuse augmentation project that will increase Aurora's water supply by 20 percent; delivering up to 9 MGD (34,000 m^3/d). The project is using return flows discharged to the South Platte River downstream of Denver. This water is recovered through a series of 17 vertical riverbank filtration wells, followed by artificial recharge and recovery (ARR), providing a retention time of approximately 30 days in the subsurface. The water is subsequently pumped to an advanced water treatment plant. The water treatment plant consists of precipitative softening, UV-AOP, biologically active carbon filtration, and granular activated carbon (GAC) filtration. At a ratio of 2:1, the final product water is blended with Aurora's current supply using mountain runoff water prior to disinfection and final distribution. Precipitative softening is employed to maintain a hardness level that is similar to Aurora's current supply. Riverbank filtration and ARR are very efficient in removing pathogens, organic carbon, trace organic chemicals, and nitrate (Hoppe-Jones et al., 2010). UV-AOP and GAC serve as an additional barrier for trace organic chemicals that might survive after the natural treatment process. The treatment scheme was selected because alternatives such as reverse osmosis with zero liquid discharge of brine or wetland treatment instead of riverbank filtration were cost-prohibitive or not viable.

SOURCE: http://www.prairiewaters.org.

ENGINEERED NATURAL PROCESSES

Natural processes in water reclamation are usually employed in combination with aboveground engineered processes and consist of managed aquifer recharge systems and natural or constructed wetlands (Figure 4-1). Natural systems can be considered as multiobjective treatment processes targeting the removal of pathogens, particulate and suspended matter, DOC, trace organic chemicals, and nutrients, either as the key treatment process or as an add-on polishing step. All natural treatment processes combine the advantage of a low carbon footprint (i.e., little to no chemical input, low energy needs) with little to no residual generation. The drawbacks of these processes are the required footprint and a suitable geology, which might not be available where the use of natural treatment systems is desired. Examples of managed natural processes in re-

use systems and the general role of environmental buffers in potable reuse projects are described in Chapter 2, but water quality improvements provided by these surface and subsurface natural systems are described in the subsequent sections.

Subsurface Managed Natural Systems

Subsurface managed natural systems can be used to enhance water quality and/or to provide natural storage for reclaimed water. These systems include surface spreading basins, vadose zone wells, and riverbank filtration wells, which take advantage of attenuation processes that occur in the vadose zone and saturated aquifer. Other processes, such as aquifer storage and recovery (ASR) and direct injection wells, introduce highly treated reclaimed water directly into a potable aquifer.

In general, subsurface treatment applications offer numerous advantages. These systems typically require a low degree of maintenance, and the energy requirements are low. The input of chemicals usually is not required, and the operation is residual free. Temperature equilibration of water is achieved during subsurface storage and excursions in water quality are buffered due to dispersion in the subsurface and dilution with native groundwater. However, subsurface applications require that a substantial aquifer be available and that it be characterized by an extensive site assessment. Although the advantages seem to outweigh the disadvantages from an operational standpoint, the lack of clear and standardized guidance for design and operation of these system limits wider establishment of managed subsurface treatment systems. Lack of process understanding can result in less-than-optimal performance or physical footprints or retention times that are larger than needed for the desired water quality improvements. Some installations might also exhibit deterioration of water quality in the recovered water due to biogeochemical reactions in the subsurface that were not anticipated.

Surface Spreading or Soil Aquifer Treatment

Surface spreading basins allow reclaimed water to infiltrate slowly through the vadose zone, where sorption, filtration, and biodegradation can enhance the water quality (also called soil aquifer treatment). Recharge basins for surface spreading operations are often located in, or adjacent to, floodplains, characterized by soils with high permeability. In some instances, excavation is necessary to remove surface soils of low permeability. For mosquito control and to maintain permeability during operation with reclaimed water, recharge basins are usually operated in alternate wet and dry cycles. As the recharge basin dries out, dissolved oxygen penetrates into the subsurface, facilitating biochemical transformation processes, and organic material accumulated on the soil surface will desiccate, allowing for the recovery of infiltration rates (Fox et al., 2001).

The removal of organic matter during SAT is highly efficient and largely independent of the level of aboveground treatment. Biodegradable organic carbon that is not attenuated during wastewater treatment represents an electron donor for microorganisms in the subsurface and is readily removed during groundwater recharge (Drewes and Fox, 2000; Rauch-Williams and Drewes, 2006). Monitoring efforts revealed consistent removal of TOC between 70 and 90 percent at full-scale SAT facilities that were in operation for several decades (Quanrud et al., 2003; Drewes et al., 2006; Amy and Drewes, 2007; Lin et al., 2008; Laws et al., 2011). The removal of easily biodegradable organic carbon in the infiltration zone usually results in depletion of oxygen and the creation of anoxic conditions. Although this transition is advantageous to achieve denitrification, it might also lead to the solubilization of reduced manganese, iron, and arsenic from native aquifer materials. If these interactions occur, appropriate post-treatment is required after recovery of the recharged groundwater.

Previous studies have characterized the transformation and removal of select trace organic chemicals during SAT for travel times ranging from ~1 day to 8 years (Drewes et al., 2003a. Montgomery-Brown et al., 2003; Snyder et al., 2004; Grünheid et al., 2005; Massmann et al., 2006; Amy and Drewes, 2007). Several studies also report efficient removal of NDMA and other nitrosamines under both oxic and anoxic subsurface conditions (Sharp et al., 2005; Drewes et al., 2006; Nalinakumari et al., 2010). A case study conducted at a facility in Southern California (Box 4-2) illustrates the efficiency of short-term SAT for the attenuation of trace organic chemicals in reclaimed water (Laws et al., 2011).

Previous studies have demonstrated that the com-

bination of filtration and biotransformation processes during subsurface treatment is very efficient for the inactivation of pathogens, especially viruses (Schijven et al., 2000, 2002; Quanrud et al., 2003; Azadpour-Keeley and Ward, 2005; Gupta et al., 2009). Attenuation of pathogens depends primarily on three mechanisms— straining, inactivation, and attachment to aquifer grains (McDowell-Boyer et al., 1986). Findings from field studies demonstrated that infiltration into a relatively homogeneous sandy aquifer can achieve up to 8 log virus removal over a distance of 30 m in about 25 days (Dizer et al., 1984; Yates et al., 1985; Powelson et al., 1990; Schijven et al., 1999, 2000). During SAT in the Dan Region Project, Israel, Icekson-Tal et al. (2003) measured 5.3 log removal of total coliform and 4.5 log removal of fecal coliform bacteria. The efficient removal of fecal and total coliform bacteria during subsurface treatment and essentially their absence in groundwater abstraction wells after SAT or riverbank filtration was confirmed by various other studies (Fox et al., 2001; Hijnen et al., 2005b; Levantesi et al., 2010). Other field studies have focused on attenuation of protozoa, and findings suggest that efficient removal occurs during passage across the surface water—groundwater interface and lesser removal is observed during groundwater transport away from this interface (Schijven et al., 1998). Further details on pathogen attenuation during SAT are provided in Chapter 7. An example of the degree of attenuation for various microbial and chemical constituents that can be achieved in SAT systems is illustrated in Tables A-7 and A-9 (Appendix A).

Nitrogen removal needs to be carefully managed when reclaimed water is applied with total nitrogen concentrations in excess of 20 mg N/L. At such high concentrations, the wetting and drying cycles of the spreading basins cannot meet the nitrogenous oxygen demand (in excess of 100 mg/L), resulting in incomplete nitrification. Ammonium is usually removed by cation exchange onto soil particles during wetting cycles, followed by nitrification of the adsorbed ammonium during drying cycles. Nitrate is not adsorbed to soils, but if sufficient carbon is present to create anoxic conditions, nitrate can be removed via denitrification during subsequent passage in the subsurface (Fox et al., 2001). Reclaimed water with nitrate concentrations in excess of 10 mg N/L can result in incomplete denitrification when applied to groundwater recharge basins because the biodegradable organic carbon usually present in a secondary or advanced-treated effluent will be insufficient to achieve complete denitrification.

For potable reuse projects, different regulatory requirements exist regarding the minimum retention time of reclaimed water in the subsurface prior to extraction. The primary intent of these regulations is to provide additional protection against pathogens in groundwater recharge projects and to provide time for corrective action in the event that substandard water is inadvertently recharged. Regulations in the state of Washington require a minimum of 6 months of hydraulic retention time in the subsurface for surface spreading operations and a minimum of 12 months for direct injection projects before the water can be recovered as a potable water source (Washington Department of Health and Washington Department of Ecology, 1997), while California's draft groundwater recharge regulations require a minimum of 2 months in the subsurface for both surface spreading and injection projects to provide time for corrective action if substandard water is inadvertently recharged (CDPH, 2011). Others have defined minimum setbacks (i.e., horizontal separation) between reclaimed water surface spreading operations and potable wells (e.g., 500 ft [150 m] in Florida; 2,000 ft [610 m] in Washington) (FDEP, 2006; Washington Department of Health and Washington Department of Ecology, 1997). However, these setbacks or minimal retention times are frequently not based on scientific findings but represent a conservative estimate to provide additional removal credits for pathogens in case of a failure in the aboveground treatment train. Reuse regulations are discussed in more detail in Chapter 10.

Riverbank Filtration

Riverbank filtration has been practiced in the United States for more than 50 years for domestic drinking water supplies utilizing streams that might have been compromised in their quality due to the discharge of wastewater effluents or other waste streams (Ray et al., 2008). Recently, water reuse projects have integrated riverbank filtration into their treatment process train to take advantage of the benefits of this natural treatment system (see Box 4-1). Aquifers used for riverbank filtration usually consist of alluvial sand and gravel deposits, with thickness ranging from 15-200 feet (5–60 m) and a hydraulic conductivity higher than 10^{-4} m/s. In riverbank filtration, constant scour forces due to streamflow prevent the accumulation of particulate and colloidal organic matter in the infiltration layer.

Biodegradation of organic matter represents a key attenuation mechanism of riverbank filtration processes (Kühn and Müller, 2000; Hoppe-Jones et al., 2010). A bioactive filtration layer forms near the water/sediment interface where dissolved oxygen concentrations are highest, which can cause significant removal of DOC during the initial phase of infiltration (first meter). Conditions can quickly transition from oxic to anoxic as the water travels with increasing distance from the river through the subsurface, although the oxidation-reduction gradient depends on site specific conditions, such as DOC and ammonia concentrations in the river (Hiscock and Grischek, 2002; Ray et al., 2008).

More than 5-log removal of pathogen surrogate microorganisms (e.g., bacteria, viruses, and parasites) has been reported in riverbank filtration under steady-state conditions, with variations of ±1-log removal efficiency associated with individual microorganism characteristics (Medema et al., 2000). Havelaar et al. (1995) reported removal in excess of 5 logs for total coliform during transport of river water over a 30-m distance from the Rhine River and over a 25-m distance from the Meuse River to a well. Total coliforms were rarely detected in riverbank-filtered waters, with 5.5- and 6.1-log reductions in average concentrations in wells relative to river water (Weiss et al., 2005). Havelaar et al. (1995) reported 3.1-log removal of protozoa surrogates during transport over a 30-m distance from the Rhine River to a well and 3.6-log removal over a 25-m distance from the Meuse River to a well. Schijven et al. (1998) measured 1.9-log removal for protozoa surrogates over a 2-m distance from a canal. This finding is consistent with field monitoring results from a riverbank filtration site in Wyoming, where Gollnitz et al. (2005) observed a 2-log removal of *Cryptosporidium* surrogates in groundwater wells characterized by flowpaths between 20 and 984 ft (6 and 300 m). At a riverbank filtration site at the Great Miami River, Gollnitz et al. (2003) reported a 5-log removal of protozoa surrogates in a production well located 98 ft (30 m) off the river.

Numerous research projects have documented the removal of trace organic compounds during riverbank filtration. For example, Ray et al. (1998) and Vers-

BOX 4-2
Montebello Forebay Groundwater Recharge Operation, California

In the United States, drinking water augmentation with reclaimed water was pioneered by the County Sanitation Districts of Los Angeles County (CSDLAC) and the Water Replenishment District of Southern California (WRD) by establishing groundwater recharge spreading operations with reclaimed water in Pico Rivera, California in the early 1960s. Laws et al. (2011) studied the fate and transport of bulk organic matter and a suite of 22 trace organic chemicals during the surface-spreading recharge operation using a smaller but well-instrumented test basin at this facility. Two monitoring wells were located at the side of the recharge basin and lysimeters were installed beneath the basin (see figure below). Based on ion signatures it appeared that all of the samples collected originated from reclaimed water that was applied to the basin; however, the samples from the deeper wells (PR 8 and 10) appeared to have been diluted by native groundwater.

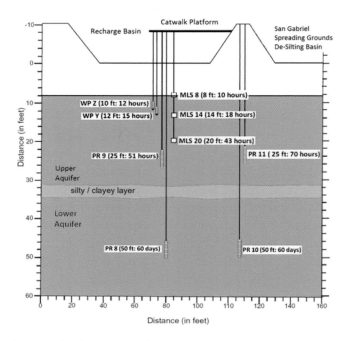

Instrumentation of groundwater recharge test basin associated with monitoring data provided in table below.
SOURCE: Laws et al. (2011).

traeten et al. (1999) reported 50 to 75 percent removal of the herbicide atrazine during riverbank filtration, although the underlying removal mechanisms were not clear. Despite the success of riverbank filtration in removing numerous compounds, certain trace organic chemicals have been regularly found in the product water of riverbank filtration systems, including urotropin (an aliphatic amine) and 1,5-naphthalindisulfonate (an aromatic sulfonate) (Brauch et al., 2000), antiepileptic drugs (e.g., carbamazepine, primidone), a blood-lipid regulator (e.g., clofibric acid), antibiotics (e.g., sulfamethoxazole), and x-ray contrast media were present

in both river water and bank-filtered water (Kühn and Müller, 2000; Schmidt et al., 2004; Hoppe-Jones et al., 2010; Maeng et al., 2010). A partial reduction in concentration was only achieved under certain redox conditions and through dilution with local groundwater.

Direct Injection

Direct injection of reclaimed water may occur in both saturated and unsaturated aquifers using wells that are constructed like regular pumping wells. In the United States, OCWD pioneered direct injection of

Over a travel time of less than three days in the upper aquifer, approximately 55 percent of the total organic carbon was removed (from 7.8 mg/L to 3.5±0.3 mg/L), and overall removal increased to 79 percent with increased travel time (60 days). Most of the observed removal occurred in the vadose zone (<2.4 m) because of its aerobic conditions. Attentuation of trace organic chemicals also occurred in the vadose zone, where concentrations decreased within the first 2.4 m (~10 hours). After 60 days travel time, the concentrations of monitored trace organic chemicals decreased further (see table below). Concentrations of primidone, carbamazepine, trimethoprim, N,N-diethyl-meta-toluamide (DEET), meprobamate, tris (2-chloroethyl) phosphate (TCEP), tris (2-chloroisopropyl) phosphate (TCPP), and triclosan were reduced less than 10 percent in the upper aquifer but contaminant attenuation increased with travel time (Laws et al., 2011).

Concentration of Select Trace Organic Chemicals in Reclaimed Water After Surface Spreading

Compound	Basin	Avg. (MLS 8-PR 11) (10-70 hrs)	Avg. (PR8, PR10) (60 days)
Atrazine	<5	5±0.2	4.0±0.1
TCEP	400	402±15	128±39
		6,483±87	
TCPP	7,200	5	797±188
Benzophenone	<1000	68±27	<50
DEET	320	238±60	50±12
Musk Ketone	<25	<MRL	<25
Triclosan	6.5	6±3	<1
Atenolol	830	31±34	<1
Atrovastatin	<10	<MRL	<0.5
Carbamazepine	330	302±28	170±0
Diazepam	<5	2±0.3	1.5±0.3
Diclofenac	24	10±2	<0.5
Fluoxetine	13	0.57±0.16	<0.5
Gemfibrozil	880	70±63	32±2.9
Ibuprofen	10	12±8	1.3±0
Meprobamate	430	375±45	132±31
Naproxen	32	6±3	2.4±0
Phenytoin	150	103±13	85±8
Primidone	150	168±45	90± 2.6
Sulfamethoxazole	460	390±129	207±12
Trimethoprim	54	58±33	3.5±2.6
Iopromide	2,700	60±41	89±18

SOURCE: Laws, et al. (2011).
NOTE: Monitoring wells (shown in the figure to the left) represent different travel times.

highly treated reclaimed water in 1976 for seawater intrusion barriers in Southern California (see Chapter 2 and Box 2-11). Direct injection wells may also be used as ASR wells where the same well serves for both injection and recovery (see also NRC, 2008c). For direct injection projects leading to drinking water augmentation, the reclaimed water is required to meet drinking water standards in addition to project-specific water quality criteria before it is injected into a potable aquifer. In these systems, the additional treatment provided in the subsurface is usually limited to temperature equilibration and blending with ambient groundwater.

Storing reclaimed water after direct injection in the subsurface may also provide additional inactivation of any remaining viruses. During a deep-well (~100 ft [300 m] below surface) injection study, Schijven et al. (2000) spiked pretreated surface water with bacteriophage (MS2 and PRD1) and observed a 6-log removal within the first 8 m of travel, followed by an additional 2-log removal during the subsequent 98 ft (30 m) of travel. The degree of water quality transformations can vary with the flow path and contact time in the subsurface. Depending on the geological conditions of the subsurface, water quality degradation is possible;

for example, redox change can result in dissolution of certain constituents from the soil matrix, including iron, manganese, or arsenic.

Infiltration rates of direct injection wells are much higher than infiltration rates in spreading basins, although direct injection wells can become clogged at the interface of the gravel envelope of a well and the aquifer. Considerable research has been conducted to understand factors that contribute to clogging and to develop approaches to evaluate the clogging potential due to biological activity and suspended solids (Asano et al., 2007). These approaches can assess the relative clogging potential of different waters, but they cannot provide an absolute prediction of clogging in injection wells. Therefore, more reliable design and operational criteria are needed for a sustainable operation. The costs of direct injection wells can also be significant where deep aquifers are used for storage, which increases the well construction costs as well as the energy costs for injecting water to maintain proper infiltration rates (Asano et al., 2007).

A high level of pretreatment is usually employed to minimize the risk of clogging and to avoid costs for redeveloping clogged injection wells. In several potable reuse applications in the United States, RO treatment is employed prior to direct injection (Table 2-3, Chapter 2). This degree of treatment, however, reduces the biodegradable organic carbon, thereby limiting the biological activities in the subsurface environments and reducing the effectiveness of the natural subsurface treatment with respect to achieving attenuation of contaminants. At OCWD in the early 2000s, low-molecular-weight compounds, such as NDMA, were present in RO permeate and persisted after direct injection during subsurface transport, presumably because co-metabolic reactions that can remove these compounds were not adequately stimulated in the aquifer (Drewes et al., 2006; Sharp et al., 2007).

Surface Managed Natural Systems

In addition to providing aesthetic benefits and providing habitat and recreational opportunities, managed natural surface water systems can provide benefits with respect to water quality. One of the main differences between surface and subsurface managed natural systems is that managers of surface water systems must frequently satisfy competing demands and multiple objectives. For example, in addition to providing water quality benefits, engineered treatment wetlands frequently serve as habitat for birds and provide recreational and educational benefits for the community. In addition, they have the potential to serve as breeding grounds for mosquitoes and other vectors.

Another important difference between surface and subsurface systems is the way in which the water flows. With the exception of fractured bedrock, the soil and groundwater systems used in managed subsurface treatment processes lead to predictable flow patterns and residence times in the subsurface. In addition, the high surface area provided by soil and geological materials provides ample surface area for microbial growth, facilitating biological attenuation processes. In contrast, managed surface systems often exhibit preferential flow and lower biological activity. As a result, a poorly managed natural system has a higher potential for providing less-effective treatment than expected, with hydraulic short-circuiting and low biological activity leading to little contaminant attenuation.

Treatment Wetlands

Treatment wetlands have been used to treat reclaimed water for nonpotable and potable reuse (see Box 2-10). Treatment wetlands are built as either subsurface-flow or surface-flow systems. Subsurface-flow wetlands consist of plants growing within a gravel bed through which reclaimed water flows whereas surface-flow systems consist of wetland plants growing in anywhere from 0.5 to 2 feet (0.15 to 0.6 m) of flowing surface water with occasional deeper areas to enhance mixing and provide habitat (Kadlec and Knight, 1996). Subsurface wetlands are more common in colder climates and in locations where there are concerns about contact with contaminants in the reclaimed water (e.g., when wetlands are used for treatment of primary effluent). With respect to water reclamation, subsurface-flow wetlands may be better suited for decentralized treatment of primary or secondary effluent (e.g., septic tank effluent) than wastewater from full-scale treatment plants.

Surface-flow wetlands are less expensive to build and maintain and provide better habitat and aesthetic benefits and are therefore more common in warmer climates. Ammonia is usually removed from reclaimed water through nitrification prior to discharge to surface-flow wetlands because ammonia toxicity affects the growth of plants and can be detrimental to resident fish that control mosquitoes.

Surface-flow wetlands frequently provide good removal of contaminants present in wastewater effluent. In particular, ample data indicate that surface-flow wetlands remove nitrate through denitrification in anoxic zones, and phosphorus through settling of particulate phosphate and uptake by growing plants (Kadlec and Knight, 1996). Wetlands also are effective in the removal of particles that settle out at the low flow velocities encountered in the wetland. As a result, wetlands provide removal of particle-associated pathogens and metals. Aerobic microorganisms living near the air–water interface and nitrate-reducing microbes below the surface also can transform organic contaminants as they metabolize decaying plants and organic matter present in the reclaimed water. Concentrations of certain trace organic chemicals, such as trihalomethane disinfection byproducts, also can decrease in treatment wetlands through volatilization (Rostad et al., 2000). Laboratory microcosm studies demonstrate the ability of microorganisms and organic compounds in wetlands to transform numerous trace organic chemicals (Gross et al., 2004; Matamoros et al., 2005; Matamoros and Bayona, 2006; Waltman et al., 2006).

Comparison of results from laboratory- or pilot-scale wetland studies with full-scale systems often indicates that hydraulic short-circuiting can result in significant decreases in treatment efficacy. For example, between 30 and 40 percent of the steroid hormones entering a pilot-scale surface treatment wetland were removed over a hydraulic residence time of approximately 2 days (Gray and Sedlak, 2005). The associated full-scale wetland, which had nearly identical plant species and a nominal hydraulic residence time of over a week should have achieved removals exceeding 90 percent, but monitoring of the inflow and outflow of the full-scale system failed to show significant removals. Presumably, this apparent discrepancy is due to hydraulic short-circuiting, which has been observed in tracer tests of the full-scale system (Lin et al., 2003). These types of findings are consistent with tracer studies of full-scale treatment wetlands that frequently show that preferential flow paths can become

dominant within wetlands, meaning that a large fraction of the flow receives little treatment (Lightbody et al., 2008). Therefore, active management of surface-flow treatment wetlands is crucial to achieving effective treatment.

Reservoirs

As mentioned previously, surface-water reservoirs frequently are managed to preserve or enhance water quality. Procedures for proper management of reservoirs that receive reclaimed water are not well established because there are a limited number of reservoirs that receive reclaimed water and the contribution of reclaimed water to the overall volume of the reservoirs is typically small. Concentrations of trace organic chemicals usually are quite low and it is difficult to assess the potential for removal from reservoirs. There is a clear research need to better understand the contribution of various attenuation processes (i.e., biotransformation, photolysis, sorption to particulate matter, and dilution) for trace organic chemicals and pathogens in surface reservoirs receiving reclaimed water.

Despite these limitations, insight into the potential importance for attenuation of contaminants in reservoirs can be made from data on reservoirs and lakes that receive discharges of wastewater effluent. For example, Poiger et al. (2001) demonstrated that the pharmaceutical diclofenac underwent photolysis in the surface of Lake Griefensee in Switzerland that receives a significant fraction of its overall flow from wastewater treatment plants. Monitoring data and models of the stratified lake demonstrated that diclofenac concentrations were significantly lower in the epilimnion of the lake because photolysis rapidly transformed the compound. Thus, for those compounds that undergo photolysis (e.g., diclofenac, sulfamethoxazole) as well as waterborne pathogens that are inactivated by sunlight, the surface-to-volume ratio of the reservoir and the depth of the drinking water plant intake both could be important to the concentration of contaminants in the water entering the treatment plant.

In recent years, there has been increasing federal attention to the impacts of nutrients on surface water ecosystems. EPA has encouraged states to develop and adopt numeric nutrient criteria for nitrogen and phosphorus, which could affect the viability of surface discharge of reclaimed water without nutrient removal.[2]

CONCLUSIONS

A portfolio of treatment options, including engineered and managed natural treatment processes, exists to mitigate microbial and chemical contaminants in reclaimed water, facilitating a multitude of process combinations that can be tailored to meet specific water quality objectives. Advanced treatment processes are capable of also addressing contemporary water quality issues related to potable reuse involving emerging pathogens or trace organic chemicals. Ways to integrate these technologies through alternative system designs that ensure water quality are discussed in Chapter 5.

Advances in membrane filtration have made membrane-based processes particularly attractive for reuse applications. Membrane advances have resulted in treatment approaches for nonpotable and potable reuse applications that are associated with a smaller space requirement, reduced labor requirement, ease of process automation, more effective pathogen removal (in particular with respect to protozoa and bacteria), consistent effluent quality, and potentially reduced chemical demand. The drawbacks of this technology are potentially higher capital costs, the limited life span of membranes, the complexity of the operation, and the potential for irreversible membrane fouling that reduces productivity. Unlike robust conventional media filters, membrane systems require a higher degree of maintenance and strategies directed to achieve optimal performance.

Environmentally sustainable and cost-effective options for brine disposal are limited in inland areas. Many of the earliest potable water reuse projects were established in coastal communities where brine concentrate from RO systems could be disposed of by ocean discharge. As a result, many coastal utilities still favor RO to mitigate salinity and risks from trace organic chemicals to produce high-quality water for potable reuse. However, limited cost-effective concentrate disposal alternatives hinder the application of membrane technologies for water reuse in inland communities. Instead, inland potable water reuse projects are increasingly relying on treatment trains that do not include RO, such as process combinations that involve managed natural treatment systems, activated carbon, ozonation, or AOP.

The lack of clear and standardized guidance for design and operation of engineered natural systems is the biggest deterrent to their expanded use, in particular for potable reuse applications. Engineered natural systems that replace certain advanced treatment unit processes are compelling from an operational standpoint, but little is known how operating conditions could be modified and retention times shortened to achieve a predictable water quality while using a smaller footprint. Additional research is needed to elucidate key attenuation processes in engineered natural systems and quantify their effects on microbial and chemical constituents of concern so that guidance for design and operation can ultimately be developed. Although each application will still require a thorough site-specific assessment, general design standards and operating procedures as well as appropriate monitoring approaches can foster a wider application of natural systems as part of reuse schemes.

---

[2] See http://water.epa.gov/scitech/swguidance/standards/criteria/nutrients/progress.cfm.

5

Ensuring Water Quality in Water Reclamation

A consistent reclaimed water quality can be achieved through appropriate treatment strategies (e.g., high-level disinfection, process redundancy), technical controls (e.g., alarm shutdowns, frequent inspection procedures), online monitoring devices (e.g., effluent turbidity, residual chlorine concentration), and/or operational controls to react to upsets and variability. Similar to drinking water practices, *quality control* in potable reuse projects is provided by monitoring and operational response plans, whereas *quality assurance* embeds the principle of establishing multiple barriers and an assessment and provision of treatment reliability. This chapter discusses the state of the science of water reuse design and operational principles to ensure water quality. Additionally, the chapter includes discussion of the role of an environmental buffer within the multiple-barrier concept. The committee then summarizes these considerations by presenting 10 steps that can be taken to ensure water quality in potable and nonpotable water reuse projects.

DESIGN PRINCIPLES TO ENSURE QUALITY AND RELIABILITY

The primary goal of any reuse project is that public health is protected continually and the finished water quality is acceptable to consumers. Four elements— monitoring, attenuation, retention, and blending—are typically embedded into the design of both nonpotable and potable reuse schemes to ensure a reclaimed water quality that is suitable for the desired use at all times (see Figure 5-1). The extent of monitoring, contami-

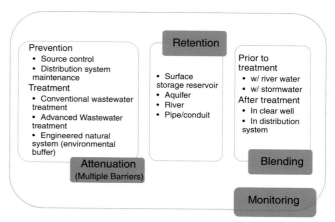

FIGURE 5-1 Four elements often used in the design of non-potable and potable reuse schemes: monitoring, attenuation, retention, and blending.

nant attenuation, retention, and blending required for a particular water reuse application (e.g., industrial, agricultural, potable) will depend on project-specific water quality objectives and the potential impacts from system failure. The following discussions focus primarily on potable reuse applications, for which rigorous quality assurance is essential, although the design concepts can be adapted to nonpotable applications as well.

Water Quality Monitoring

As with conventional drinking water supplies, water quality monitoring for potable water reuse is composed of a combination of online monitoring devices (e.g., filter effluent turbidity, chlorine residual, pH) and discrete measurements using grab or composite samples (e.g., ammonia, nitrate, dissolved organic carbon [DOC], *Eschericia coli*) to ensure the quality of the finished product water. These practices usually follow standards and protocols similar to those applied in drinking water treatment. Although these monitoring controls can fail, the acknowledged imperfection of the monitoring technology is comparable to that of drinking water treatment facilities. In some states potable reuse systems are required to include water retention after discharge from the treatment plant (e.g., in surface or subsurface storage of the product water). In theory, this retention allows time for additional contaminant attenuation and for water to be diverted from the distribution system if water quality problems are detected. However, significant water retention is often not cost-effective for potable reuse projects. Additionally, past experience with water reuse has demonstrated that unanticipated contaminants can be detected in final product water, even when state-of-the-art treatment and monitoring programs are employed (e.g., see Box 3-2 on NDMA).

An idealized monitoring program would measure critical microbial and chemical contaminants in real time in the finished product water before it leaves the reclamation plant. The availability of instantaneous monitoring techniques could allow significant reduction of required reclaimed water retention times. Water quality goals would need to be well defined, and measuring techniques would need to be selected with sensitivity suitable for confirming that water treatment goals have been achieved. Although several new techniques to monitor pathogens and a diverse set of chemicals in real time have recently been proposed (Panguluri et al., 2009; Cahill et al., 2010; Puglisi et al., 2010), significant additional research is required to develop reliable and appropriate approaches to real-time monitoring that are suitable for water reclamation settings. Also, to be truly protective of public health, such monitoring programs would need to be comprehensive enough to include all potential contaminants that pose significant risks in the anticipated reuse applications. Real-time monitoring techniques that are both sufficiently comprehensive and sensitive are unlikely to be available in the next decade. Thus, in the meantime, alternative approaches to quality assurance are needed to address shortcomings in real-time monitoring of contaminants.

The problem of ensuring the quality of an ongoing production operation is not new. The food and drinking water industries have faced it for some time, particularly where pathogens are concerned. Where drinking water is concerned, this need has been addressed by a three-part strategy: (1) characterizing critical elements that control the performance of unit processes in removing specific contaminants, (2) identifying parameters that can be reliably monitored and used to confirm that these elements are in place and that the processes are performing as expected, and (3) routine analysis of certain constituents in samples taken from the finished water to confirm that the previous measures are reliable.

Recently, a monitoring approach with similar components has been proposed for management of trace organic chemicals in potable reuse schemes (Drewes et al., 2008). This approach combines the monitoring of bulk parameters (i.e., surrogates) and a select number of indicator chemicals to ensure proper performance of unit processes. In this work, performance indicators and surrogate parameters are defined as follows:

• *Indicator*—"An indicator compound is an individual chemical occurring at a quantifiable level, that represents certain physicochemical and biodegradable characteristics of a family of trace organic constituents that are relevant to fate and transport during treatment. It provides a conservative assessment of removal." (Drewes et al., 2008).

• *Surrogate*—"A surrogate parameter is a quantifiable change of a bulk parameter that can measure the performance of individual unit processes or operations in removing trace organic compounds" (Drewes et al., 2008). Surrogates can often be used in real time.

As an analogy, the measurement of indicators plays a similar role to the measurement of *E. coli* in drinking water, and the monitoring of surrogates plays a role similar to the monitoring of chlorine residual and contact time. This analogy makes it clear that the indicators and surrogates concept can be extended to address virtually any constituent targeted by a treatment train.

In 2010, an independent scientific advisory panel appointed by the California State Water Resources Control Board endorsed this concept to ensure proper performance of water reclamation processes that remove trace organic chemicals. The panel suggested a combination of appropriate surrogate parameters and performance-based and health-based indicator chemicals for monitoring reclaimed water quality of surface spreading operations (i.e., soil aquifer treatment) and direct injection projects in California (Anderson et al., 2010). Indicator chemicals were selected with a range of properties in an attempt to account for unknown chemicals and newly developed compounds that may be released to the environment in the future, provided they fall within the range of chemical properties covered. This committee encourages further development of this concept.

Monitoring requirements usually become more stringent (e.g., more frequent sampling and more constituents to be monitored) as the potential for human contact with the reclaimed water increases. Municipal wastewater can contain thousands of chemicals originating from consumer products (e.g., household chemicals, personal care products, pharmaceutical residues), human waste (e.g., natural hormones), industrial and commercial discharges (e.g., solvents, metals), or chemicals that are generated during water treatment (e.g., transformation products; see also Chapter 3). Thus, it is appropriate for monitoring programs for reclaimed water used for potable applications to be more comprehensive than programs commonly used for monitoring water quality for conventional drinking water supplies.

Attenuation

Attenuation of microbial and chemical contaminants of concern can be achieved by establishing multiple barriers. A reuse scheme usually is composed of a combination of *treatment barriers* that are suitable to reduce the concentrations of compounds of concern and *preventive measures* that control exposure to certain contaminants, although the actual number of barriers differs among different reuse projects (Drewes and Khan, 2010). Tailored source control programs that limit the discharge from industrial activities to a municipal sewer system or the maintenance of a reclaimed water distribution system are examples of preventive barriers. Attenuation of water quality constituents of concern can occur through conventional wastewater treatment, advanced water treatment, or engineered natural systems.

Multiple barriers are an important concept in

ensuring that performance goals are met. Multiple barriers accomplish this objective in two ways: (1) by expanding the variety of contaminants the process train can effectively address (i.e., robustness) and (2) by improving the degree to which the process can be relied upon to remove any one of them (i.e., reliability, or the extent of consistent performance of a unit process to attenuate a contaminant). These principles are illustrated in Figure 5-2. Multiple barriers can also provide redundancy (defined as a series of unit processes that is capable of attenuating the same type of contaminant) so that if one process fails another is still in the line (Haas and Trussell, 1998; NRC, 1998). Additionally, even when true redundancy is not provided, multiple barriers can reduce the consequences of a failure when it does occur (Olivieri et al., 1999; Crittenden et al., 2005).

Given the nature of the associated risk, the performance criteria of multiple barriers are generally different for pathogens, which can cause acute (sudden and severe) health effects, as compared with organic chemicals, which can cause chronic health effects after prolonged or repeated exposures in drinking water scenarios (see also Chapter 6). Acute health effects from exposure to organic chemicals in drinking water

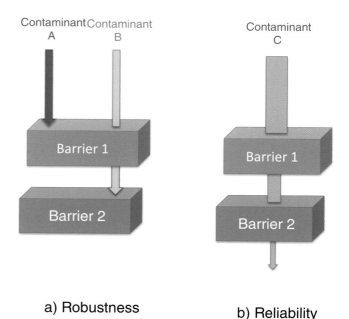

a) Robustness   b) Reliability

FIGURE 5-2 Multiple barriers function in two ways: (a) robustness—increasing the variety of contaminants addressed and (b) reliability—decreasing the likelihood that any one contaminant will fail to be removed, in this example by incorporating redundancy.

or reclaimed water are highly unlikely absent cross connections or backflows. From a public health standpoint, disinfection, which addresses acute risks, is the process element that requires the highest degree of reliability for applications involving significant human contact. In the case of pathogens in potable reuse projects, the performance expectation is that the overall objective for pathogen reduction needs to be met even if a single treatment barrier fails (NRC, 1998). The level of redundancy applied to address microbial constituents is typically not applied in the same way to multiple barriers for chemicals, because of the long-term exposure associated with significant elevated risks for most chemical constituents. Instead, multiple barriers for organic contaminants are designed to encompass a sequence of different processes capable of targeting classes of chemicals with different physicochemical properties, given the wide range of different chemicals present in reclaimed water (Drewes and Khan, 2010). For example, multiple barriers for chemical contaminants might consist of an advanced oxidation process followed by granular activated carbon, where GAC is attenuating chemicals that are not amendable to oxidation.

Retention

Within a water reuse context, retention time may serve two purposes: (1) to allow additional opportunities for attenuation of contaminants and (2) to provide time to respond to system failures or upsets. Retention time can be provided by storing reclaimed water in a surface storage reservoir, storing it in an engineered storage tank, recharging it to an unconfined or confined aquifer, releasing it into a segment of a river, or conveying it through a pipeline system. Proper documentation should be provided of how the water provider would be able to respond to specific types of upsets, including strategies for diverting compromised product water to avoid contaminated water reaching consumers and to ensure that the desired retention time is actually provided.

Blending

Blending of reclaimed water with a water source other than wastewater (e.g., surface water, stormwater, native groundwater) may occur prior to treatment of reclaimed water in engineered processes or after treatment prior to a distribution system. For advanced treatment processes that demineralize reclaimed water and remove trace chemicals, it may be necessary to balance the water chemistry by blending after treatment for public health concerns (e.g., absence of magnesium and calcium), to enhance taste, to prevent downstream corrosion (e.g., calcium saturation index), and to minimize damage to soils (e.g., sodium adsorption ratio) and crops (e.g., magnesium deficiency) (Tchobanoglous et al., 2011). Blending with traditional sources can also ensure some degree of contaminant dilution if a treatment system failure occurs. It is noteworthy that in many cases the blending water might actually represent a lower quality source. Therefore, a careful evaluation of the water quality prior to and after blending is warranted to avoid any degradation of the final product water.

Balancing Monitoring, Attenuation, Retention, and Blending

The need for using retention and/or blending to ensure water quality is dependent on the reliability and robustness of the measures taken for attenuation and monitoring. Early projects using limited technologies for attenuation and monitoring depended heavily on retention and blending. In the future, as more advanced technologies are used for attenuation that address a broader variety of contaminants with greater reliability and as these technologies are supported by improved techniques for monitoring and control, retention and blending will have less significance. However, an overarching comprehensive monitoring program tailored to the specific barriers and local conditions of a reuse scheme is necessary in all water reuse systems to ensure proper performance of each barrier.

Role of the Environmental Buffer in the Multiple-Barrier Concept

Up to the present time, the environmental buffer has often been a core element of the multiple-barrier concept in potable reuse projects. As discussed in Chapter 2, an environmental buffer is a water body or aquifer that is perceived by the public as natural and serves to eliminate the connection between the water and its past history. It also may provide some or all of the following design elements discussed in the previous section: (1) attenuation of contaminants of concern, (2) provision of retention time, and (3) blending (or dilution). The performance of various environmental buffers is discussed in Chapter 4.

Attenuation of contaminants can occur in certain environmental buffers (e.g., wetlands, soil aquifer treatment, riverbank filtration). In this function, an engineered natural treatment system can be used before or after an aboveground water reclamation plant. However, the role of environmental buffers in attenuation of contaminants is not well documented. As detailed in Chapter 4, contaminant attenuation has been reported for some environmental buffers. However, considering site-specific differences, environmental buffers are likely to exhibit some variability in performance with respect to contaminant attenuation.

There is no widely accepted standard for retention time in environmental barriers for potable reuse systems. The retention provided by various examples discussed in this report varies from days to more than 6 months. Retention is particularly uneven where de facto reuse is concerned. Additionally, relying on environmental buffers as the only means of lengthening response times is questionable, especially in systems with short hydraulic residence times.

For potable reuse projects implemented through groundwater recharge, blending or dilution of reclaimed water with water deemed not to be of wastewater origin can occur before application or in aquifers. For surface water augmentation, blending typically occurs in a raw drinking water reservoir. The extent of dilution varies with the different natural systems, and can range from substantial dilution (<1 percent reclaimed water) to minimal dilution (>50 percent reclaimed water). As mentioned before, the need for blending depends heavily on the nature of the process train employed for attenuation.

Currently, the use and application of an environmental buffer for potable reuse is based on regulatory guidance and current practice rather than specific scientific evidence. Sufficient science does not currently exist to determine if current guidance is, in fact, appropriately protective, overprotective, or underprotective of public health. From a public outreach perspective, environmental buffers have often been perceived as important for gaining public acceptance as they create the perception of a "natural" system and provide time

to respond to potential problems should they arise (Ruetten et al., 2004). NRC (1998) described a "loss of identity" that occurs in an environmental buffer, although the committee noted that "loss of identity is an issue that seems more relevant to public relations than public health protection" (NRC, 1998).

During the past decade, extensive research on the performance of reuse operations using modern engineered systems (Ternes et al., 2003; Drewes et al., 2003b; Snyder et al., 2006c; Bellona et al., 2008) as well as those using environmental buffers (Fox et al., 2001; Laws et al., 2011; Maeng et al., 2011) has demonstrated some engineered systems can perform equally well as some existing environmental buffers in diluting and attenuating contaminants, and the proper use of indicators and surrogates in the design of reuse systems offers the potential to address many concerns regarding quality assurance (Drewes et al., 2008). This committee concludes that the practice of classifying potable reuse projects as indirect and direct based on the presence or absence of an environmental buffer is not meaningful to an assessment of the final product water quality because it cannot be demonstrated that such "natural" barriers provide any public health protection that is not also available by other means. Moreover, the science required to design for uniform protection from one environmental buffer to the next is not available.

Accordingly, although the committee does view environmental buffers as useful elements of design that should be considered along with other processes and management actions in formulating potable water reuse projects, the committee does not consider environmental buffers to be an essential element of potable reuse projects. Rather than relying on environmental buffers to provide public health protection that is poorly defined, the level of quality assurance required for public health protection needs to be better defined so that potable reuse systems can be designed to provide it, with or without environmental buffers. A more quantitative understanding of the protections provided by different environmental buffers will allow engineered natural systems to be more effectively designed and operated.

Case Studies for System Design

The role of the design elements mentioned earlier (monitoring, attenuation, retention, blending) can be illustrated using three case studies that practice drinking water augmentation. The three case studies employ different treatment processes including engineered unit processes as well as engineered natural treatment systems providing attenuation of contaminants, and provide a final quality of drinking water that is considered safe by public health agencies and accepted by the public. It is noteworthy that the sequence and location of the individual treatment barriers within the potable reuse scheme also differs.

- Case Study 1 describes a groundwater recharge project favoring direct injection of reclaimed water into a potable aquifer after advanced treatment (Figure 5-3). This case study is similar to the practice of groundwater recharge established by the Orange County Water District (see also Box 2-11) or West Basin Municipal Water District in California.
- Case Study 2 (Figure 5-4) illustrates a groundwater recharge project employing surface spreading followed by soil aquifer treatment. The case study is similar to the groundwater recharge operation in the Montebello Forebay operated by the County Sanitation Districts of Los Angeles County and the Water Replenishment District of Southern California (see also Box 4-2).
- Case Study 3 (Figure 5-5) represents a groundwater recharge scenario using a combination of engineered natural treatment systems with advanced engineered unit processes for drinking water augmentation, similar to that established by the Prairie Waters Project of the City of Aurora, Colorado (see also Box 4-1).

For each case study, the key processes that provide attenuation of contaminants are highlighted, the retention process is identified, and the role of blending in these projects is characterized. These examples reveal that multiple combinations and sequences of treatment processes can be selected for a potable reuse scheme resulting in comparable qualities of finished drinking water.

OPERATIONAL PRINCIPLES TO ASSURE QUALITY AND RELIABILITY

Treatment plant reliability is defined as the probability that a system can operate consistently over extended periods of time (Olivieri et al., 1987). In the case of a water reclamation plant, reliability might be defined as the likelihood of the plant achieving an effluent that matches or is superior to predetermined water reuse quality objectives. Traditional drinking water treatment plants consider reliability in their operations, but even greater attention to reliability is necessary in water reclamation facilities that supply water for potable reuse or other applications with significant human exposures. Failure of wastewater reclamation treatment processes could result in exposure of the population served by nonpotable or potable reuse applications to considerable health risk, particularly from acute illnesses caused by microbial pathogens (see Chapters 3 and 6). It is therefore important to minimize the probability of failure, or, in other words, to increase reliability. Although appropriate design is necessary to ensure reliable delivery of a product such as reclaimed water (as discussed in the previous section), it is also necessary to maintain an operational protocol to cope with intrinsic variability and react to process and conveyance upsets.

Some definitions of reliability only encompass the variability associated with treatment processes and assume that the plant is properly designed, operated, and maintained. Expansion of the definition of reliability to include the probability that the plant will be nonfunctional at any given time requires an evaluation of plant operational reliability, separate from reclaimed water quality variability. Operational reliability is affected by mechanical, design, process, or operational failures, which may be triggered by a wide range of causes, including human error or severe weather events. Previous sections of this chapter discuss ways to incorporate reliability into project design.

Reliability analysis can also be used to reveal weak points in the process so that corrections and/or modifications can be made. Even a well-maintained, well-operated plant is not perfectly reliable, and some variation will necessarily be inherent in any system (e.g., variations in influent flow and quality can lead to variation in effluent characteristics). Other factors, including power outages, equipment failure, and operational (human) error also affect plant reliability and need to be incorporated into the reliability analysis (Olivieri et al., 1987). There are a number of formal techniques for assessing reliability by looking at historical performance of individual components (e.g., pumps, valves, electric

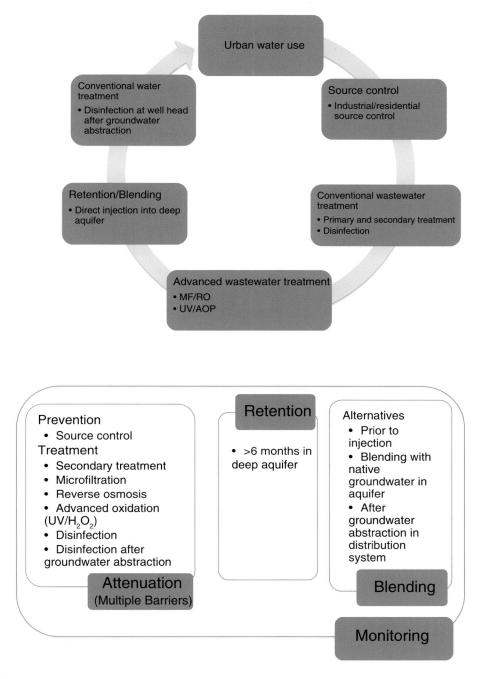

FIGURE 5-3 Case Study 1: Potable reuse design elements (including attenuation, retention, and blending) used for groundwater recharge of reclaimed water.
NOTE: Residential source control could include voluntary programs to reduce the discharge of potentially problematic chemicals.

supply) and the potential for various hazards (e.g., storms, wind, earthquakes) to occur. By using historical data of these individual events (including data from individual components in other applications), failure

or event trees can be constructed (Rasmussen, 1981; Kumamoto and Henley, 1996), and the probability distribution of consequences of different levels of severity can be illustrated.

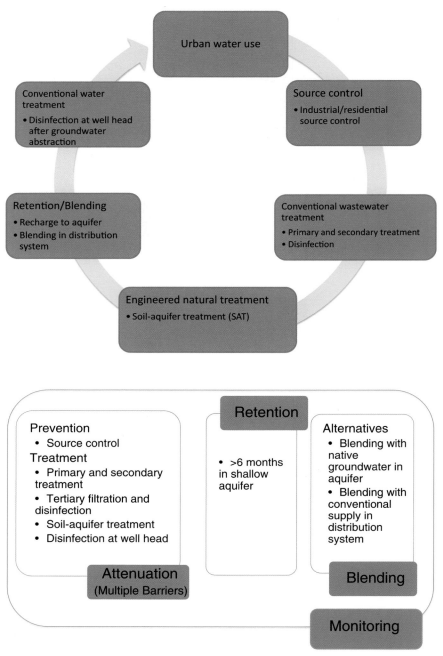

FIGURE 5-4 Case Study 2: Potable reuse design elements (including attenuation, retention, and blending) used for surface spreading of reclaimed water followed by soil aquifer treatment.

Strategies for Incorporating Reliability into System Operation

No matter how well designed a treatment system is, there will be inevitable fluctuations in performance due to intrinsic variability of processes, variability in the influent stream, equipment failures, and human error. Therefore, systems delivering potable reclaimed water need to incorporate deliberate strategies to ensure reliable operation. The centrality of the operational plan in ensuring water quality has been emphasized by the World Health Organization (WHO, 2005) in its concept of water safety plans.

One formal approach for ensuring operational reliability is the hazard analysis and critical control points (HACCP) framework. HACCP was developed

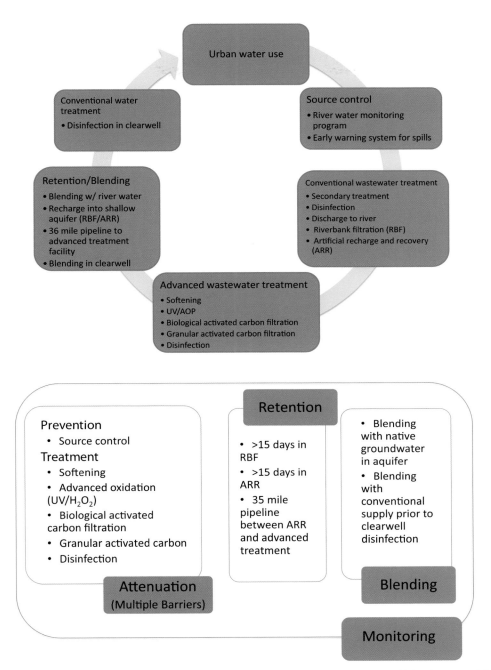

FIGURE 5-5 Case Study 3: Potable reuse design elements (including attenuation, retention, and blending) used for riverbank filtration of reclaimed water followed by softening, advanced oxidation, and carbon adsorption.

in the late 1950s to ensure adequate food quality for the nascent National Aeronautics and Space Administration program. HACCP was further developed by the Pillsbury Corporation and ultimately codified by the National Advisory Committee on Microbiological Criteria for Foods (NACMC, 1997). The ultimate framework consists of a seven step sequence outlined in Box 5-1. These principles are important parts of

the international food safety protection system. The development of HACCP broke reliance on the use of testing of the final product as the key determinant of quality, but rather emphasized the importance of understanding and control of each step in a processing system (Sperber and Stier, 2009).

Havelaar (1994) was one of the first to note that the drinking water supply, treatment, and distribution

---

BOX 5-1
Steps of the HACCP Framework and Application to Potable Reuse

The following seven steps represent the key components of the HACCP framework (adapted from NACMC, 1997), which was originally developed for food safety but has been applied to other areas, including drinking water quality.

1. Conduct a hazard analysis. Under HACCP, hazards are chemical or microbial constituents likely to cause illness if not controlled.

2. Determine the critical control points (CCPs). Defined originally for the food sector, a critical control point is "any point in the chain of food production from raw materials to finished product where the loss of control could result in unacceptable food safety risk" (Unnevehr and Jensen, 1996).

3. Establish critical limit(s). Critical limits are performance criteria—specific maximum or minimum values of biological, chemical, or physical parameters that are readily measurable—that must be attained in each process (at the CCPs) to prevent occurrence of a hazard or reduce it to an acceptable level. These parameters will be process-specific and determined through experimentation, computational models, quantitative risk analysis, or a combination of such methods (Havelaar, 1994; Notermans et al., 1994).

4. Establish a system for monitoring the CCPs.

5. Establish the corrective action(s) that will be taken when monitoring signals that a CCP is not under control.

6. Establish verification procedures to confirm that the HACCP system is working effectively.

7. Document all procedures and records relevant to these HAACP principles and their application.

Application to Potable Reuse

As an illustration of the use of HACCP, the committee developed the following set of steps that might be followed to implement this framework in the potable reuse context using an example of managing risks from pathogenic organisms:

1. Identify the critical organisms of interest, considering the type of source water used, that are likely to cause illness if not controlled. Determine the overall log reductions needed after treatment, given the nature of an incoming water to achieve the targeted final acceptable risk level and allocate these reductions among individual treatment processes.

2. Enumerate CCPs for water reclamation, considering each particular treatment process in the treatment train as well as the overall treatment method.

3. Given criteria in the finished reclaimed water, determine the minimum performance criteria for each treatment process. Note that these performance criteria should be based on easily measurable parameters (e.g., surrogates, residual chlorine) that can be used for operational control.

4. Establish a monitoring system to track the identified performance criteria at the critical control points. The finished product of a reclamation system is only acceptable for utilization when the performance criteria are all within the acceptable bounds.

5. Establish an operational procedure for implementing appropriate corrective actions at a particular installed process should a performance criterion be outside acceptable limits. These actions might include additional holding time, recirculating the water to allow for additional treatment, or some other measure. These procedures would also include actions to protect public health in the case of systemwide failure (e.g., natural disaster leading to extended power failure).

6. Establish a quality assurance process for periodic validation and auditing (e.g., by an independent third-party organization) to assess that the procedures are working effectively.

7. Document all procedures and records.

---

chain has a formal analogy to the food supply, processing, transport, and sale chain and that HACCP could be applicable to water treatment. The development of the U.S. Surface Water Treatment Rule under the Safe Drinking Water Act (SDWA; 40 CFR Parts 141-142) and subsequent amendments incorporate a HACCP-like process. Under this framework, an implicitly acceptable level of viruses and protozoa in treated water was defined. Based on this, specific processes operated under certain conditions (e.g., filter effluent turbidity for granular filters) were "credited" with certain removal efficiencies, and a sufficient number of removal credits needed to be in place depending on an initial program of monitoring of the microbial quality of the supply itself. The use of treatment technique in drinking water regulation is an option when it is not "economically or technically feasible to set an MCL" (SDWA § 1412(b)(7)(A)). HAACP has also been used as a framework for the Australian Drinking Water Guidelines (NHMRC, 2004), which have been expanded to address potable reuse [see Box 5-2; NRMMC/EPHC/NHMRC, 2008]). Box 5-1 highlights an example of how the HACCP approach might be applied in the context of reclaimed water to ensure operational reliability.

STEPS TO ENSURE WATER QUALITY IN WATER REUSE

In the following section, the committee identifies reasonable steps that can and should be taken to

ensure water quality in potable and nonpotable water reuse projects. These steps address potential public health impacts from microbial pathogens and chemical contaminants found or likely to be found in reclaimed water and include considerations of reliability and quality assurance, and therefore merit careful consideration from designers and managers of reuse projects. The extent of each activity will depend on the type of reuse (nonpotable vs. potable) and degree of exposure:

1. Implement and maintain an effective source control program.

2. Utilize the most appropriate technology in wastewater treatment that is tailored to site-specific conditions.

3. Utilize multiple, independent barriers, especially for the removal of microbiological and organic chemical contaminants.

4. Employ quantitative reliability assessments to monitor and assess performance including major and minor process failures (i.e., both process control and final water quality monitoring and assessment as well as assessment of mechanical reliability).

5. Establish a trace organic chemical monitoring program that goes beyond currently regulated contaminants.

6. Document a strategy to provide retention time necessary to allow time to respond to system failures or upsets (e.g., this could be based, in part, on turnaround time to receive water quality monitoring results).

7. Provide for alternative means for diverting the product water that does not meet required standards.

8. Avoid "short-circuiting" in environmental buffers to ensure maintenance of appropriate retention times within the buffers (i.e., groundwater, wetlands, reservoir).

9. Train and certify operators of advanced water reclamation facilities regarding the principles of operation of advanced treatment processes, and educate them on the pathogenic organisms and chemical contaminants likely to be found in wastewaters and the relative effectiveness of the various treatment processes in reducing microbial and chemical contaminants concentrations. This is important because, in general, operators at water reclamation facilities have not received training on the operation of advanced water treatment

---

BOX 5-2
Australian Potable Reuse Guidelines

The Australian Guidelines for Water Recycling: Augmentation of Drinking Water Supplies (Phase 2) (NRMMC/EPHC/NHMRC, 2008), were developed to complement the Australian Drinking Water Guidelines (NHMRC, 2004).The approach to risk management for potable reuse is modeled on the approach developed for the Australian Drinking Water Guidelines and incorporates a generic framework applicable to any system that is reusing water based on 12 elements focusing on ensuring safety and reliability, rather than verification monitoring. The framework incorporates HACCP principles, based on a risk management approach designed to assure water quality at the point of use. The guidelines also provide detailed information on topics such as setting health-based targets for microorganisms and chemicals, the effectiveness of various treatment processes, CCPs, and monitoring.

In the Australian potable reuse guidelines, approaches for calculating contaminant guideline values based on toxicological data and specific guideline values for individual contaminants, as outlined in the Australian Drinking Water Guidelines, are applied to potable reuse. Microbial risk is evaluated using disability adjusted life years (DALYs), performance targets, and reference pathogens (based on WHO, 2008; see also Box 10-4). The tolerable microbial risk adopted in the potable reuse guidelines is 10^{-6} DALYs per person per year, which is roughly equivalent to 1 diarrheal illness per 1,000 people per year. The approach adopted in these guidelines for chemical parameters is based on approaches and guideline values outlined in the Australian Drinking Water Guidelines. The potable reuse guidelines also describe an approach, using thresholds of toxicological concern, for addressing chemicals without guideline values or those that lack sufficient toxicological information for guideline derivation (see also Chapter 6 for further discussion of this and other methods).

processes or the public health aspects associated with drinking water.

10. Institute formal channels of coordination between water reclamation agencies, regulatory agencies, and agencies responsible for public water systems. This will, for example, allow for rapid communication and immediate corrective action(s) to be taken by the appropriate agency (or agencies) in the event that the reclaimed water does not meet regulatory requirements.

CONCLUSIONS AND RECOMMENDATIONS

In both nonpotable and potable reuse schemes, monitoring, contaminant attenuation processes, post-treatment retention time, and blending can be effective tools for achieving quality assurance. Today, most reuse projects find it necessary to employ all these elements. Attenuation can be achieved through the establishment of multiple barriers (consisting of treatment and prevention approaches) to minimize public health risks. Over the last 15 years, several potable reuse projects of significant size have been developed in the United States. Although these projects share the design principle of multiple barriers, the type and sequence of water treatment processes employed in these schemes differ significantly. All these schemes have demonstrated that different configurations of unit processes can achieve similar levels of water quality and reliability. **In the future, as new technologies improve capabilities for both monitoring and attenuation, it is expected that retention and blending requirements currently imposed on many potable reuse projects will become less significant in quality assurance.**

Reuse systems should be designed with treatment trains that include reliability and robustness. Redundancy strengthens the reliability of contaminant removal, particularly important for contaminants with acute affects, while robustness employs combinations of technologies that address a broad variety of contaminants. Reuse systems designed for applications with possible human contact should include redundant barriers for pathogens that cause waterborne diseases. Potable reuse systems should employ diverse processes that can function as barriers for many types of chemicals, considering the wide range of physicochemical properties of chemical contaminants.

Reclamation facilities should develop monitor-

ing and operational plans to respond to variability, equipment malfunctions, and operator error to ensure that reclaimed water released meets the appropriate quality standards for its use. Redundancy and quality reliability assessments, including process control, water quality monitoring, and the capacity to divert water that does not meet predetermined quality targets, are essential components of all reuse systems. Particularly in potable reuse, systems need to be designed to be "fail-safe." The concept of HACCP, water safety plans, or their equivalent may be used as a guide for such operational plans. A key aspect involves the identification of easily measureable performance criteria (e.g., surrogates), which are used for operational control and as a trigger for corrective action.

Natural systems are employed in most potable water reuse systems to provide an environmental buffer. However, it cannot be demonstrated that such "natural" barriers provide public health protection that is not also available by other engineered processes. Environmental buffers in potable reuse projects may fulfill some or all of three design elements: (1) provision of retention time, (2) attenuation of contaminants, and (3) blending (or dilution), although the extent of these three factors varies widely across different environmental buffers. In some cases engineered natural systems, which are generally perceived as beneficial to public acceptance, can be substituted for engineered unit processes. However, the science required to design for uniform protection from one environmental buffer to the next is not available.

The potable reuse of highly treated reclaimed water without an environmental buffer is worthy of consideration, if adequate protection is engineered within the system. Historically, the practice of adding reclaimed water directly to the water supply without an environmental buffer—a practice referred to as direct potable reuse—has been rejected by water utilities, by regulatory agencies in the United States, and by previous National Research Council committees. However, research during the past decade on the performance of several full-scale advanced water treatment operations indicates that some engineered systems can perform as well or better than some existing environmental buffers in diluting (if necessary) and attenuating contaminants, and the proper use of indicators and surrogates in the design of reuse systems offers the potential to address many concerns regarding quality assurance. Environmental buffers can be useful elements of design that should be considered along with other processes and management actions in formulating the composition of potable water reuse projects. However, environmental buffers are not essential elements to achieve quality assurance in potable reuse projects. Additionally, the classification of potable reuse projects as indirect (i.e., includes an environmental buffer) and direct (i.e., does not include an environmental buffer) is not productive from a technical perspective because the terms are not linked to product water quality.

6

Understanding the Risks

Although people commonly ask whether the actions they take are "safe," with an implication that safety poses no risk of harm to human health, it is impossible to demonstrate such a definition of safety or indeed to achieve zero risk. It has previously been recommended (NRC, 1998) "that water agencies considering potable reuse fully evaluate the potential public health impacts from the microbial pathogens and chemical contaminants found or likely to be found in treated wastewater through special microbiological, chemical, toxicological, and epidemiological studies, monitoring programs, risk assessments, and system reliability assessments." In other words, an evaluation of the adequacy of public health and ecological protection rests upon a holistic assessment of multiple lines of evidence, such as toxicology, epidemiology, chemical and microbial analysis, and risk assessment.

Major research efforts have attempted to refine our understanding of the human health risks of water reuse, particularly the risks of potable reuse, through toxicological and epidemiological studies (see Boxes 6-1 and 6-2; NRC, 1998).[1] In the context of reclaimed water

projects, epidemiological analyses of health outcomes are an imprecise method to quantify chronic health risks at levels generally regarded as acceptable. This is especially true when interpreting negative study results, which typically do not have the statistical power to detect the level of risks considered significant from a population-based perspective (e.g., an additional lifetime cancer risk of 1:10,000 to 1:1,000,000). Although epidemiology is invaluable as part of an evaluative suite of analytical tools assessing risk, epidemiology may be most useful at bounding the extent of risk, rather than actually determining the presence of risk at any level. Direct toxicological methods (Box 6-2) are intriguing, as indeed was noted in the National Research Council report on *Issues in Potable Reuse* (NRC, 1998), yet there remains insufficient development and knowledge for these methods to be broadly applied.

There will always be a need for human-specific data, and epidemiological studies will remain important to assessing and monitoring the occurrence of health impacts. However, today's decisions as to health and environmental protection remain grounded in the measurement of chemical and microbiological parameters and the application of the formal process of risk assessment. Risk can be identified, quantified, and used by decision makers to assess whether the estimated likelihood of harm—no matter how small—is socially acceptable or whether it may be justified by other benefits. Risk assessment provides input to the overall decision process, which also includes consideration of financial costs and social and environmental benefits (discussed in Chapter 9).

The focus of this chapter is to present risk assess-

---

[1] Toxicological studies expose animals or organisms to a series of doses or dilutions of a single contaminant, complex mixtures, or actual concentrates of reclaimed water to predict adverse health effects (e.g., mortality, morphological changes, effects on reproduction, cancer occurrence). Toxicological tests on mammals often are used to identify doses associated with toxicity, and these dose-response data are subsequently used to estimate human health risks. Potential adverse human health effects are more difficult to predict based on studies in nonmammalian species or microorganisms; however, observed effects are considered cause for further investigation. Epidemiological studies examine patterns of human illness (morbidity) or death (mortality) at the population level to assess associated risks of exposure.

BOX 6-1
Water-Reuse–Specific Epidemiological Information

NRC (1998) provided a comprehensive review of six toxicological and epidemiological studies of reuse systems. The epidemiological study findings from potable reuse applications are briefly summarized in this box. The results from several toxicology studies are summarized in Box 6-2.

Windhoek, Namibia, is the first city to have implemented potable reuse without the use of an environmental buffer (sometimes called direct potable reuse; see Box 2-12). It has been doing so since 1968, especially during drought conditions, and the plant provides up to 35 percent of the potable water supply during normal periods. Epidemiological evaluations of the population have found no relationships between drinking water source and diarrheal disease, jaundice, or mortality (Isaacson et al., 1987; Isaacson and Sayed, 1988).

Three sets of studies have been conducted for the Montebello Forebay Project in Los Angeles County, California: (1) a 1984 Health Effects Study, which evaluated mortality, morbidity, cancer incidence, and birth outcomes for the period 1962–1980; (2) a 1996 RAND study, which evaluated mortality, morbidity, and cancer incidence for the period 1987–1991; and (3) a 1999 RAND study, which evaluated adverse birth outcomes for the period 1982–1993. The first studies looked at two time periods (1969–1980 and 1987–1991) and characterized census tracts into four or five categories by 30-year average percentage of reclaimed water in the water supply. The annual maximum percentage of reclaimed water ranged from less than 4 percent to between 20 and 31 percent. The studies included 21 and 28 health outcome measures, respectively, including health outcomes related to cancer, mortality, and infectious disease incidence. Although some outcomes were more prevalent in the census tracts with a higher percentage of reclaimed water in the water supply, neither study observed consistently higher rate patterns or dose-response relationships (Frerichs et al., 1982; Frerichs, 1984; Sloss et al., 1996). Sloss et al. (1996) identified reclaimed water use and control areas so that comparisons could be made. Compared with the control areas, reclaimed water use areas had some statistically higher as well as lower rates of disease. After evaluating the overall patterns of disease, the authors concluded that the study results did not support the hypothesis of a causal relationship between reclaimed water and cancer, mortality, or infectious disease. Although assessment of a dose-response relationship was possible in the study design, none was identified for the excesses of disease seen.

Since the NRC (1998) report, there have been only a few additional epidemiological studies of human health impacts of wastewater reuse. The largest and most comprehensive study was the third continuation of the Montebello Forebay study (Sloss et al., 1999). Sloss et al. (1999) included a health assessment utilizing administrative health data from 1987–1991 and birth outcomes from 1982–1993. They found some differences between study groups but saw no pattern and concluded that the rates of adverse birth events were similar between the control group and the region receiving reclaimed water.

The most recent study (Sinclair et al., 2010) compared the health status of residents in two housing developments: one with dual plumbing to support nonpotable reuse and a nearby development using a conventional water supply. The study assessed the rates that residents consulted with primary care physicians for gastroenteritis, respiratory complaints, and dermatological complaints (conditions that could be related to reclaimed water exposure) as well as two conditions unrelated to water reuse or waterborne disease exposure. Sinclair et al. (2010) reported no differences in consultation rates between the two groups. There were slight differences in the ratios of specific consultations (i.e., dermal versus respiratory), but the seasonal reporting patterns did not match the timing of reclaimed water exposure.

Population-based studies, also called ecological studies, such as these face significant challenges such as short study periods for chronic disease outcomes, changing exposures over time, nonspecific disease outcomes with unknown attributable risks, and the inability to know actual water consumption rates. Their use for quantitative risk assessment is extremely limited. Such studies simply cannot have the statistical power to achieve detection of the risk expectations established in public water supply regulatory standards such as 10^{-5} or 10^{-6} lifetime cancer risk. Population-based studies are probably best viewed as "scoping" or hypothesis-forming exercises. They cannot prove that there is no adverse effect from the reuse of water in these areas (indeed no study can do so), but they can suggest an upper bound on the extent of the impact if one did exist.

Two alternative study approaches could be considered for assessing the effects of reclaimed water on public health. Blinded-design household intervention studies could be used in which all households in the study receive point of use (POU) "treatment devices," although the control group receives sham devices, and the occurrence of acute gastroenteritis illness is tracked. Most health concerns related to chemical exposures are chronic diseases that may take years to appear. To avoid the need for long observation periods, the household intervention approach could use human tissue chemical biomarkers rather than disease occurrences. Another methodology that is more passive but holds promise for assessing the health impacts of reclaimed water consumption is the "opportunistic natural experiment," epidemiologically characterized as a community intervention study. These studies assess the incidence of acute gastrointestinal illness before and after scheduled changes in water sources or treatment processes. An example of such a study is a 1984–1987 Colorado Springs study of water reuse for public park irrigation. Three different sources of water (potable, nonpotable water of wastewater origin, and nonpotable water of runoff origin) were used to irrigate municipal parks, and randomly selected park users were surveyed for the occurrence of gastrointestinal disease. Wet grass conditions and elevated densities of indicator bacteria, but not exposure to nonpotable irrigation water per se, were associated with an increased rate of gastrointestinal illness. Increased levels of disease and symptoms were observed when several different bacterial indicators exceeded 500/100 mL. These levels occurred most commonly with the nonpotable water of runoff origin (Durand and Schwebach, 1989). A well-designed case control study can also be used in select populations. Such studies in the context of ordinary potable water have been conducted by a number of authors (Payment et al., 1997; Aragón et al., 2003; Colford et al., 2005).

BOX 6-2
Potable Reuse Toxicological Testing

In 1982, the National Research Council Committee on Quality Criteria for Reuse concluded that the potential health risks from reclaimed water should be evaluated via chronic toxicity studies in whole animals (NRC, 1982). Early studies in laboratory animals, most notably the Denver and Tampa Potable Water Reuse Demonstration Project studies, which used rats and mice exposed to concentrates of reclaimed water, failed to identify adverse health effects when tested in subchronic, reproductive, developmental, and chronic toxicity studies (Lauer et al., 1990; CH2M Hill, 1993; Condie et al., 1994; Hemmer et al., 1994; see also more comprehensive descriptions in NRC, 1998). The absence of adverse effects following repeated, long-term exposure to concentrates of reclaimed water was also confirmed in mice chronically exposed to 150 and 500× concentrates of reclaimed water from a Singapore reclamation plant (NEWater Expert Panel, 2002). Although data from the 24-month tests were planned for completion in 2002, the Singapore Water Reclamation Board did not reconvene the NEWater Expert Panel to evaluate the results or issue an updated final report.

The Orange County Water District conducted online biomonitoring of Japanese Medaka fish exposed to effluent-dominated Santa Ana River water over 9 months and found no statistically significant differences in mortality, gross morphology, reproduction, or gender ratios (Schlenk et al., 2006). The Singapore Water Reclamation Board also exposed Japanese Medaka fish (*Oryzias latipes*) to reclaimed water over multiple generations and identified no estrogenic or carcinogenic effects in fish (Gong et al., 2008). However, the relevance of these findings to human health remains unclear.

In addition to the in vivo studies described above, a number of in vitro genotoxicity studies have been conducted on samples of reclaimed water and/or concentrates of reclaimed water sampled from sites in Montebello, California, Tampa, Florida, San Diego, California, and Washington, DC (summarized in C. Rodriguez et al., 2009). These studies have identified a small number of positive results—a few tests showed mutagenic effects in the Ames assay in *Salmonella typhimurium*—although most in vitro and in vivo genotoxicity assays (e.g., mammalian cell transformation, 6-thioguanine resistance, micronucleus, Ames, and sister chromatid assays) have been negative (Nellor et al., 1985; Thompson et al., 1992; Olivieri et al., 1996; CSDWD, 2005). Although in vitro assays are useful for identifying specific bioactivity and chemical modes of action, they are not likely to be used in isolation for the determination of human health risk. Such bioassays provide a high degree of specificity of response, but they generally cannot represent the actual situation in animals that includes metabolism, multicell signaling, and plasma protein binding, among others. In addition, some chemicals can be rapidly degraded during digestion and metabolism, whereas others are transformed into more toxic metabolites. At the same time, many limitations also plague the current in vivo testing paradigm in that interspecies and intraspecies variability can obfuscate the interpretation of animal testing results when applied to humans. For this reason, uncertainty factors are applied in an attempt to provide a conservative estimate of human health risk from animal models.

The U.S. Environmental Protection Agency (EPA) and the National Toxicology Program continue to investigate modern in vitro, genomic, and proteomic methods for rapid screening of chemicals and mixtures and to better deduce the complex pathways leading to disease (NRC, 2007; Collins et al., 2008). Although high-throughput screening using in vitro tools will increase the knowledge on various modes of toxicity of chemicals, in vivo testing will remain an integral part of evaluation of human health consequences from chemical exposure. However, a powerful approach to screening waters can involve a battery of bioassays, each with different toxicological endpoints (Escher et al., 2005).

---

ment methods for chemical and microbial contaminants that can be used to quantify health risks associated with water reuse applications. In Chapter 7, these methods are applied in a comparative analysis of several reuse scenarios compared to a conventional drinking water source commonly viewed as safe.

INTRODUCTION TO THE RISK FRAMEWORK

With the limitations of toxicological testing and population-level epidemiological studies, quantitative risk assessment methods become a critically important basis for assessing the acceptability of a reclaimed water project (NRC, 1998; Asano and Cotruvo, 2004; C.

Rodriquez et al., 2007b, 2009; Huertas et al., 2008). Quantitative methods to assess potential human health risks from chemical and microbial contaminants in reclaimed water have evolved over the past 30 years and are still being refined. Although EPA has extensive health effects data on regulated contaminants, potable reuse and de facto reuse involve some level of exposure to minute quantities of contaminants that are not regulated. Many of these classes of constituents may require innovative approaches to assess health risks. Challenges associated with assessing risks posed by such contaminants include incomplete toxicological datasets, uncertainties associated with concomitant low-level exposures to multiple chemical and biological materials that may share similar modes of action; and deficiencies in analytical methods to accurately identify and quantify the presence of these contaminants in reclaimed water (Snyder et al., 2009, 2010a; Drewes et al., 2010).

The contribution of water-associated risks to the total U.S. disease burden is estimated to be relatively small. However, water that is not treated to the appropriate level for the end use can pose significant human health risks. These include chronic effects, such as cancer or genetic mutations, or acute effects, such as neurotoxicity or infectious diseases. These adverse outcomes may be caused by different agents, such as inorganic constituents, organic compounds, and infectious agents. The impact of an agent may be a function of the route of exposure (e.g., oral, dermal, inhalation, ocular). Rarely can an observed outcome be ascribed to a particular agent and exposure route in a particular vehicle (such as reclaimed water). In water reuse considerations, there will invariably be multiple substances, types of effects, and modes of exposure that may be relevant.

Historically, the paradigm for risk analysis has been divided into risk assessment (based on objective technical considerations) and risk management, wherein more subjective aspects (e.g., cost, equity) are considered. Risk characterization served as the conduit between the two activities, as introduced in NRC (1983; also known as the "Red Book"). However, evolution in the use of risk to regulate human exposure has resulted in substantial evolution of the framework.

Early in 2009, an updated risk framework, encapsulated in Figure 6-1, was developed (NRC, 2009b). This updated framework has a number of important revisions that are of particular relevance to the problem under consideration in this report. This framework shares a number of similarities with the 1983 Red Book framework with respect to the central tasks of risk assessment (i.e., hazard characterization, dose-response assessment, exposure assessment, and risk characterization). However, it formally introduces several new aspects to the risk analysis and management process that are particularly germane to assessing and managing health risks from reclaimed water:

• **Problem formulation:** At the outset, there should be a problem formulation and scoping phase in which the risk management question(s) to be answered should be explicitly framed, and the nature of the assessment activities—with respect to agents, consequences, routes, and methodologies—should be outlined. The nature of the management question to be addressed should drive the nature of the assessment activities. Examples of potential scoping questions relevant to water reuse include what is the risk from using groundwater that has been mixed with reclaimed water as a supplement to an existing surface water supply, or what is the human health risk from the application of undisinfected secondary effluent to fruit crops?

• **Stakeholder involvement:** At all stages, there should be well-understood processes available for involvement of internal and external stakeholders. This is an important consequence of the fact that risk assessment per se involves a number of trans-scientific assumptions (Crump, 2003), and the involvement of stakeholders at all stages promotes transparency to the process and, it is hoped, greater acceptance of the ultimate risk management decision.

• **Evaluation:** Within the assessment phase itself, there is an explicit evaluation step to determine whether the computations have produced results of sufficient utility in risk management and of the nature contemplated in problem formulation and scoping. If this is not the case, further developed assessments should be conducted. This recognizes that there are various levels of complexity that can be used in risk assessment with a tradeoff between time and resources required for the assessment and degree of uncertainty in the results. If a risk management question can be addressed satisfactorily with a less intensive assessment process, such an approach would be favorable inasmuch as it would enable a decision to be reached more expeditiously with less resource expenditure.

There is also more explicit recognition (NRC 2009b) that risk management decisions will involve consideration not only of the risk assessment results, but of issues of economics, equity, and law, which are discussed in Chapters 9 and 10.

In the following sections, four core components of risk assessment are discussed with regard to a range of water reuse applications:

1. *hazard identification*, which includes a summary of chemical and microbiological agents of concern;
2. *exposure assessment*, which explains the route and extent of exposure to contaminants in reclaimed water;

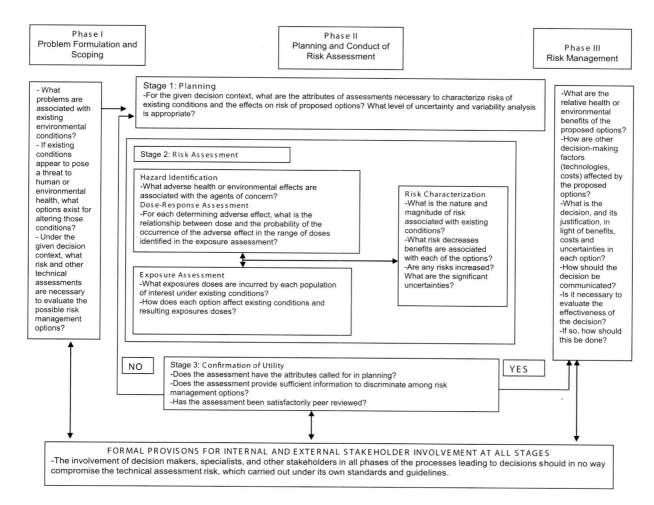

FIGURE 6-1 Consensus risk paradigm.
SOURCE: NRC (2009b)

3. *dose-response assessment*, which explains the relationship between the dose of agents of concern and estimates of adverse health effects, and

4. *risk characterization*, in which the estimated risk under different scenarios is compiled. This may include a determination of relative risk (via the route under consideration, e.g., reclaimed water) versus risks from the same contaminants via other routes (e.g., alternative supplies).

CONTEXT FOR UNDERSTANDING WATERBORNE ILLNESSES AND OUTBREAKS

As noted in Chapter 2, the early 20th century brought significant public health improvements due to the implementation of constructed water treatment and supply systems as well as wastewater collection and treatment systems. Despite much success across the developed world to consistently deliver safe water, diseases associated with microorganisms in water continue to occur. Epidemiological investigations have resulted in estimates of between 12 million and 19.5 million waterborne illnesses per year in the United States (Reynolds et al., 2008). Such illnesses are caused by exposure to bacteria, parasites, or viruses (Barzilay et al., 1999).

Fortunately, in the United States these illnesses rarely result in death. On the other hand, death due to acute gastrointestinal illness, especially in the vulnerable young, is all too common in the developing world. Most obvious to the public are the reported outbreaks of acute gastrointestinal illness largely due to pathogens in the water supply (Mac Kenzie et al., 1994).

Epidemiologists have been conducting surveillance for waterborne outbreaks for nearly 100 years and keeping statistics since 1920. The epidemiological investigation of these events has helped identify the vulnerabilities in our drinking water delivery systems and led to many system improvements. From 1991 to 2002, an annual average of 17 waterborne outbreaks were reported and investigated in the United States compared with an annual average of 23 during 1920–1930 (Craun et al., 2006), while over the same period, the U.S. population increased by a factor of over 2.5. From 1991 to 2000 there were 155 outbreaks recorded in the national epidemiological surveillance system. In 39 percent of the reports, no causative agent was identified, and in 16 percent, the cause was a chemical. These studies suggest that the epidemiology of waterborne disease is complex and that outbreak surveillance is far from complete, with significant underreporting. Analyses from recent years have identified that deficiencies in the water distribution system rather than failure in the treatment process are increasingly the cause of outbreaks (Craun et al., 2006; NRC, 2006). Thus, water may be free of contamination when it leaves the municipal water treatment plant but becomes recontaminated by the time it reaches the household tap. The adequacy of the distribution system may therefore provide a limit to the degree of risk reduction even though treatment becomes more stringent. This also heightens the need for monitoring at the point of exposure (i.e., the tap) rather than relying solely on monitoring immediately after treatment. Data collected by the Centers for Disease Control and Prevention's Surveillance for Waterborne Diseases and Outbreaks indicated that *Escherichia coli*, norovirus, and unidentified microbial pathogens (likely viral) are the common causes of the waterborne disease outbreaks (Blackburn et al., 2004; Liang et al., 2006; Yoder et al., 2008). Cases of meningitis and other infectious diseases also were reported during water recreation in virus-contaminated coastal waters (Begier et al. 2008).

The record of waterborne disease outbreaks, however, is only the tip of the iceberg. Large numbers of waterborne infectious diseases are undocumented. The level of background endemic diseases associated with water and water supplies is not well understood. There is no estimate of waterborne diseases by specific region or community or by water utility treatment modalities.

A review of 33 studies of incidence and prevalence of acute gastrointestinal illness from all exposure sources ranged from 0.1 to 3.5 episodes per adult per year, with child estimates higher (Roy, 2006). Roy (2006) estimated 0.65 episode per person per year in the United States. Health effects from marine recreational exposures to microbial pathogens in water receiving treated wastewater discharge (e.g., eye infection, ear and nose infections, wound infections, skin rashes) are also underreported (Turbow et al., 2003, 2008).

As illustrated above, many human illnesses have the potential to be transmitted via water exposure. There are few if any waterborne pathogens that are distinct to reclaimed water, as opposed to other modes of introduction into the potable or nonpotable aquatic environments. Sometimes these other modes can result in large waterborne outbreaks. For example, an estimated 400,000 cases of *Cryptosporidium* illness occurred in Milwaukee in 1993 caused by a failure in a filtration process at a water treatment plant (Mac Kenzie et al., 1994), and an acute gastrointestinal illness outbreak in Ohio affected over 1,500 people from microbial contamination of a groundwater supply (Fong et al., 2007). Therefore, although this chapter focuses on the risks of water reuse, potential waterborne hazards should be considered in the context of the full suite of possible exposure routes.

HAZARD IDENTIFICATION

The first step in any risk assessment (microbial or chemical) is hazard identification, defined as "the process of determining whether exposure to an agent can cause an increase in the incidence of a health condition" (NRC, 1983) such as cancer, birth defects, or gastroenteritis, and whether the health effect in humans or the ecosystem is likely to occur.[2] Hazards of reclaimed water may depend on factors such as its composition and source water (industrial and domestic sources), varying removal effectiveness of different treatment processes, the introduction of chemicals, and the creation of transformation byproducts during the water treatment process (NRC, 1998). It is important to remember that risk is a function of hazard and exposure, and where there is no exposure, there is no risk.

---

[2] http://www.epa.gov/oswer/riskassessment/human_health_toxicity.htm

Chemical and microbial contaminants constitute two types of agents that may cause a spectrum of adverse health impacts, both acute and/or chronic. *Acute* health effects are characterized by sudden and severe illness after exposure to the substance. Acute illnesses are common after exposure to pathogens, but acute health effects from exposure to regulated or unregulated chemical contaminants found in drinking water or reclaimed water are highly unlikely under anything but aberrant conditions due to system failures, chemical spills, unrecognized cross connections with industrial waste streams, or accidental overfeeds of disinfection agents. *Chronic* health effects are long-standing and are not easily or quickly resolved. They tend to occur after prolonged or repeated exposures over many days, months, or years, and symptoms may not be immediately apparent. There is recently recognized concern for effects arising via an epigenetic route wherein an agent alters aspects of gene translation or expression; such effects can be manifested in a variety of end points (Baccarelli and Bollati, 2009).

Chemical Hazards and Risks

Health hazards from chemicals present in reclaimed water (discussed in Chapter 3) include potential harmful effects from naturally occurring and synthetic organic chemicals, as well as inorganic chemicals. Some of these chemicals, including the carcinogens *N*-nitrosodimethylamine (NDMA; see Box 3-2), and trihalomethanes (EAO, Inc., 2000), may be produced in the course of various treatment processes (e.g., disinfection), rather than arising from the source water itself. Among the most studied of this latter class of chemicals are the chlorination disinfection byproducts, which have been associated with cancer as well as adverse birth outcomes. Because of the need to disinfect wastewater, which may have comparatively higher organic content than typical drinking water sources, such treatment-related contaminants may be problematic in some reclaimed waters.

Multiple studies in the scientific literature have described associations between chemical contaminants in drinking water and chronic disease such as cancer, chronic liver and kidney damage, neurotoxicity, and adverse reproductive and developmental outcomes such as fetal loss and birth defects (NRC, 1998). Most toxic

chemicals that are relevant to water reuse pose chronic health risks, where long periods of exposure to small doses of potentially hazardous chemicals can have a cumulative adverse effect on human health (Khan, 2010; see Chapter 10 for discussion of regulation of drinking water contaminants). As noted in Box 6-1, epidemiological studies are seldom able to determine which of the many chemicals typically present in the water over time are associated with the chronic health effects described. Box 6-3 provides a list of the biologically plausible diseases investigated in the literature for associations with water exposures as well as the organ systems most vulnerable to the contaminants present in wastewater (Sloss et al. 1996; NRC, 1998).

As noted in Chapter 3, a large array of chemicals are present at low concentrations in the nation's source waters and drinking water, including pharmaceuticals and personal care products (see Table 3-3; Kolpin et al., 2002; Weber et al., 2006; Rodriquez et al., 2007a,b; Snyder et al., 2010b; Bull et al., 2011). There is a growing public concern over potential health impacts from long-term ingestion of low concentrations of trace organic contaminants (Snyder et al., 2009, 2010b; Drewes

---

BOX 6-3
Biologically Plausible Possible Health Outcomes from Exposures to Chemicals Found in Wastewater

Cancer

Bladder[a]	Liver
Colon[a]	Pancreas
Esophagus	Rectum[a]
Kidney	Stomach

Reproductive and Development Outcomes

Spontaneous abortion[a]	Birth defects[a]
Low birth weight	Preterm birth

Target Organ Systems

Gastrointestinal organs	Cardiovascular organs
Kidney	Cerebrovascular organs
Liver	

[a]Most consistently increased in epidemiological studies, especially those of trihalomethane disinfection byproducts.

et al., 2010). In contrast to well-documented adverse health effects associated with exposure to specific disinfection byproducts (such as trihalomethanes) in municipal water systems, health hazards posed by long-term, low-level environmental exposure to trace organic contaminants in reclaimed water or from de facto reuse scenarios are not well characterized, nor are their subsequent health risks known (NRC, 2008a; Khan, 2010; Snyder et al., 2009, 2010b). Although chemicals currently regulated in drinking water have comparatively robust toxicological databases, many more chemicals present in water are unregulated and are missing critical toxicological data important to understanding low-level chronic exposure impacts (Drewes et al., 2010). These same agents can be present in treated wastewater in concentrations not otherwise encountered in most public water supply sources.

To date, epidemiological analyses of adverse health effects likely to be associated with use of reclaimed water have not identified any patterns from water reuse projects in the United States (Khan and Roser, 2007; NRC, 1998; see Box 6-1). In laboratory animals and in vitro studies, there is a mixed picture, with more recent studies on genotoxicity, subchronic toxicity, reproductive and developmental chronic toxicity, and carcinogenicity showing negative results (summarized in Nellor et al., 1985; Lauer et al., 1990; Condie et al., 1994; Sloss et al., 1999; Singapore Public Utilities Board and Ministry of the Environment, 2002; R. A. Rodriguez et al., 2009; see also Box 6-2). Collectively, while these findings are insufficient to ensure complete safety, these toxicological and epidemiological studies provide supporting evidence that if there are any health risks associated with exposure to low levels of chemical substances in reclaimed water, they are likely to be small.

Microbial Hazards

Most waterborne infections are acute and are the result of a single exposure. Disease outcomes associated with infection from waterborne pathogens include gastroenteritis, hepatitis, skin infections, wound infections, conjunctivitis, and respiratory infections. Microbial infection rates are determined by the survival ability of the pathogen in water; the physicochemical conditions of the water, including the level of treatment; the

pathogen infectious dose; the virulence factor; and the susceptibility of the human host.

Bacterial pathogens in general are more sensitive to wastewater treatment than are viruses and protozoa; thus, few survive in disinfected water for reuse (see Chapter 3, Table 3-2). Most bacterial pathogens (e.g., Vibrios) also have a high median infectious dose, which requires ingestion of many cells for a likely establishment of infection in healthy adults (Nataro and Levine, 1994). Other bacteria, such as *Salmonella*, can constitute a likely human infection with 1 to 10 cells if consumed with high-fat-content food (Lehmacher et al., 1995). Toxigenic *E. coli* O157:H7 with two potent toxins is also suspected of having a low median infectious dose (Teunis et al., 2004).

In comparison with bacterial pathogens, protozoan cysts and viruses are more resistant to inactivation in water. Protozoan cysts are resistant to low doses of chlorine, and high infection rates in water are associated with suboptimal chlorine doses. Viruses can pass the filtration system in water treatment plants because of their small size. Some viruses are also resistant to ultraviolet disinfection (see Chapter 4). Because they have a low median infectious dose, viruses have the potential to present a concern in water reuse applications.

In addition to microbial characteristics, human host susceptibility plays an essential role in microbial hazards. Microbial agents that are benign to a healthy population can lead to fatal infections in a susceptible population. The growing numbers of immunocompromised individuals (e.g., organ transplant recipients, those infected with HIV, cancer patients receiving chemotherapy) are especially vulnerable to such infection. Because of their clinical status, infection is difficult to treat and often becomes chronic. Infectious-agent disease can also lead to chronic secondary diseases, such as hepatitis and kidney failure, and can contribute to adverse reproductive outcomes. The exacerbating factors are not unique to water reuse but apply to all exposure to infectious microorganisms via water, food, and other vehicles. Table 3-1 lists the microbial agents that have been associated with waterborne disease outbreaks and also includes some agents in wastewater thought to pose significant risk.

WATER REUSE EXPOSURE ASSESSMENT

For the purpose of human health risk assessments, exposure is defined as contact between a person and a chemical, physical, or biological agent. The amount of exposure (or dose) is a product of two variables: concentration of a substance in a medium (e.g., the concentration of trihalomethanes in reclaimed water) and the amount of that medium to which an individual is exposed (e.g., via ingestion or inhalation). For an ingested contaminant, the dose is the concentration in water multiplied by the amount of water ingested. Accurately assessing exposure to reclaimed water is a critically important aspect of assessing health risks, because the likelihood of harm from exposure distinguishes risk from hazard.

Influence of Water Treatment on Potential Exposures

Reclaimed wastewater that has undergone varying degrees of water treatment will have different levels of microbial and chemical contamination (see Table 3-2 and Appendix A). As discussed in Chapter 2, the appropriate end use of reclaimed water is dependent on the level of water treatment, with greater intensity of treatment more effectively reducing or removing microbial and chemical contaminants as needed by particular applications (EPA, 2004; de Koning et al., 2008). The treatment and conveyance of waters of different qualities is not novel and dates to the Roman imperial times (Robins, 1946).

Over the course of time, a unit volume of water undergoes changes in quality (illustrated conceptually in Figure 6-2). With use, a deterioration in quality occurs that may be reversed with treatment. Depending on the desired use, water may be abstracted at different locations along this continuum (i.e., at the right-hand side, increasing degrees of treatment will produce reclaimed wastewater suitable for increasingly stringent usages).

Reclaimed water that has undergone secondary treatment (biological oxidation or disinfection) has numerous nonpotable uses in applications with minimal human exposure potential, such as industrial cooling and nonfood crop irrigation (see Chapter 2). Secondary effluent that has undergone further treatment (e.g., chemical coagulation, disinfection, microfiltration,

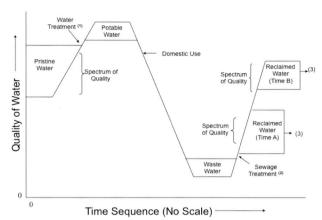

FIGURE 6-2 Continuum of water quality with use and treatment.
NOTES:
(1) Typical processes include coagulation-flocculation, sedimentation, filtration, and disinfection.
(2) Processes include secondary treatment and disinfection.
(3) Effluent discharged to environmental receiving water or reused.
SOURCE: Adapted from McGauhey (1968); T. Asano, personal communication, 2010).

reverse osmosis, high-energy ultraviolet light with hydrogen peroxide) is suitable for a greater number of nonpotable or potable uses, including uses that have a higher degree of human exposure to the constituents in reclaimed water, such as food crop irrigation and groundwater recharge. In contrast, wastewater that has only undergone primary treatment (sedimentation only), has no use as reclaimed water in the United States because of the likely chemical and microbial contamination. It should be recognized that more extensive treatment generally is more cost- and energy-intensive, may have greater potential for byproducts to occur, and may have greater environmental footprints. Different applications of reclaimed water are also associated with different exposure scenarios, discussed in more detail later in this section.

Influence of Different Exposure Circumstances and Routes of Exposure on Dose

Exposure to contaminants in reclaimed water occurs not only through the ingestion of water that has been designed for potable reuse applications but also from food, skin and eye contact, accidental ingestion during water recreation, and inhalation in other reuse applications (Gray, 2008). Exposure can also result from improper use of reclaimed water, improper operation of a reclaimed water system, or inadvertent cross connections between a potable water and a nonpotable water distribution system (see Box 6-4). This illustrates that regardless of the intended use, the assessment of risk should consider unintended but foreseeable plausible inappropriate uses of the reclaimed water.

A key component of a human health risk assessment is the estimation of an individual's average daily dose (ADD) of a chemical. The ADD of a chemical in reclaimed water represents the sum of the ADDs

for ingestion, dermal contact, and inhalation of that chemical from reclaimed water. To assess the likelihood that adverse health effects may occur, the ADD can be compared with a daily dose determined to be acceptable over a lifetime of exposure. See Appendix B for equations used for calculating each of these terms.

Ingestion of Reclaimed Water

Ingested volumes of tap water vary with gender, age, pregnancy status (Burmaster, 1998; Roseberry and Burmaster, 1992), ethnicity (Williams et al., 2001), climate, and likely other factors. Also, the concentration of contaminants in reclaimed water, which affects

---

BOX 6-4
Cross Connections

Several cross connections between nonpotable reclaimed water and potable water lines have been reported in the United States and elsewhere (e.g., Australia). Some of the cross connections existed for 1 year or longer prior to detection. Only a few cross connections involving reclaimed water have resulted in reported illnesses, and fewer still have been medically documented. Most cross connections that occur are accidental, although some are intentional by homeowners or others.

Some examples of cross connection incidents reported in the literature are provided below:

• In 1979, several people reportedly became ill as a result of a cross connection between potable water lines and a subsurface irrigation system that supplied reclaimed water for irrigation at a campground. Based on a survey of 162 persons who camped at the site, at least 57 campers reported symptoms of diarrheal illness (Starko et al., 1986).

• In 2004, a cross connection in a large residential development with a dual-distribution system reportedly affected approximately 82 households (Sydney Water, 2004). The cross connection resulted from unauthorized plumbing work during construction of a house in the development.

• A meter reader discovered a cross connection in 1996 when he noticed that a water meter at a private residence was registering backwards, which indicated that reclaimed water was flowing into the public potable water system (University of Florida TREEO Center, 2011). The reclaimed water service had recently been connected to an existing irrigation system at the residence. The irrigation system had previously been supplied with potable water and was still connected to the potable system. A backflow prevention device was not installed at the potable water service connection, and it was estimated that about 50,000 gallons of reclaimed water backflowed into the public potable water system.

• Homeowners reported illnesses (diarrhea and digestion and intestinal problems) resulting from a cross connection that occurred in 2002 between a reclaimed water line supplying reclaimed water to a golf course and a potable water line supplying water to more than 200 households. Contractors failed to sever a potable water line that previously provided irrigation water, which created a cross connection between the potable line and the reclaimed water irrigation system. Pressures in the reclaimed and potable systems were comparable, and when a higher demand was created on the potable system, water from the nonpotable reclaimed system was siphoned into the potable system (Bloom, 2003).

• A cross connection between reclaimed (nonpotable) and drinking water lines was discovered at a business park in 2007. It was determined that occupants in 17 businesses at the business park had been drinking and washing their hands with reclaimed water for 2 years. The cross connection was found after the water district increased the percentage of reclaimed water in the nonpotable water line from 20 percent (the remaining 80 percent being potable water) to 100 percent, and occupants complained that the water tasted bad and had a odor and a yellowish tint (Krueger, 2007).

Detailed information on cross-connection control measures is available in manuals published by the American Water Works Association (AWWA, 2009) and the EPA (2003a). Regulations often address cross-connection control by specifying requirements that reduce the potential for cross connections (see Box 10-5). However, effective as such programs are, 100 percent compliance has not been achievable.

end-user exposure, will differ according to the source water, the level of treatment (see Chapter 4), and the extent of dilution with other water sources. If the water is treated to levels intended for nonpotable uses, but it is inadvertently ingested (e.g., after a cross connection of the delivery pipes), the exposure might be much greater than the ingestion of water intended for potable consumption, depending on the level of treatment of the reclaimed water (see Box 6-4). In terms of potential health risks, ingestion of reclaimed water is of greater importance than other reclaimed water uses because exposure and estimation of potential health risks are assessed on the basis of the consumption of drinking water, which most governments (including EPA and countries such as Australia) assume to be 2 L/d (NRMMC/EPHC/NHMRC, 2008).

Aside from the consumption of reclaimed water for drinking water, other sources of ingestion exposure of reclaimed water—primarily from incidental exposures—would be less. Although more data are needed to define the variability of such exposures, Tanaka et al. (1998) provide useful benchmarks for reclaimed water ingestion exposures (see Table 6-1). Indirect exposure pathways through ingestion of contaminants in reclaimed water could potentially occur when reclaimed water is used for food crop irrigation, for fish or shellfish growing areas, or in recreational impoundments that are used for fishing. In these cases, exposure may occur from the accumulation of chemicals within the particular food. Some compounds that occur in wastewater such as nonylphenol (Snyder et al., 2001a) and perfluorinated organic compounds (Plumlee et al., 2008) have been shown to bioconcentrate in animals as the result of water exposures. The potential for bioaccumulation of chemicals and pathogenic microbes can occur, as well as decay of chemicals or microbes during product cultivation. With long-term use of reclaimed water on agricultural land, attention should be paid to accumulation in food crops of persistent substances such as perfluorinated chemicals and metals from repeated application of reclaimed water containing these substances. Limited data have suggested that certain compounds potentially present in reclaimed water may be detectable in irrigated food crops (Boxall et al., 2006; Redshaw et al., 2008). Thus, more research is needed to assess the importance of these indirect pathways of exposure.

Inhalation and Dermal Exposures

Household uses of water can result in inhalation and dermal exposure to chemicals from showering (Xu and Weisel, 2003) and by volatilization (for volatile substances) from other water uses in household appliances, such as clothes washers and dryers (Shepherd and Corsi, 1996). Experimental studies in humans and in vitro test systems using skin samples indicate that certain classes of chemicals can be absorbed into the body following inhalation or dermal exposure to water following bathing or showering. Research has examined dermal and inhalation exposures to neutral, low-molecular-weight compounds, such as water disinfection byproducts present in conventional water systems, including trihalomethanes (e.g., chloroform, bromoform, bromodichloromethane, dibromochloromethane) and haloketones, (e.g., 1,1-dichloropropanone, 1,1,1-trichloropropanone) (Weisel and Wan-Kuen, 1996; Baker et al., 2000; Xu and Weisel, 2005). Levels of these chemicals are not known to be higher in reclaimed water than in conventional water systems (see Appendix A). As reliance on membrane processes in reclaimed water increases (see Chapter 4), there will be a need to assess the potential exposure to neutral, low-molecular-weight organic compounds that could be present, such as 1,4-dioxane and dichloromethane.

Use of reclaimed water in ornamental fountains, landscape irrigation, and ecological enhancement may

TABLE 6-1 Illustration of Differential Water Ingestion Rates from Different Reclamation Uses

Application Purposes	Risk Group Receptor	Exposure Frequency	Amount of Water Ingested in a Single Exposure, mL
Scenario I, golf course irrigation	Golfer	Twice per week	1
Scenario II, crop irrigation	Consumer	Everyday	10
Scenario III, recreational impoundment	Swimmer	40 days per year—summer season only	100
Scenario IV, groundwater recharge	Groundwater Consumer	Everyday	1000

SOURCE: Tanaka et al. (1998).

result in inadvertent exposure via aerosolization, dermal contact, or ingestion from hand-to-mouth activity. Although these have not been studied with respect to reclaimed waters, there have been outbreaks or expressions of concern from many of these exposure pathways fed by other waters (Benkel et al., 2000; Fernandez Escartin et al., 2002), and therefore the potential for such effects cannot be neglected.

In instances where there is adequate information and justification to assess exposure following dermal and/or inhalation exposure to a contaminant in reclaimed water, an average daily dose for dermal and inhalation exposures can be computed analogously to that for ingestions as shown in Appendix B.

Recreational Exposures

The storage of reclaimed water in recreational impoundments or the conveyance through rivers used for recreational purposes may result in exposure via all three routes: oral, dermal, and inhalation. Frequently, for swimming, it is assumed that ingestion of 10–100 mL per incident occurs (Tanaka et al., 1998; Heerden et al., 2005), although direct estimation of this ingestion rate is not common (Schets et al., 2008).

DOSE-RESPONSE ASSESSMENTS

Dose-response assessment is "the process of characterizing the relation between the dose of an agent administered or received and the incidence of an adverse health effect in the exposed populations and estimating the incidence of the effect as a function of human exposure to the agent" (NRC, 1983). The assessment includes consideration of factors that influence dose-response relationships such as age, illness, patterns of exposure, and other variables, and it can involve extrapolation of response data (e.g., high-dose responses extrapolated to low-doses animal responses extrapolated to humans) (NRC, 1994a,b). Dose-response relationships form the basis for the risk assessments used for establishing drinking water regulatory standards. To protect public health, drinking water standards are established at levels lower than those associated with known adverse health effects following analysis of a chemical's dose-response curve

and cost-benefit analysis. These standards are intended to protect against adverse health effects such as cancer, birth defects, and specific organ toxicity, that occur after prolonged exposures and are generally established using various margins of safety or acceptable risk levels to protect humans, including sensitive subpopulations (e.g., children, immunocompromised persons).

Chemical

Dose-response assessment and the subsequent estimation of health risk from exposure to chemicals has traditionally been performed in two different ways: linear methods to address cancer effects and nonlinear (or threshold) methods to address noncancer health effects. These different approaches have been used historically because cancer and noncancer health effects were thought to have different modes of action. Cancer was thought to result from chemically induced DNA mutations. Because a single chemical-DNA interaction in a single cell can cause a mutation that leads to cancer, it has generally been accepted that any dose of chemical that causes mutations may carry some finite risk. Thus, in the absence of additional data on the mode-of-action, cancer risk is typically estimated using a linear, nonthreshold dose-response method. In contrast nonlinear, threshold dose-response methods are typically used to estimate the risk of noncancer effects becausemultiple chemical reactions within multiple cells have been thought to be involved.

Dose-response assessment for chemicals is a two-step process. The first step involves an assessment of all available data (e.g., in vitro testing, toxicology experiments using laboratory animals, human epidemiological studies) that document the relationship(s) between chemical dose and health effect responses over a range of reported doses. In the second step, the available observed data are extrapolated to estimate the risk at low doses, where the dose begins to cause adverse effects in humans (EPA, 2010c; WHO, 2009). Upon considering all available studies, the significant adverse biological effect that occurs at the lowest exposure level is identified as the critical health effect for risk assessment (Barnes and Dourson, 1988). If the critical health effect is prevented, it is assumed that no other health effects of concern will occur (EPA, 2010c).

For both carcinogens and noncarcinogens, it is common practice to also include uncertainty factors to account for the strength of the underlying data, interspecies variation, and intraspecies variation. The effect of these factors may be several orders of magnitude in the estimated effect/no-effect level.

Noncarcinogens/Threshold Chemicals

Chemicals that cause toxicity through mechanisms other than cancer are often thought to induce adverse effects through a threshold mechanism. For these chemicals, it is generally thought that multiple cells must be injured before an adverse effect is experienced and that an injury must occur at a rate that exceeds the rate of repair. For chemicals that are thought to induce adverse effects through a threshold mechanism, the general approach for assessing health risks is to establish a health-based guidance value using animal or human data. These health-based guidance values, known as reference dose (RfD), acceptable daily intake (ADI), or tolerable daily intake (TDI), are generally defined as a daily oral exposure to the human population (including sensitive subgroups) that is likely to be free of appreciable health risks over a lifetime (see Box 6-5 for the derivation of RfDs). For pharmaceuticals, maximum recommended therapeutic doses (MRTDs) are generally derived from doses employed in human clinical trials, and are estimated upper dose limits beyond which a drug's efficacy is not increased and/or undesirable adverse effects begin to outweigh beneficial effects. For a number of drug categories (e.g., some chemotherapeutics and immunosuppressants), a clinical effective dose may be accompanied by substantial adverse effects (Matthews et al. 2004). Matthews et al. (2004) analyzed FDA's MRTD database and found that the overwhelming majority of drugs do not demonstrate efficacy or adverse effects at a dose approximately 1/10 the MRTD.

Carcinogens/Nonthreshold Chemicals

A dose-response assessment for a carcinogen comprises a weight-of-evidence evaluation relating to the potential of a chemical to cause cancer in humans, considering the mode of action (EPA, 2005a). For chemicals that can cause tumors by inducing mutations within a cell as well as chemicals whose mode of action is unknown, the dose response is assumed to be linear, and the potency is expressed in terms of a cancer slope factor (CSF, expressed in units of cancer risk per dose; see Box 6-6). Cancer risk then is assumed to be linearly proportional to the level of exposure to the chemical, with the CSF defining the gradient of the dose-response relationship as a straight line projecting from zero exposure–zero risk (Khan, 2010).

Tumors that arise through a nongenotoxic mechanism and exhibit a nonlinear dose-response are quantified using an RfD-like method. Ideally, the risk is evaluated on the basis of a dose-response relationship for a precursor effect considering the mode of action leading to the tumor (EPA, 2005a; Donohue and Miller, 2007). In the absence of specific mechanistic information relating to how chemical interaction at the target site is responsible for a physiological outcome or pathological event, nonthreshold and threshold approaches are generally employed when analyzing dose-responses for carcinogens and noncarcinogens, respectively.

Microbiological

Microbiological dose-response models serve as a link between the estimate of exposed dose (number of organisms ingested) and the likelihood of becoming infected or ill. Infectivity has been used as an end point in drinking water disinfection because of the potential for secondary transmission (Regli et al., 1991; Soller et al., 2003).

From deliberate human trials ("feeding studies"), such as for cryptosporidium (Dupont et al. 1995), rotavirus (Ward et al. 1986), and other organisms, mechanistically derived dose response relationships (exponential and beta-Poisson) have been developed (Haas, 1983). It has also been possible to use outbreak data to develop dose response information, as in the case of E. coli O157:H7 (Strachan et al., 2005); however, this will likely only be possible with agents in foodborne outbreaks where exposure concentration data are available.

In some cases, dose-response relationships relying on animal data must be used. It has generally been found that the ingested dose in animals from a single

BOX 6-5
Derivation of Reference Doses

RfDs, ADIs, and TDIs can be derived from no-observed-adverse-effect levels (NOAELs) or lowest-observed-adverse-effect levels (LOAELs) in animal or human studies, or from benchmark doses (BMDs) that are statistically estimated from animal or human studies. The overall process associated with derivation of an RfD, ADI, or TDI is illustrated in the figure below, and the detailed equation is

$$RfD = \frac{(NOAEL_{Critical\ Effect}\ or\ LOAEL_{Critical\ Effect}\ or\ BMD_{Critical\ Effect})}{UF_H \times UF_A \times UF_S \times UF_L \times UF_D}$$

Where

NOAEL = The highest exposure level at which there are no biologically significant increases in the frequency or severity of adverse effect between the exposed population and its appropriate control. Some effects may be produced at this level, but they are not considered adverse or precursors of adverse effects.

LOAEL = The lowest exposure level at which there are biologically significant increases in frequency or severity of adverse effects between the exposed population and the appropriate control group.

BMD = A dose that produces a predetermined change in response rate of an adverse effect (called the benchmark response) compared with background.

UF_H = A factor of 1, 3, or 10 used to account for variation in sensitivity among members of the human population (intraspecies variation).

UF_A = A factor of 1, 3, or 10 used to account for uncertainty when extrapolation from valid results of long-term studies on experimental animals to humans (interspecies variation).

UF_S = A factor of 1, 3, or 10 used to account for the uncertainty involved in extrapolating from less-than-chronic NOAELs to chronic NOAELs.

UF_L = A factor of 1, 3, or 10 used to account for the uncertainty involved in extrapolating from LOAELs to NOAELs.

UF_D = A factor of 1, 3, or 10 used to account for the uncertainty associated with extrapolation from the critical study data when data on some of the key toxic end points are lacking, making the database incomplete (Donohue and Miller, 2007).

Both the NOAEL approach and BMD approach involve use of uncertainty factors (UFs), which account for differences in human responses to toxicity, uncertainties in the extrapolation of toxicity data between humans and animals (if animal data are used), as well as other uncertainties associated with data extrapolation.

The underlying basis of calculating an RfD, ADI, or TDI is the dose-response assessment, where critical health effects are identified for each species evaluated across a range of doses. The critical effect should be observed at the lowest doses tested and demonstrate a dose-related response to

exposure presents the same risk as ingesting the same dose in humans; thus, there is not a need for interspecies "correction." This has been shown, for example, for *Legionella* (Armstrong and Haas, 2007), *E. coli* O157:H7 (Haas et al., 2000), and *Giardia* (Rose et al., 1991).

While the one-time exposure to a pathogen carries the possible risk of an adverse effect, multiple exposures (e.g., exposures on successive days) may enhance the risk. Very little is known about the description of risk from multiple exposures to the same agent. As a default, multiple exposures are modeled as independent events (Haas, 1996), although it is biologically plausible that

either positive deviations (due to sensitization) or negative deviations (due to immune system inactivation) could occur. Dose response experiments using multiple dose protocols would be necessary to further inform this assessment.

Depending on the agent, effects from exposure to pathogens can produce a spectrum of illnesses, from mild to severe, either with acute or chronic effects. For some agents, particularly in sensitive subpopulations, mortality can occur. To determine public health consequences, it is necessary to integrate across the spectrum of effects. This can be done using disability adjusted life years (DALYs) or quality adjusted life years (QALYs) (see Box 10-4).

support the conclusion that the effect is due to the chemical in question (Donohue and Miller, 2007; Faustman and Omenn, 2008).

The RfD, ADI, TDI, or MRTD can then be used as the basis for deriving an acceptable level of chemical contaminant in reclaimed water, using the following equation:

$$\text{Acceptable Level in Reclaimed Water}_{\text{Noncarcinogen/Threshold Chemical}} = \frac{Rfd \times Body\ Weight \times RSC}{Drinking\ Water\ Intake}$$

where drinking water intake is assumed to equal 2 L/d, and the relative source contribution (RSC) equals the portion of total exposure contributed by reclaimed water (default is 20 percent).

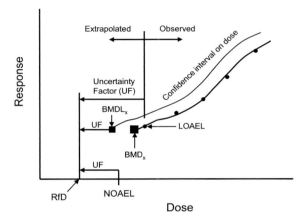

Example RfD derivation for noncarcinogens or chemicals with a threshold effect. This figure shows graphically how various dose-response data are converted to an RfD, considering confidence intervals and various uncertainty factors.

SOURCE: Adapted from Donohue and Orme-Zavaleta (2003).

RISK CHARACTERIZATION

Risk characterization is the last stage of the risk assessment process in which information from the preceding steps of the risk assessment (i.e., hazard identification, dose response assessment, and exposure assessment) are integrated and synthesized into an overall conclusion about risk. "In essence, a risk characterization conveys the risk assessor's judgment as to the nature and existence of (or lack of) human health or ecological risks" (EPA, 2000). Ideally, a risk characterization outlines key findings and identifies major assumptions and uncertainties, with results that are transparent, clear, consistent, and reasonable.

When estimates or measures of exposure and potency (i.e., dose-response relationships) exist, risk can be formally characterized in terms of expected cases of types of illness (with uncertainties) resulting under a given scenario. For example, for a nonthreshold chemical or microbial agent that has a linear dose-response relationship, the characterized risk from a uniform exposure is the simple product of the potency multiplied by the dose. The process is illustrated in Chapter 7. There are a variety of summary measures of risk that can be used (e.g., RfD, ADI, TDI, risk quotient [RQ; i.e., the level of exposure in reclaimed water divided by the risk-based action level, such as the maximum contaminant level or MCL]).

BOX 6-6
Derivation of Cancer Slope Factors

CSFs can be derived using a multistage model of cancer (available through EPA's Benchmark Dose Modeling software), where the quantal relationship of tumors to dose is plotted. A point of departure, or dose that falls at the lower end of a range of observation for a tumor response, is estimated, and a straight line is plotted from the lower bound to zero. The below figure illustrates a linear cancer risk assessment (Donohue and Orme-Zavaleta, 2003). The CSF is the slope of the line (cancer response/dose) and is the tumorigenic potency of a chemical.

The CSF can be used as the basis for deriving an acceptable level of chemical contaminant in reclaimed water, using the following equation:

$$\text{Acceptable Level in Reclaimed Water } (\mu g/L) = \frac{\text{Acceptable Risk Level} \times \text{Body Weight} \times \text{CSF}}{\text{Drinking Water Intake}}$$

where the acceptable risk level generally equals 10^{-6}, and drinking water intake is assumed to be 2 L/d.

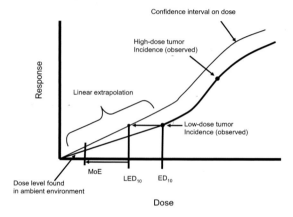

Example cancer risk extrapolation, using the linear dose-response model. The CSF is the slope of the line (i.e., cancer response/dose) and represents the tumorigenic potency of a chemical.

NOTES: MoE = margin of exposure; ED_{10} = effective dose at 10 percent response; LED_{10} = lower 95th confidence interval of ED_{10}.
SOURCE: Adapted from Donohue and Orme-Zavaleta (2003)

Risk Characterization Given Lack of Data

For many chemicals, dose-response information is unavailable. Nonetheless, communities still need to make decisions on water reuse projects in the absence of such data. In this section, frameworks for providing information on risk in absence of dose-response data are discussed.

Numerous organic and inorganic chemicals have been identified in reclaimed water and waters that receive wastewater effluent discharges, and only a limited number of these chemicals are actually regulated in water supplies. Current regulatory testing protocols

address only one chemical at a time, leaving a gap in our understanding of the potential adverse effects of chronic, low-level exposure to a complex mixture of chemicals. A mixture of chemicals may result in toxicity that is additive (i.e., reflecting the sum of the toxicity of all individual components), antagonistic (i.e., toxicity is less than that of an individual component), potentiated (i.e., toxicity is greater than that of an individual component), or synergistic (i.e., with toxicity that is greater than additive). Of particular concern are chemicals that are mutagenic or carcinogenic and share similar modes of action. As with other types of exposures, in the case of reclaimed water, multiple chemicals may be present at the same time for prolonged exposure periods, and they may have a synergistic relationship.

Due to the absence of a federal risk assessment paradigm for evaluating health risks from trace contaminants in reclaimed water, private associations as well as states (particularly California) have embarked upon their own programs to use existing screening paradigms to assess health risks of contaminants in reclaimed water (e.g., Rodriquez et al., 2007a; Bruce et al., 2010; Drewes et al., 2010; Snyder et al., 2010b; Bull et al., 2011). Techniques to conduct such water quality evaluations and subsequently perform exposure and risk assessments are summarized in Khan (2010).

Rodriquez et al. (2007a,b, 2008) and Snyder et al. (2010b) used these screening health risk assessment approaches to evaluate potential health risks from chemicals in reclaimed water in Australia and the United States, respectively. In both evaluations, potential health impacts of chemical contaminants were evaluated using a combination of approaches based on extrapolating health risks using actual health effects data on a specific contaminant, as well as chemical class-based evaluation approaches in the absence of contaminant-specific data. For regulated chemicals, EPA MCLs, Australian drinking water guidelines, or WHO drinking water guideline values were used as benchmark risk values (or risk based action levels, RBALs), from which risk quotients can be evaluated (see also example in Appendix A). RBALs for unregulated chemicals with existing risk values can be based upon EPA reference doses (RfDs), WHO acceptable daily intakes (ADIs), lowest therapeutic doses for pharmaceuticals, or EPA cancer slope factors (CSFs), among other risk values. If existing risk values have not been derived, it is possible to derive risk values for noncarcinogens or carcinogens using human or laboratory animal datasets on the chemical under consideration using methods described in Boxes 6-5 and 6-6. The selection of one risk value over another (e.g., RfD vs. ADI) or selection of a specific epidemiological or toxicological dataset used to derive a RBAL generally should be based upon the critical health effect(s) identified for the specific chemical in the most sensitive species.

Potential health risks from the presence of a chemical in reclaimed water can be assessed by dividing a chemical's RBAL by the concentration of that chemi-

cal in reclaimed water. This risk quotient is known as a Margin of Safety (MOS), with values >1 indicating that the presence of a chemical in reclaimed water is unlikely to pose a significant risk of adverse health effects. This is exampled in Chapter 7 for 24 organic contaminants in reclaimed water.

Benchmarks for unregulated chemicals without complete epidemiological or toxicological datasets or risk values were evaluated by Rodriquez et al. (2007a,b) and Snyder et al. (2010b) using class-based risk assessment approaches, including the Threshold of Toxicological Concern (TTC), FDA's Threshold of Regulation (TOR; see Box 6-7), or EPA's Toxicity Equivalency Factor (TEF) approach. Rodriquez et al. (2007a,b) used the TTC approach for both unregulated noncarcinogens and carcinogens without available toxicity information, while Snyder et al. (2010b) used TTC for noncarcinogens and nongenotoxic carcinogens. The Toxicity Equivalency Factor (TEF)/Toxicity Equivalents (TEQ) approach was used by Rodriquez et al. (2008) to assess potential health risks from dioxin and dioxin-like compounds in Australian reclaimed water used to augment drinking water supplies, based

---

BOX 6-7
Threshold of Regulation (TOR)

One class-based approach is the Threshold of Regulation, which was developed as a method to evaluate the potential toxicity of carcinogens extracted from food contact substances. The TOR is a concentration of chemicals unlikely to pose a significant risk of adverse health effects, including cancer risk (10^{-6}) over a lifetime (FDA, 1995; Rulis, 1987, 1989). The FDA derived a threshold value of 0.5 ppb for carcinogens in the diet based on carcinogenic potencies of 500 substances from 3500 experiments of Gold et al.'s (1984, 1986, 1987) Carcinogenic Potency Database. The distribution of chronic dose rates that would induce tumors in 50 percent of test animals (TD50s) was plotted. This distribution was extrapolated to a Virtually Safe Dose (10^{-6} lifetime risk of cancer) in humans and is equal to 0.5 μg chemicals/kg of food, or 1.5 μg/person/day (based on 3 kg food/drink consumed/day). This value can be extrapolated to a concentration in water intended for ingestion, as follows:

$$\text{TOR: } 0.5 \text{ μg/kg food/day} \times (3 \text{ kg food/day}) / (2 \text{ L water/day}) = 0.75 \text{ μg/L.}$$

BOX 6-8
Thresholds of Toxicological Concern (TTCs)

For carcinogens, distributions of chronic dose rates from lifetime animal cancer studies were statistically evaluated for more than 700 carcinogens to identify an extrapolated threshold value in humans unlikely to result in a significant risk of developing cancer over a lifetime of exposure (Cheeseman et al., 1999; Kroes et al., 2004; Barlow, 2005). This threshold value is equal to 1.5 µg/person/day. For noncarcinogens, analyses have been performed to identify human exposure thresholds for chemicals falling into certain chemical classes. One of the best known TTC evaluations is Munro et al. (1996)'s evaluation of 613 organic chemicals that had been tested in noncancer oral toxicity studies in rodents and rabbits, where chemicals are grouped into three general toxicity classes based on the Cramer classification scheme (Cramer et al., 1978):

• Class I—Simple chemicals, efficient metabolism, low oral toxicity
• Class II—May contain reactive functional groups, slightly more toxic than Class I
• Class III—Substances that have structural features that permit no strong initial presumption of safety or may even suggest significant toxicity

Human exposure thresholds (TTCs) of 1800, 540, and 90 µg/person/day (30, 9, and 1.5 µg/kg body weight/day, respectively) were proposed for class I, II, and III chemicals using the 5th percentile of the lowest No Observed Effect Level for each group of chemicals, a human body weight of 60 kg, and a safety/uncertainty factor of 100 (Munro et al., 1996). Using the above TTC human exposure thresholds, an acceptable level of each chemical in reclaimed water can be derived as follows:

$$\text{Acceptable Level In Reclaimed Water (µg/L)} = \frac{[X] \text{ µg/person/day} \times \text{RSC}}{2 \text{ L/person}}$$

Where X = 1800 µg/day for class I compounds, 540 µg/day for class II compounds, and 90 µg/day for class III compounds; Relative Source Contribution (RSC) = 0.2 (assumed default), and drinking water intake = 2 L/day. Therefore, the TTC approach assigns acceptable levels for these three classes of chemicals in reclaimed water as follows:180 µg/L for Class I compounds, 54 µg/L for Class II compounds, and 9 µg/L for Class III compounds.

on TEFs developed by the WHO. (For details on calculation of TEQs, see EPA, 2010c.)

Although newer than traditional risk assessments, which are based upon chemical-specific data, these class-based values are widely used by regulatory authorities to assess health risks in the absence of complete substance-specific health effects datasets. The TTC approach is used by the World Health Organization's Joint Expert Commission on Food Additives (JECFA) to assess health risks from food additives present at low levels in the diet, and the U.S. Food and Drug Administration (FDA) uses the TOR approach when assessing health risk from indirect food additives (such as chemicals in food contact articles; Box 6-7).

The TTC approach has evolved over the past 20 years, starting from the FDA's TOR concept (Rulis, 1987, 1989) and more recently developing into a tiered appear, where different threshold doses are established based on chemical structure and class (Munro, 1990; Munro et al., 1996; Kroes et al., 2004). The TTC approach is based on the existence of a threshold for a toxic effect (e.g., cancer or a systemic toxicity endpoint such a liver toxicity), which is usually identified through animal experiments. TTC values are statistically derived by analyzing toxicity data for hundreds of different chemicals, where doses in animal studies are extrapolated to doses that are unlikely to cause adverse health effects in humans. TTC values have been derived for carcinogens and noncarcinogens (see Box 6-8).

Despite the utility of TTC, there are multiple classes of chemicals that cannot be screened using the TTC approach, such as heavy metals, dioxins, endocrine active chemicals, allergens, and high potency carcinogens, which instead must be evaluated using different risk assessment approaches (Kroes et al., 2004, Barlow, 2005, SCCP, 2008). Reasons for this are primarily public health protective and include the following factors:

• Heavy metals and dioxins may bioaccumulate, and safety factors used in derivation of TTC values may not be large enough to account for differences in elimination of such chemicals in the human body compared to laboratory animals. In addition, the original databases used to develop TTC threshold values may not have included structurally similar chemicals.

• Endocrine active chemicals have limited datasets relating at lower doses.

• Allergens don't always display a clear threshold, and may elicit adverse effects even at extremely low doses.

• High potency carcinogens, such as aflatoxin-like, N-nitroso and azoxy compounds, are toxic even at low levels.

The TTC approach is meant solely as a method to derive relatively rapid conservative estimation of risk for compounds without detailed risk assessment or with limited datasets. The screening approach was not intended for detailed regulatory decision making. This tool also provides a means to prioritize attention to chemicals where complete toxicological relevance data are absent. The screening value also provides a means for analytical chemists to target meaningful method reporting limits based on health, rather than simply relying on absolute maximum instrumental and method sensitivity.

Results of Screening-Level Analyses

Rodriquez et al. (2007b) evaluated a total of 134 chemicals, including volatile organic compounds, disinfection byproduct, metals, pesticides, hormones and pharmaceuticals, in water that had undergone advanced treatment (microfiltration or ultrafiltration followed by reverse osmosis) at the Australian Kwinana Water Reclamation Plant (KWRP). Calculated risk quotients (RQ) were 10 to 100,000 times below 1 for all volatile organic compounds and all pharmaceuticals except cyclophosphamide (RQ=0.5). Risk quotients <1 indicate that there is unlikely to be a significant health risk associated with exposure to a specific chemical. RQs for all metals were also <1. Rodriquez et al. (2007a) concluded that there were no increased health risks from the KWRP reclaimed water destined for indirect potable reuse as evidenced by levels of contaminants being well below benchmark values.

Soller and Nellor (2011a,b) performed quantitative relative risk assessments of two different water reuse projects in Southern California: the Montebello Forebay and Chino Basin Groundwater Recharge projects. In each project, water samples from wells that contained "relatively high proportions" of reclaimed water were analyzed for chemical contaminants and compared against water samples from control wells containing little or no reclaimed water. Health risks from contaminants of potential health concern were estimated, and the datasets were compared. For both types of groundwater samples, hazard indices were calculated representing the sum of potential noncancer effects from exposure to the identified chemicals; cancer risks were assessed by estimating lifetime cancer risks associated with drinking water exposure to the chemicals present in wells. For both projects, hazard indexes in the reclaimed and control water samples were below the threshold for potential health effects (i.e., <1). In the Chino Basin Groundwater Recharge Project, noncancer and cancer risks were judged to be equivalent among the reclaimed water wells compared with the control wells. In the Montebello Forebay Groundwater Recharge Project, noncancer and cancer risks were equivalent among the reclaimed water wells compared to control wells, with the exception of risks associated with arsenic. An analysis by the authors indicates that arsenic concentrations in water do not appear to be influenced by reclaimed water content, but rather are caused by naturally occurring arsenic.

CONSIDERATION OF UNCERTAINTY

Many elements going into a risk characterization contain elements of uncertainty and/or variability. These terms are defined as (NRC, 2009b):

Uncertainty: Lack or incompleteness of information. Quantitative uncertainty analysis attempts to analyze and describe the degree to which a calculated value may differ from the true value; it sometimes uses probability distributions. Uncertainty depends on the quality, quantity, and relevance of data and on the reliability and relevance of models and assumptions.

Variability: Variability refers to true differences in attributes due to heterogeneity or diversity. Variability is usually not reducible by further measurement or study, although it can be better characterized.

The inputs to a risk characterization may have a number of sources of uncertainty and variability, and therefore, the final risk characterization has inevitable uncertainty as well. Some of these sources of uncertainty and variability are

• uncertainty from the use of animal species to derive effects data,
• uncertainty from effects data based on single contaminants rather than mixtures,
• variability in occurrence of contaminants and performance of treatment processes,[3]
• variability in response of populations (susceptibility),[3]
• variability in exposure to water with contaminants (e.g., ingestion rates, inhalation rates),[3] and
• uncertainty in models (e.g., contaminant transport; dose-response).

Given the variability and uncertainty in the inputs to a risk characterization that may arise in both exposure assessment and dose-response assessment, any final characterization can never be known with absolute precision and certainty. Therefore, the uncertainty in the risk assessment should be characterized. In speaking about the level of analysis with which these facets are considered, NRC (2009b) makes the following statement in the context of EPA decision making:

> The characterization of uncertainty and variability in a risk assessment should be planned and managed and matched to the needs of the stakeholders involved in risk-informed decisions. In evaluating the tradeoff between the higher level of effort needed to conduct a more sophisticated analysis and the need to make timely decisions, EPA should take into account both the level of technical sophistication needed to identify the optimal course of action and the negative impacts that will result if the optimal course of action is incorrectly identified. If a relatively simple analysis of uncertainty (for example, a non-probabilistic assessment of bounds) is sufficient to identify one course of action as clearly better than all the others, there is no need for further elucidation. In contrast, when the best choice is not so clear and the consequences of a wrong choice would be serious, EPA can proceed in an iterative manner, making the analysis more and more sophisticated until the optimal choice *is* sufficiently clear. (NRC, 2009b)

Depending on the preferences of the decision makers and stakeholders (recalling that the objective of risk characterization is to provide information in a form useful to these groups), uncertainty can be captured and described in different ways (Patè-Cornell, 1996). The use of uncertainty or "safety" factors is perhaps the simplest. The quantitative factors contributing to uncertainty (footnoted in the list above) can be characterized by probability distributions, and a Monte Carlo analysis can be performed to present the characterized risk as a probability distribution (Burmaster and Anderson, 1994). As the most intensive alternative, a second-order or two-dimensional Monte Carlo analysis (Burmaster and Wilson, 1996) may be performed in which elements of uncertainty (that could be reducible if more information were obtained) are separated from elements of variability (reflecting the intrinsic heterogeneity of the scenario). For the sake of conciseness, the details of these various methods (beyond the use of uncertainty factors) are not detailed in this report. However, formal uncertainty analysis can often be useful to decision makers (Finkel, 1990). Although safety factors and simple Monte Carlo analyses have been performed in the context of reuse, the committee is not aware of use of the second-order methods in this context.

An uncertainty analysis can also be used to assess the risks involved in excursions from usual process performance, accidents, or failure of one or more processes. Essentially the likelihood of such deviations and the impact on removal of contaminants are combined to assess the impact on overall risk on a per day (or per year) basis. However, to perform such analyses, more data are needed on the process variability (including in the distribution system) and the risk of failure under long-term operations. The risk of a cross connection in distribution systems (Box 6-4) is a special type of such risk that also should be considered, although a strong quantitative database to estimate the frequency and impact of such occurrences is lacking.

CONCLUSIONS AND RECOMMENDATIONS

Health risks remain difficult to fully characterize and quantify through epidemiological or toxicological studies, but well-established principles and processes exist for estimating the risks of various water reuse applications. Absolute safety is a laudable goal of society; however, in the evaluation of safety, some

---

[3] Quantitative factors contributing to uncertainty.

degree of risk must be considered acceptable (NAS, 1975; NRC, 1977). To evaluate these risks, the principles of hazard identification, exposure assessment, dose-response assessment, and risk characterization can be used. Although risk assessment will be an important input in decision making, it forms only one of several such inputs, and risk management decisions incorporate a variety of other factors, such as cost; equitability; social, legal, and regulatory factors; and qualitative public preferences.

The occurrence of a contaminant at a detectable level does not necessarily pose a significant risk. Instead, only by using dose-response assessments (the second step of risk assessment), can a determination be made of the significance of a detectable and quantifiable concentration.

Risk assessment screening methods enable estimates of potential human health effects for circumstances where dose-response data are lacking. Approaches such as the threshold of toxicological concern and the toxicity equivalency factor may useful in this regard, although additional research in such approximate methods and assessment of their performance is needed.

A better understanding and a database of the performance of treatment processes and distribution systems are needed to quantify the uncertainty in risk assessments of potable and nonpotable water reuse projects. Failures in reliability of a water reuse treatment and distribution system may cause a short-term risk to those exposed, particularly for acute contaminants where a single exposure is needed to produce an effect. Although there are many sources of uncertainty and variability, by using well-understood methods in risk assessment, the impact of such sources of variability and uncertainty on estimated human health risk can be determined. To assess the overall risks of a system, the performance (variability and uncertainty) of each of the steps needs to be understood. Although a good understanding of the typical performance of different treatment processes exists, an improved understanding of the duration and extent of any variations in performance at removing contaminants is needed.

When assessing risks associated with reclaimed water, the potential for unintended or inappropriate uses should be assessed and mitigated. If the risk is then deemed unacceptable, some combination of more stringent treatment barriers or more stringent controls against inappropriate uses would be necessary if the project is to proceed. Inadvertent cross connection of potable and nonpotable water lines represents one type of unintended outcome that poses significant human health risks from exposure to pathogens. To significantly reduce the risks associated with cross connections, particularly from exposure to pathogens, nonpotable reclaimed water distributed to communities via dual distributions systems should be disinfected to reduce microbial pathogens to low or undetectable levels. Enhanced surveillance during installation of reclaimed water pipelines may be necessary for nonpotable reuse projects that distribute reclaimed water that has not received a high degree of treatment and disinfection.

Guidance and user-friendly risk assessment tools would improve the understanding and application of these risk assessment methods. Although risk assessment is a useful tool to help prioritize efforts to protect public health in the face of uncertainty, conducting a chemical or microbial risk assessment is complex and resource-intensive. As the extent of water reclamation increases around the United States, it may be desirable and appropriate for regulatory authorities (e.g., state, federal) to prepare guidance or reference materials to facilitate understanding of these methods for water reuse applications and to develop user-friendly tools for the use of more advanced assessment methods that can be used by a greater number of utilities and stakeholders.

7

Evaluating the Risks of Potable Reuse in Context

In this chapter, the committee summarizes the findings of previous National Research Council (NRC) committees as they examined the question of the safety of water reuse. Building on the risk assessment methodologies presented in Chapter 6, the committee then presents a comparative analysis of potential health risks of potable reuse in the context of the risks of the common contemporary circumstance of a conventional drinking water supply derived from a surface water that receives a small percentage of treated wastewater (see Chapter 2). By means of this analysis, the committee compares the estimated risks of a drinking water source generally perceived as safe (i.e., de facto potable reuse) against the estimated risks of two other potable reuse scenarios.

PREVIOUS NRC ASSESSMENTS OF REUSE RISKS

The 1982 Committee on Quality Criteria for Water Reuse, citing the many unknowns in making an assessment of the health effects of potable reuse, adopted the view that "the quality of reused water could be compared to that of conventional drinking water supplies, which are assumed to be safe" (NRC, 1982). That committee was providing advice for an extensive testing program being undertaken by the U.S. Army Corps of Engineers on the treatment of an effluent-dominated Potomac River as part of an evaluation of the future water supply for Washington, D.C., and it outlined specific testing procedures for evaluating the treated water based on the state of science in 1982. A second NRC committee reviewed the results of the Corps' testing program and found them inconclusive, primar-

ily because the Corps had not included all the tiers of toxicological testing recommended (NRC, 1984).

In the years that followed, more extensive toxicological tests were conducted for other proposed potable reuse projects, particularly in Tampa (CH2M Hill, 1993) and Denver (Lauer and Rogers, 1996). In reviewing those data (see Box 6-2), which did not demonstrate any adverse health effects, a third NRC committee concluded that such tests provide a database too limited to draw general conclusions about the safety of potable reuse (NRC, 1998). NRC (1998) also pointed out new concerns that had arisen since the earlier NRC reports and outlined new testing techniques that had been developed. The committee also recommended that new toxicity tests be conducted, particularly long-term fish exposure testing, which were partially implemented in subsequent evaluations in Orange County and Singapore. Advice on testing methods will continue to evolve as science advances, but based on the progression of this research, it is evident that, although such testing might be used to show evidence of potential health risk, it cannot be used to establish the absence of risk.

Examining this history from the vantage point of 2011, the most profound contribution of the 1982 committee was the idea that the quality of the water in potable reuse scenarios should be compared with the quality of conventional drinking waters, which are assumed safe. Advice on specific tests that might be useful in making these comparisons will continue to follow developments in the underlying science, but it is unlikely that any laboratory test will ever establish the absence of health risk in a drinking water from any source. This chapter builds on that foundation, comparing the estimated health risk of water from potable reuse projects with conventional supplies using established tools of risk assessment.

The 1982 committee went on to say that the comparison should be made with the highest quality water that can be obtained from that locality even if that source may not be in use. In a similar vein, the 1998 committee, after concluding that planned potable reuse is viable, suggested that planned potable reuse should be, "an option of last resort—to be adopted only if all the alternatives are technically or economically infeasible" (NRC, 1998). All three committees (NRC, 1982, 1984, and 1998) also took the view that U.S. drinking water regulations were not intended to protect public health when raw water supplies were heavily contaminated with municipal and industrial wastewater.

In the committee's judgment, current circumstances call for a reassessment of those views. First, as shown in Chapter 1, the United States has been operating near the limit of its water supply for several decades since about the time of the first study. As a result of further stress from continued population growth and climate change, this report is being written with a view to providing useful advice to the nation as it comes to terms with this new world where pristine water is ever less abundant, even as the domestic wastewater from an increasing population is discharged into the nation's waterways. Second, as demonstrated in Chapter 2, the committee concludes that de facto reuse (i.e., when a drinking water source consists of some significant percentage of treated wastewater effluent from an upstream discharger) is becoming increasingly common in the United States. Finally, it has become evident to the committee that, in many communities, today's drinking water regulations are already being employed to address the quality of drinking water prepared from water supplies that have substantial wastewater content (see also Chapter 10 for a discussion of regulations). Although this fact does not imply that the regulations are adequate for that charge, it does reflect a notable shift in perspectives since the prior NRC reuse reports were written.

THE RISK EXEMPLAR

Under these conditions, the committee judges that it is appropriate to compare the risk associated with potable reuse projects with the risk associated with de facto reuse scenarios that are representative of the supplies that are widely experienced today. The committee chose to construct a "risk exemplar" to examine how these comparisons might be made. The analysis in this exemplar uses the quantitative risk assessment methods originally proposed for organic chemicals by the NRC (1983) as expanded for microbial contaminants (Haas et al., 1999) and more recently updated (NRC, 2009b). Other methods recently developed to address pharmaceuticals, personal care products, and other anthropogenic contaminants (Rodriquez et al., 2007a,b; Snyder et al., 2008a; Bull et al., 2011) are also used in the analysis to address the risk of classes of contaminants for which rigorous toxicological data are lacking. In the committee's judgment, these risk assessment techniques represent the best means available at this time for estimating the relative risk in such circumstances (see Chapter 6) and offer a method for evaluating the relative merits of various options for managing health risks from chemical and microbial contaminants in reclaimed water.

The committee chose to develop an exemplary comparison of risks associated with various potable reuse scenarios, including de facto potable reuse, modeled upon circumstances currently encountered in the United States today. Based on the discussion in Chapter 2, the committee concluded that it would be appropriate to compare the quality of the water in potable reuse scenarios with the quality of a de facto reuse scenario where a conventional water supply has an average annual wastewater content of 5 percent. This situation is commonly found among current surface water supplies (see Box 2-4). As shown in the figure in Box 2-4, there are many circumstances where de facto reuse exceeds 5 percent, and the committee discussed at length the appropriate wastewater content for use in the exemplar. In the end, 5 percent was selected as a wastewater content that can be reasonably viewed as commonplace and not exaggerated. Swayne et al. (1980) reported that more than 24 million people of the 76 million people surveyed were using drinking water supplies with a wastewater content of 5 percent or more in low-flow conditions (see figure in Box 2-4). Although no data exist, anecdotal evidence based on the population growth in urban areas suggests that wastewater content is often higher today.

The comparative risk approach used in this analysis was designed to examine the presence of selected pathogens and trace organic chemicals in final product waters from de facto reuse and two potable reuse scenarios. Contaminant occurrence data, compiled from several sources, were critically evaluated for each scenario. The data were then analyzed to assess whether there are likely to be significantly greater human health concerns from exposure to contaminants in these hypothetical reuse scenarios, compared with a common de facto reuse scenario. For the chemicals in each of the scenarios, a risk-based action level was used, such as the U.S. Environmental Protection Agency's maximum contaminant levels (MCLs), Australian drinking water guidelines, or World Health Organization drinking water guideline values. Also, a margin of safety is reported, defined as the ratio between a risk-based action level (such as an MCL) and the actual concentration of a chemical in reclaimed water. The resulting ratio between these two values (i.e., the margin of safety) can be used to characterize potential health risks associated with exposure to a chemical (Illing, 2006). For microbials, the dose-response relationships were used to compute risk from a single day of exposure. Additional underlying assumptions are described below.

It was beyond the means of the committee to conduct an analysis of every possible contaminant in reclaimed water. In addition, certain assumptions were made to simplify the analysis. The committee focused on four pathogens commonly of concern in reuse applications and selected 24 chemicals representing different classes of contaminants (i.e., nitrosamines, disinfection byproducts (DBPs), hormones, pharmaceuticals, antimicrobials, flame retardants, and perfluorochemicals), for which occurrence and toxicological data were available in the published literature.

Potable Reuse Scenarios Considered in the Exemplar

Three hypothetical scenarios were evaluated to compare the relative risk from exposure to pathogens and trace organic chemicals in the conventional water supply and potable reuse scenarios. Scenario 1, the conventional water supply scenario, considers de facto reuse with a conventional drinking water treatment plant drawing water from a supply that receives a 5 percent contribution of disinfected wastewater effluent upstream of the intake for the drinking water plant (Figure 7-1a). The two reuse cases describing drinking water supplies derived from planned potable reuse projects include one with groundwater recharge to a potable aquifer via surface spreading basins with subsequent soil aquifer treatment (SAT; Scenario 2, Figure 7-1b) and one with groundwater recharge to a potable aquifer by direct injection of reclaimed water that has received advanced water treatment (Scenario 3, Figure 7-1c).

Scenario 1: De Facto Reuse (Common Surface Water Supply)

In Scenario 1, a surface water supply that serves as a drinking water source receives discharge from secondary treated wastewater effluent that is disinfected and dechlorinated prior to discharge to meet a standard of 200 fecal coliform/100 mL. The surface water is assumed to be free of pathogens with no measurable trace organic chemicals prior to effluent discharge.

Attenuation of contaminants after wastewater discharge can vary widely as a function of distance between discharge point and raw drinking water withdrawal (i.e., retention time), streamflow geometry (i.e., depth, mixing), and environmental conditions such as temperature, ultraviolet light penetration, particulate matter, biological activity. In this scenario, a worst case is assumed where no inactivation or attenuation of pathogens or chemicals occurs in the surface water body. The wastewater discharged constitutes 5 percent of the flow in the source at the point where water is abstracted for the drinking water treatment plant. Subsequently, this water is extracted by a conventional drinking water plant employing coagulation/flocculation, followed by granular media filtration with disinfection designed to meet the requirements of the Long Term-2 Surface Water Treatment Rule (EPA, 2006a).

Scenario 2: Soil Aquifer Treatment

In Scenario 2, a potable aquifer is augmented with reclaimed water via groundwater recharge by surface spreading. Advanced wastewater treatment in this exemplar assumes secondary treatment, followed by nitrification, partial denitrification and granular media filtration but no disinfection. The reclaimed water is

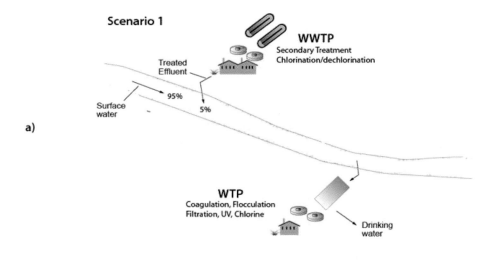

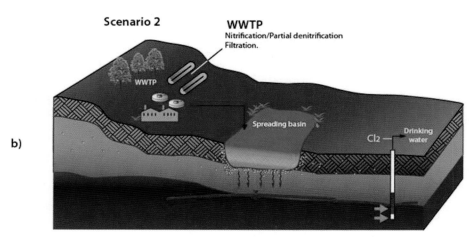

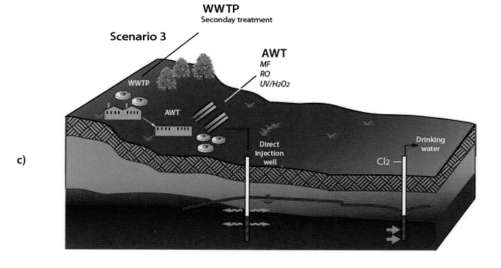

FIGURE 7-1 Summary of scenarios examined in the risk exemplar. (a) Scenario 1—A conventional water plant drawing from a source that is 5 percent treated wastewater in origin; (b) Scenario 2—A deep well in an aquifer fed by reclaimed water via a soil aquifer treatment system and (c) Scenario 3—A deep well drawing from an aquifer fed by injection of reclaimed water from an advanced water treatment (AWT) plant.

applied to surface spreading basins with subsequent SAT (see Chapter 4). It is assumed that the water remains in the subsurface for 6 months with no dilution from native groundwater. The assumption of no dilution is a worst case, being more conservative than the real-world hydrogeological characteristics of typical subsurface systems. This condition was selected to assign removal credits only to physicochemical and biological attenuation processes occurring during SAT. Subsequently, the water is abstracted and disinfected with chlorine at the wellhead prior to consumption, assuming that no blending with other source waters occurs in the distribution system. This assumption describes a scenario where all the drinking water that is consumed originates from the reclaimed water source. This assumption is conservative, given that most existing potable reuse projects blend their product water with other sources, providing additional dilution.

Scenario 3: Microfiltration, Reverse Osmosis, Advanced Oxidation, Groundwater Recharge

In Scenario 3, a potable aquifer is augmented with advanced-treated wastewater followed by groundwater recharge by direct injection. The advanced treatment train includes secondary treatment with chloramination, followed by microfiltration, reverse osmosis, and advanced oxidation using UV irradiation in combination with hydrogen peroxide (UV/H_2O_2). It is assumed that the water remains in the subsurface for 6 months with no dilution from native groundwater. Again, this scenario assumes that any attenuation of pathogens and trace organic chemicals in the aquifer is achieved only by physicochemical and biological processes rather than dilution. Subsequently, the groundwater is abstracted and disinfected at the wellhead with chlorine prior to consumption. Again, this case describes a scenario where 100 percent of the drinking water consumed originates from reclaimed water after advanced water treatment and direct injection into a potable aquifer. This assumption is also conservative, given that most existing potable reuse projects blend their product water with other sources, providing additional dilution.

Other Scenarios Considered

The construction of an additional scenario, similar to Scenario 2, but with disinfection similar to the 450 mg-min/L requirement in California's Title 22 regulations was considered, but was not included because sufficient data could not be obtained to estimate the impact of disinfection on contaminant concentrations. From a qualitative perspective, this scenario would result in significantly lower levels of microbial exposure, particularly for *Salmonella*, but higher levels of DBPs would be present—trihalomethanes and haloacetic acids if free chlorine is used or *N*-nitrosodimethylamine (NDMA) if combined chlorine is used. A review of the literature shows that these DBPs are typically removed during the SAT, particularly NDMA (Kaplan and Kaplan, 1985; Yang et al., 2005; LACSD, 2008; Zhou et al., 2009; Nalinakumari et al., 2010; Patterson et al., 2011), but vigilance is called for when disinfected effluent is used for SAT because as shown in the following section, the margin of safety is smaller with DBPs than with most other trace organic chemicals.

Further details for Scenarios 1 through 3 are provided in Appendix A with respect to water quality characteristics, attenuation, and generation of contaminants during various treatment steps.

Contaminants Considered in the Risk Exemplar

The committee considered a broad cross section of common pathogenic bacteria, viruses, and protozoa, as well as regulated and nonregulated trace organic chemicals that have been reported in reclaimed water or surface water receiving wastewater discharge, to determine the contaminants to be considered in the risk exemplar. The contaminants that were ultimately selected met the following criteria: (1) sufficient information was available on their occurrence, health effects, fate and transport, and behavior in treatment systems such that reasonable calculations could be made for each scenario; and (2) they are either recognized to be of concern based on possible health effects or they are of interest to the public.

Four pathogens were selected: adenovirus, norovirus, *Salmonella*, and *Cryptosporidium*. All of these organisms are transmitted by the fecal-to-oral route, and they all play an important role in waterborne illness in the United States. All four pathogens have been studied in effluents, and, for each, dose-response data are available. *Salmonella* is a classic bacterial pathogen associated with both food- and waterborne disease, while the significance of the other three pathogens has only become clear in recent decades. Toxigenic *Escherichia coli* was originally considered as well, but sufficient dose-response data were not available.

Potential adverse health effects associated with trace organic chemicals in drinking water are an important concern among stakeholders and the public. As noted in Chapter 3, reclaimed water can contain chemicals originating from consumer products (e.g., personal care products, pharmaceuticals), human waste (e.g., natural hormones), and industrial and commercial discharges (e.g., solvents). The reclaimed water, itself, can also contain compounds that are generated during water treatment (e.g., DBPs). Collectively, the number of potential compounds present in reclaimed water is in the thousands. For the risk exemplar, 24 chemicals were selected (see Box 7-1) that represent different classes of these contaminants (i.e., DBPs, including nitrosamines; hormones; pharmaceuticals; antimicrobials; flame retardants; and perfluorochemicals).

The starting concentrations for the microbial and chemical contaminants were selected on the basis of a review of contaminant occurrence data in the scientific literature. More details are provided in Appendix A.

BOX 7-1
Chemicals Selected for Evaluation in the Risk Exemplar

Disinfection Byproducts

Bromate
Bromoform
Chloroform
Dibromoacetic acid (DBCA)
Dibromoacetonitrile (DBAN)
Dibromochloromethane (DBCM)
Dichloroacetic acid (DCAA)
Dichloroacetonitrile (DCAN)
Haloacetic acid (HAA$_5$)
Trihalomethanes (THMs)
N-Nitrosodimethylamine (NDMA)

Hormones and Pharmaceuticals

17β-Estradiol
Acetaminophen (paracetamol)
Ibuprofen
Caffeine
Carbamazepine
Gemfibrozil
Sulfamethoxazole
Meprobamate
Primidone

Others

Triclosan
Tris(2-chloroethyl)phosphate (TCEP)
Perfluorooctanesulfonic acid (PFOS)
Perfluorooctanoic acid (PFOA)

Assumptions Concerning Fate, Transport, Removal, and Estimates of Risk

The assumptions in the exemplar concerning the fate, transport, and removal of the pathogens and chemicals considered in each scenario of the exemplar are discussed in detail in Appendix A. Literature references are provided for the sources of the data that make up each scenario, including expected densities or concentrations following the various treatment steps (including both engineered treatment systems and engineered natural systems), to characterize the fate of contaminants from the initial water or wastewater source to the product water at the consumer's tap. Quantitative microbial risk assessment methodologies are described that are used to estimate the risk of disease that results. Chemical risk assessment techniques used in the exemplar are also described that detail methods to derive risk-based action levels for chemicals in reclaimed water.

Exemplar Results

The goal of the exemplar exercise is to illustrate the relative risk among the scenarios. In Appendix A, the qualities of the surface and reclaimed water and the final water qualities are described for each case as well as the rest of the assumptions behind the exemplar. Scenario 1 represents a scenario to which the public is already commonly exposed in many locations throughout the United States and which is generally regarded as safe, whereas Scenarios 2 and 3 represent planned potable reuse projects. Because of the nature of the risk characterization tools employed, risks from pathogens are displayed in a different form than the risks from chemicals. The pathogen risks are calculated as an estimate of the risk of increased gastroenteric illness. These data also can be usefully displayed as a relative risk—the risks of the potable reuse Scenarios 2 and 3 relative to the risks of Scenario 1—de facto reuse (Figure 7-2).

Figure 7-2 presents a summary of the relative risk of illness from exposure to norovirus, adenovirus, *Salmonella*, and *Cryptosporidium* as a result of drinking water from each of the three scenarios. All of the risks have been normalized to Scenario 1, the de facto potable reuse example. As shown, both potable reuse scenarios have reduced risks, especially where viruses

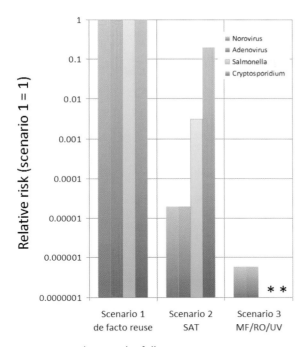

FIGURE 7-2 Relative risk of illness (gastroenteritis) to persons drinking water from each of the reuse scenarios relative to de facto reuse (Scenario 1). The smaller the number, the lower the relative risk of the reuse applications for each organism. For example in Scenario 2, the risk of illness due to *Salmonella* is estimated to be less than 1/100th of the risk due to *Salmonella* in Scenario 1.
NOTES: *The risks for *Salmonella* and *Cryptosporidium* in Scenario 3 were below the limits that could be assessed by the model.

are concerned with the SAT supply, and with all four organisms where the microfiltration/reverse osmosis/ UV supply is concerned. In the latter instance, the densities of *Salmonella* and *Cryptosporidium* are estimated to be reduced to such low levels that the model was unable to calculate a risk. On the basis of these calculations the committee concludes that microbial risks from these potable reuse scenarios are much less than those from de facto reuse.

Table 7-1 summarizes the estimates of the margin of safety for each of the 24 organic compounds studies in the exemplar. These are also displayed graphically in Figure 7-3.

Findings of the Risk Exemplar

The results of the risk assessment (Table 7-1, Figures 7-2 and 7-3) can be used to ascertain whether the

particular process trains produce water of an acceptable quality. Note that these assessments were based on an ingestion scenario. For other end uses, such as showering, some modification in the analyses would need to be made.

For the pharmaceuticals, triclosan, and TCEP, the margin of safety ranges from 1000 to 1,000,000 for all three scenarios (a margin of safety lower than 1 poses potential concern). The perfluorinated chemicals (PFOA and PFOS) have lower margins of safety, but have margins of safety exceeding 1 for all three scenarios. With one exception, the DBPs are shown to have margins of safety above 1. For NDMA, the data for all three scenarios show that it was below the limit of detection, but the detection limit (2 ng/L) exceeds the 10^{-6} lifetime cancer risk level used for the risk-based action level for this compound in the exemplar (0.7 ng/L). As a result the margin of safety for NDMA can only be established as greater than 0.35 for all three scenarios. These results have not identified any chemical that presents a health risk of concern in any of the scenarios studied, although further research is warranted to ensure confidence in these assessments (see Chapter 11). Despite uncertainties inherent in the analysis, these results demonstrate that following proper diligence and employing appropriately designed treatment trains (see Chapter 5), potable reuse systems can provide protection from trace organic contaminants comparable to what the public experiences in many drinking water supplies today. As a general rule, DBPs and perfluorinated chemicals deserve continued scrutiny in all drinking water supplies.

For microbial agents, if one illness or infection/10,000 persons per year is used as a benchmark, it is apparent that the risks from bacterial and protozoan exposure are below this benchmark for all the scenarios, with the exception of Scenario 1, the de facto reuse example (see Appendix A, Table A-6). In this particular instance, it is likely that the risks for the viruses are overestimated, perhaps as a result of the conversion of the genome copy density to the density of infectious units (IU) and/or because predation and die-off in the stream was neglected. In any case, the consistent use of conservative assumptions throughout all three scenarios assures that the assessment of the relative risk of one scenario over the other is robust. The relative analysis makes it clear that the potable reuse scenarios exam-

TABLE 7-1   Summary of Margin of Safety (MOS) Estimates for the Three Scenarios Analyzed by the Committee

Chemical	Risk-Based Action Level[a]	MOS Scenario 1, de Facto Reuse	MOS Scenario 2, SAT, No Disinfection	MOS Scenario 3 MF/RO/UV
Nitrosamines				
NDMA	0.7 ng/L	>0.4	>0.4	>0.4
Disinfection byproducts				
Bromate	10 µg/L	N/A	N/A	> 2
Bromoform	80 µg/L	27	160	>160
Chloroform	80 µg/L	16	80	16
DBCA	60 µg/L	>60	>60	>60
DBAN	70 µg/L	>54	>140	N/A
DBCM	80 µg/L	>80	N/A	>160
DCAA	60 µg/L	12	>60	>60
DCAN	20 µg/L	>20	>20	N/A
HAA5	60 µg/L	6	12	12
THM	80 µg/L	2.7	16	8
Pharmaceuticals				
Acetaminophen	350,000,000 ng/L	>350,000,000	>350,000,000	>35,000,000
Ibuprofen	120,000,000 ng/L	>120,000,000	56,000,000	>280,000,000
Carbamazepine	186,900,000 ng/L	10,000,000	1,200,000	>190,000,000
Gemfibrozil	140,000,000 ng/L	8,600,000	2,300,000	>140,000,000
Sulfamethoxazole	160,000,000 ng/L	>80,000,000	720,000	>160,000,000
Meprobamate	280,000,000 ng/L	17,000,000	8,800,000	>930,000,000
Primidone	58,100,000 ng/L	10,000,000	450,000	>58,000,000
Others				
Caffeine	70,000,000 ng/L	3,500,000	>70,000,000	>23,000,000
17-β Estradiol	3,500,000 ng/L	>35,000,000	>35,000,000	>35,000,000
Triclosan	2,100,000 ng/L	>3,500,000	840,000	>2,100,000
TCEP	2,100,000 ng/L	>84,000	5,800	>210,000
PFOS	200 ng/L	17	4	>200
PFOA	400 ng/L	36	19	>80

NOTES: > indicates that the assumed concentration was below detection, and only an upper limit on the risk calculation was determined. See Appendix A for further detail. [a]Sources of the risk-based action limits are provided in Table A-11 of Appendix 11.

ined here represent a reduction in microbial risk when compared with the de facto scenario that has become a common occurrence throughout the country.

It should be emphasized that the committee presents these calculations as an exemplar. This should not be used to endorse certain treatment schemes or determine the risk at any particular site without site-specific analyses. For example, the presence of a chemical manufacturing facility in the service area of a wastewater utility being used for potable reuse would dictate scrutiny of chemicals that might be discharged to the sanitary sewer. In addition, the various inputs and assumptions of this risk assessment contain sources of variability and uncertainty. Good practice in risk assessment would require full consideration of these factors, such as by a Monte Carlo analysis (Burmaster and Anderson, 1994).

CONCLUSIONS

It is appropriate to compare the risk from water produced by potable reuse projects with the risk associated with the water supplies that are presently in use. The committee conducted an original comparative analysis of potential health risks of potable reuse in the context of the risks of a conventional drinking water supply derived from a surface water that receives a small percentage of treated wastewater. By means of this analysis, termed a risk exemplar, the committee compared the estimated risks of a common drinking water source generally perceived as safe (i.e., de facto potable reuse) against the estimated risks of two other potable reuse scenarios.

The results of the committee's exemplar risk assessments suggest that the risk from 24 selected chemical contaminants in the two potable reuse sce-

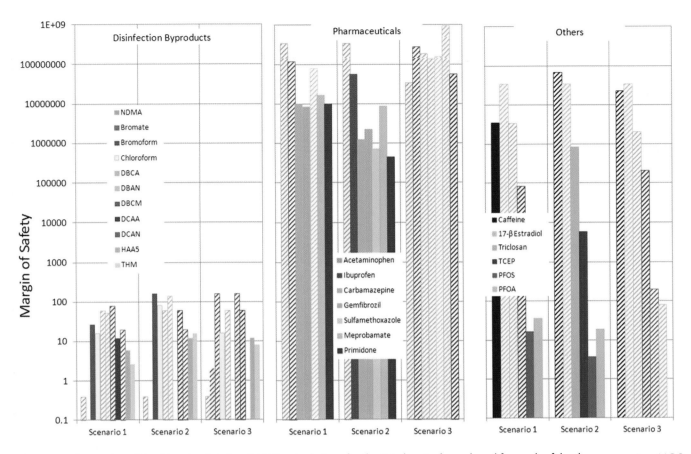

FIGURE 7-3 Display of the Margin of Safety (MOS) calculations for the 24 chemicals analyzed for each of the three scenarios. MOS <1 is considered a potential concern for human health.
NOTE: Bars with diagonal stripes are for MOS values represent the lower limit of the actual value, considering that the concentration of the contaminant was below the detection limit.

narios does not exceed the risk in common existing water supplies. The results are helpful in providing perspective on the relative importance of different groups of chemicals in drinking water. For example, DBPs, in particular NDMA, and perfluorinated chemicals deserve special attention in water reuse projects because they represent a more serious human health risk than do pharmaceuticals and personal care products, given their lower margins of safety. Despite uncertainties inherent in the analysis, these results demonstrate that following proper diligence and employing tailored advanced treatment trains and/or natural engineered treatment, potable reuse systems can provide protection from trace organic contaminants comparable to what the public experiences in many drinking water supplies today.

With respect to pathogens, although there is a great degree of uncertainty, the committee's analysis suggests the risk from potable reuse does not appear to be any higher, and may be orders of magnitude lower than currently experienced in at least some current (and approved) drinking water treatment systems (i.e., de facto reuse). State-of-the-art water treatment trains for potable reuse should be adequate to address the concerns of microbial contamination if finished water is protected from recontamination during storage and transport and if multiple barriers and quality assurance strategies are in place to ensure reliability of the treatment processes (see Chapter 5). The committee's analysis is presented as an exemplar (see Appendix A for details and assumptions made) and should not be used to endorse certain treatment schemes or determine the risk at any particular site without site-specific analyses.

8

Ecological Enhancement via Water Reuse

Rivers, lakes, and streams provide many recreational activities and benefits, as well as important ecosystem services such as nutrient cycling, wildlife habitat, and flood mitigation. With the increasing demand in urban and agricultural areas for freshwater, few options are available to ensure that aquatic systems maintain their respective ecohydrological requirements (Neubauer et al., 2008). Environmental applications of water reuse include river and wetland habitat creation and augmentation of existing water sources for the express purpose of improving conditions for aquatic biota. The Florida Everglades, for example, are at risk due to a decrease of incoming freshwater (see Box 8-1). For areas such as the Everglades and others, water reuse for ecological enhancement may be a beneficial option because reclaimed water could be used to augment streamflow, restore wetlands, and/or enhance water quality (Wintgens et al., 2008; Carey and Migliaccio, 2009). In addition to ecological benefits, there may also be economic benefits (e.g., increased tourism, hurricane protection) from such projects (Carvalho, 2007; Costanza et al., 2008; see also Chapter 9).

Reclaimed water may have potential for augmenting existing surface water systems and creating new habitats. In most instances, reclaimed water used for the purpose of ecological enhancement will meet or exceed local wastewater discharge standards. Nevertheless, the ecological risk of such planned applications needs to be considered to ensure that the level of risk to the environment is acceptable relative to the benefits. The level of acceptable ecological risk in these projects will likely vary between reuse scenarios; for example,

the acceptable level of risk in a newly constructed wetland may be different than in a pristine system such as the Everglades. The level and cost of the assessment will also vary depending on the scenario.

Based on these considerations, the purpose of this chapter is to (1) present what is known about risks associated with the purposeful reuse of treated wastewater for habitat restoration and creation, (2) describe methods for assessing ecological risks from a historical and state-of-the-science perspective, and (3) recommend future research needs in the area of water reuse and ecological risk assessment.

POTENTIAL CONCERNS ABOUT ENVIRONMENTAL APPLICATIONS

As presented in Chapter 2, treated wastewater is routinely discharged to the nation's rivers as part of the wastewater disposal process, with nearly 99 percent of wastewater discharges receiving secondary or greater treatment (see Table 2-1). The quality of reclaimed water used for ecological applications would be no lower than that of traditional wastewater discharge, and may be treated to higher levels. Therefore, available data on the ecological effects from the chemical, physical, and biological stressors in treated wastewater effluent discharged to rivers and lakes provide a worst-case scenario of effects that could occur in ecological enhancement water reuse projects.

Typical wastewater discharges contain a mixture of microbes, inorganic chemicals, and organic chemicals, some of which may cause adverse ecological effects in

---

BOX 8-1
Proposed Reuse Projects to Expand Environmental Water Supply in the Everglades

The Comprehensive Everglades Restoration Plan (CERP) was envisioned as a multidecadal effort to achieve ecological restoration by reestablishing the hydrological characteristics of the historic Everglades ecosystem, where feasible, and to create a water system that simultaneously serves the needs of both the natural and human systems of South Florida (NRC, 2010). The conceptual plan (USACE/SFWMD, 1999) included 68 different project components focused on restoring the quantity, quality, timing, and distribution of water in the ecosystem. The largest component of the budget for this $13 billion project is devoted to water storage, including conventional surface water storage reservoirs, in-ground reservoirs, aquifer storage and recovery, and seepage management. To provide sufficient water supply to meet anticipated future environmental, urban, and agricultural water demands in South Florida, the comprehensive plan included two water reuse projects in Miami-Dade County, which together would treat more than 200 million gallons per day (MGD; 760 million m³/d). In the preliminary project concept, the reclaimed water would be used for aquifer recharge to enhance urban water supplies and reduce seepage out of the Everglades. Additionally, reclaimed water could be provided to Biscayne Bay National Park to help meet freshwater flows to support ecosystem needs. However, the plan acknowledged the high costs of such treatment to support ecological needs and noted that other potential sources of water would be investigated before water reuse was pursued.

Pilot projects were planned to assess the "cost effectiveness and environmental feasibility of applying reclaimed water to sensitive natural areas" and to "identify treatment targets consistent with preventing degradation to natural area," among other objectives. A pilot plant was constructed by Miami-Dade County that included several different wastewater reclamation treatment trains (e.g., with and without reverse osmosis; ozone vs. ultraviolet/advanced oxidation processes), and trace organic chemical data were collected for several months. However, the pilot project was halted in 2011 before the planned toxicity testing was initiated because of general concern about the economic feasibility of the larger ecological restoration project (Jim Ferguson, Miami Dade County Water and Sewer Department, personal communication, 2011).

---

receiving water bodies. However, the level of toxicant exposure and dilution within the receiving systems are key considerations when assessing toxicity. The individual constituents may arise from industrial, household, or wastewater treatment plant applications.

For instance, chlorine is often used as a disinfection chemical to reduce pathogen load and disease risk in wastewater. Low levels of chlorine may cause toxicity in the receiving stream or form chlorinated byproducts capable of causing ecotoxicity. Organic chemicals in wastewater have the potential to deplete the receiving aquatic system of oxygen, thus impacting aquatic life. Suspended solids from wastewater can block sunlight, thus reducing the photosynthetic capability of aquatic plants. Reduction in sunlight penetration may reduce plant life, as well as vertebrate and invertebrate populations. All of these stressors singularly or in combination may affect aquatic life, which includes macroinvertebrates, fish, plants, and amphibians (Sowers et al., 2009; Brix et al., 2010; Slye et al., 2011). Ecological assessments of wastewater effluent-dominated surface waters have shown that aquatic life can be sustained in these types of waters; however, site-specific factors may influence the aquatic life in various locations (Brooks et al., 2006; Slye et al., 2011).

Many studies associated with municipal effluents have been focused on standard measures of water quality, such as pH, temperature, total nitrogen and phosphorus, dissolved oxygen, and the impact of the effluent on the receiving system (Howard et al., 2004; Kumar and Reddy, 2009; Odjadjare and Okoh, 2010). Regulatory agencies, such as the U.S. Environmental Protection Agency (EPA), have developed guidance documents and criteria for many of these water quality parameters on a site-specific or ecoregion basis. Further, the EPA created the National Pollutant Discharge Elimination System to prevent aquatic life impacts associated with these traditional forms of wastewater pollution. As information on new classes of environmental contaminants arise, standard methods for assessing risk (e.g., whole effluent toxicity [WET] testing) may be unable to detect the subtle changes associated with these compounds. For instance, there have been recent reports of treated wastewater causing severe lesions and developmental alterations in amphibians, which are not common sentinel testing organisms in the WET testing paradigm (Sowers et al., 2009; Keel et al., 2010; Ruiz et al., 2010).

Because stressors may be different between each reuse scenario, basic information on the effects of potential ecological stressors in treated wastewater are described in this chapter.

Nitrogen and Phosphorus

Nutrients represent one of the historical problems with direct discharge of wastewater effluent, although the nutrient discharge concentrations are highly dependent on the type of wastewater treatment provided (Carey and Migliaccio, 2009; see Box 8-2). EPA has recently focused increasing attention to the impacts of nutrients on surface water ecosystems and has encouraged states to develop and adopt numeric nutrient criteria for nitrogen and phosphorus (EPA, 2011).[1] Excess nutrients to an aquatic ecosystem can be problematic, because they cause an increase in the primary productivity of the ecosystem, known as eutrophication. Eutrophication can lead to changes in dissolved oxygen concentrations, algal blooms, decreases in submerged aquatic vegetation, and fish-kills. Increases in the limiting nutrient (i.e., the nutrient needed for plant growth but which typically occurs in small quantities) will accelerate eutrophication. Typical levels of nitrate in effluents receiving secondary treatment with disinfection are between 5 and 20 mg nitrogen (N)/L. Typical levels of phosphorus in effluents receiving conventional activated sludge (i.e., secondary) treatment are 4–10 mg/L, and these concentrations can be lowered to 1–2 mg/L with biological nutrient removal (BNR) (see Table 3-2).

Ammonia is particularly toxic to aquatic organisms, with the toxicity dependent on pH and temperature. The roles of pH and temperature relate to the amount of un-ionized ammonia (NH_3) in the water body. The acute and chronic criteria for ammonia (pH 8 at 25°C) are 2.9 and 0.26 mg/L, respectively (EPA, 2009a). Typical levels of ammonia in secondary effluents with disinfection are 1–10 mg/L and 1–3 mg/L with BNR (Asano et al., 2007).

Metals

Trace metals (cadmium, copper, etc) are common regulated contaminants in wastewater discharges. The toxicity of metals in aquatic systems is complex and is often related to the amount of dissolved or free metal in the water. Water quality parameters, such as hardness, pH, and organic matter, can greatly affect toxicity. When considering copper, for instance,

---

[1] See also http://water.epa.gov/scitech/swguidance/standards/criteria/nutrients/strategy/index.cfm.

BOX 8-2
Santa Clara Valley Water District (SCVWD) Stream Augmentation Project

SCVWD proposed a pilot project to augment the flows of Coyote Creek with advanced-treated reclaimed water from the San Jose/SCVWD treatment plant for the purpose of ecological enhancement. Reclaimed water would be discharged into Upper Silver Creek 2 km upstream from its confluence with Coyote Creek in San Jose and released from May to October at a flow rate of 1 to 2 cubic feet per second (cfs) (2,400 to 4,900 m³/d). Baseline studies were conducted prior to the project to monitor water quality parameters (e.g. nutrients, oxygen, temperature) and algal biomass (Hopkins et al., 2002). Hopkins et al. (2002) concluded that augmentation to Coyote Creek could result in increased nutrient and ammonia levels, as well as algal biomass. Analysis of advanced-treated wastewater (from treatment plants using dual-media filtration followed by disinfection by either chlorination or chloramination) indicated that it contained measurable levels of perfluorooctane sulfonate (PFOS) and perfluorooctanoate (PFOA) at total concentrations ≤ 470 ng/L (Plumlee et al., 2008). The bioaccumulation and biomagnifacation factors for PFOS and PFOA that were used in the ecological risk assessment of Coyote Creek were based on data obtained from the Great Lakes. Because the Great Lakes and Coyote Creek are disparate water bodies, there were higher levels of uncertainty in the analysis of the risks of PFOS and PFOA in Coyote Creek. Nonetheless, the detection of these chemicals placed this project on hold in an attempt to understand the meaning of these findings.

---

a low pH increases the most toxic form (i.e., Cu^{2+}) of copper. Hardness and copper toxicity are inversely proportional, whereby elevated water hardness leads to decreased copper toxicity (Erickson et al., 1996). Organic matter forms complexes with copper and reduces toxicity (Hollis et al., 1997). EPA has national water quality guidelines to protect aquatic life for most metals, but site-specific parameters may need be considered for ecological applications of reclaimed water in sensitive ecosystems, particularly in areas with little dilution of the wastewater discharge in the ecosystem (EPA, 2009a).

The impact of silver- and titanium-based nanoparticles in the aquatic environment is an emerging topic of research interest. Fabrega et al. (2011) reported that concentrations of silver nanoparticles as low as a few nanograms per liter can affect fish and invertebrates, although mechanisms of toxicity, nanoparticle fate in wastewater treatment and the environment, and ecological risk in the environment remain poorly understood.

Salinity

Changes in salinity may occur with the use of reclaimed water. Typical levels of salt (measured as total dissolved solids [TDS]) in effluents receiving secondary treatment with disinfection are 270–860 mg/L (Asano et al., 2007). Although the TDS of treated wastewater is not expected to be significantly greater than that of many surface waters, ecological applications should consider the TDS of the native water before introducing reclaimed water into existing ecosystems. Currently, no federal TDS aquatic life criterion exists (Soucek et al., 2011). However, site-specific criteria have been advocated. For example, in certain regions of Alaska, a TDS criterion of 500 mg/L has been suggested for periods of salmon spawning, while a TDS criterion of 1,500 mg/L has been suggested for nonspawning periods (Brix et al., 2010).

Temperature and Dissolved Oxygen

Changes in water temperature may be associated with the use of reclaimed water for environmental purposes. Temperature can influence aquatic community structure and productivity of microbes to fish. For instance, water temperature has been shown to influence factors that affect growth in aquatic organisms (e.g., metabolic rate, respiration), which may alter community structure and trophic interactions (i.e., predator-prey dynamics) within a water body (Sobral and Widdows, 1997; Abrahams et al., 2007; Hoekman, 2010). Further, temperature can alter aquatic habitat influencing species composition and biodiversity (Jones et al., 2004). Typically, the temperature of treated wastewater discharge is in the normal range of the receiving environment.

Dissolved oxygen is an important parameter for aquatic life and is related to various water quality parameters including temperature. As temperature increases in a water body, dissolved oxygen decreases. Dissolved oxygen can also be reduced by algal blooms spurred by high nutrient concentrations. National and site-specific dissolved oxygen criteria have been developed to protect aquatic life (EPA, 1986, 2000). For

instance, dissolved oxygen acute mortality criteria for non-embryo/early-life-stage freshwater fish is 3 mg/L (EPA, 1986). An increase in organism mortality and/or growth, in addition to changes in species composition, may be observed if dissolved oxygen levels fall below the developed criterion.

Boron

Boron, in the form of borates, is released into the environment from anthropogenic sources (i.e., wastewater treatment plant discharge), as well as from weathering of sedimentary rocks (Frick, 1985; Howe, 1998; Dethloff et al., 2009; see also Chapter 3). Boron in reclaimed water is generally less than 0.5 mg/L, while concentrations in surface waters are generally ≤1.0 mg/L (Butterwick et al., 1989, Asano et al., 2007). Fish, amphibian, invertebrate, and plant effects associated with boron exposure generally occur in the low to mid milligram-per-liter range (Powell et al., 1997; Howe, 1998; Laposata and Dunson, 1998; Davis et al., 2002; Dethloff et al., 2009). The concentration of boron that affects fish, amphibians, invertebrates, and plants, including landscape plants, are typically above the concentrations observed in reclaimed water (Wu and Dodge, 2005).

Trace Organic Chemicals

As discussed in Chapter 3, trace organic contaminants (e.g., pharmaceuticals and personal care products, and flame retardants) have been detected in municipal wastewater effluent and in the nation's surface waters, creating concerns for both human and aquatic systems (Daughton and Ternes, 1999; Kolpin et al., 2002; see also Appendix A). The presence of these chemicals (e.g., carbamazepine, triclosan, brominated diphenyl ethers) is associated with normal human use of trace organic compounds. When considering the sensitivity of human and aquatic organisms to trace organic compounds detected in reclaimed water, it is important to note that aquatic organisms are generally as sensitive or more sensitive than humans to these chemicals (Table 8-1). Further, the potential toxicity for many of these compounds may be heavily influenced by water quality parameters (e.g., pH), thus complicating the risk assessment process described below (Valenti et al.,

TABLE 8-1 Comparison of Human Monitoring Trigger Levels for Potable Reuse and Aquatic Predicted No Effect Concentrations for Selected Chemicals in Reclaimed Water

Chemical	Example Occurrence in Secondary/ Advanced-Treated Water (ng/L)[a,b]	Human Monitoring Trigger Levels (ng/L)[b]	Predicted No Effect Concentrations (PNECs) for Aquatic Ecosystems (ng/L)[c]
Ethinyl Estradiol	≤1	280	0.35
Carbamazepine	400	1,000	250,00
Fluoxetine	31	10,000	1,400
PFOS	90	200	1,200
Triclosan	485	350	69
DEET	1,520	2,500	7,700
Atenolol	1,780	70,000	1,800
Nonylphenol	161	500,000	1,700

[a]Defined as the 90th percentile average occurrence in secondary or advanced-treated wastewater, representative of water quality required by California's Title 22 regulations for urban irrigation (Drewes et al., 2010).

[b]Calculated from risk-based acceptable daily intakes (ADIs; see Chapter 6 and Box 6-5) in the California State Water Resources Control Board (SWRCB) Science Advisory Panel Report (Drewes et al., 2010).

[c]Derived by methods outlined under Single Chemical Risk Assessment in this chapter using Brooks et al. (2003); Cleuvers (2003); EPA (2005b); Beach et al. (2006); Costanzo et al. (2007); Caldwell et al. (2008); Capdevielle et al. (2008); Küster et al. (2010).

2009). Human health impacts related to trace organic contaminants are discussed in Chapters 6 and 7.

Natural and synthetic chemicals have the ability to mimic endogenous hormones and alter the endocrine system in aquatic organisms. Chemicals that alter the endocrine system may ultimately cause reproductive dysfunction and population-level decline of organisms. While there are a myriad of chemicals that may interact and disrupt the endocrine system (e.g., bisphenol-A, cadmium), one of the best-studied endocrine disruptors is the birth control contraceptive 17-α ethinyl estradiol (EE2; see Box 8-3; Lange et al., 2001; Maunder et al., 2007). The science is still developing with respect to biological assay for rapid detection of endocrine disruptors (discussed later in this chapter).

One of the current limitations in evaluating the ecological risk of trace organics relates to the amount of ecotoxicity data available. For many trace organics, few data are available to make a reliable assessment of risk. With respect to pharmaceuticals, for instance, the recent improvement in the European Medicines Agency guidelines for the environmental risk assessment of pharmaceuticals should reduce these data gaps.

> **BOX 8-3**
> **17-α Ethinyl Estradiol: A Case in Ecological Endocrine Disruption**
>
> Natural and synthetic chemicals have the ability to mimic endogenous hormones, alter the endocrine system, and lead to reproductive dysfunction in aquatic organisms. In particular, numerous studies have focused on the toxicity associated with the birth control contraceptive 17-α ethinyl estradiol (EE2) in fish (Lange et al., 2001). Fish reproduction is the most sensitive end point associated with EE2, with a laboratory predicted no effect concentration of 0.35 ng/L (Caldwell et al., 2008). A whole Canadian lake study was conducted with EE2, where lakes treated with 5–6 ng/L EE2 caused population declines in fathead minnows and other organisms (Kidd et al., 2007). Although these data supported the laboratory findings of EE2, the levels are higher than those normally expected in the environment (Hannah et al., 2009).

To date, few field studies have evaluated the impact that water reuse and associated trace organics may have on the environment. In addition, few studies are available linking the relationship of laboratory endocrine and reproductive responses to effects in natural systems. Although endocrine disruption is a major scientific research thrust, the detection and risk of endocrine disruptors may be different depending on the reuse scenario. Atkinson et al. (2009) and Slye et al. (2011) investigated surfactants along a 100-mile gradient on the Trinity River spanning the Dallas-Fort Worth Metroplex to Palestine, Texas, where in some areas in the summer months >95 percent of the flow comes from municipal wastewater effluent from multiple inputs. No risk to aquatic organisms could be attributed to surfactants associated with this effluent-dominated river. These two studies represent examples for how geographic information systems (GIS) and chemical and biological monitoring can be incorporated to evaluate an ecosystem dominated with effluent.

APPROACHES FOR ASSESSING ECOLOGICAL RISKS OF RECLAIMED WATER

Many questions remain about the risk of trace organic chemicals to the environment because of the lack of associated environmental fate and effects information. Historically, most chemicals have been tested one chemical at a time. However, mixtures of bioactive trace organic chemicals are often present in water, for which new techniques need to be developed and refined to better understand their risk to the environment. As described in Chapter 6, a mixture of chemicals may result in toxicity that is equal to, less than, or greater than the sum total of the toxicity of the individual components. Using chemicals with the same mode of action (e.g., environmental estrogens), it has been demonstrated that the combined toxicity could be predicted based on the toxicity of the individual chemicals (Thorpe et al., 2003). However, it is much more difficult to model mixture responses when the modes of action of the individual chemicals within the mixture are different. This section discusses historical as well as newer techniques that can be used to assess ecological risk even in the absence of chemical-specific data.

The ecological risk assessment (ERA) process is adapted from and is not dissimilar to the human health risk assessment process described in Chapter 6. An ERA consists of four phases: problem formulation, characterization of exposure, characterization of effects, and risk characterization (EPA, 1998a). Following the risk characterization phase, the information can be used by risk managers to determine the course of action for the particular action or question. Furthermore, the data can be used to prioritize which chemicals are of greatest concern and deserve further research.

An ERA is typically conducted to evaluate the likelihood of adverse effects in the environment associated with exposure to chemical, biological, or physical stressors (EPA, 1998a). In addition, the ERA is designed to accommodate mixtures of stressors on aquatic life and habitat. In this respect, it can be used as the foundation for determining potential adverse effects of using reclaimed water for ecological purposes. Key factors in the ERA are the end point to be evaluated (e.g., habitat, endangered species) and the sensitivity of the ecosystem, which may be different for each reuse scenario.

Once the end points of concern have been identified, an understanding of the magnitude of exposure and response to the stressors will ultimately determine the level of risk. One of the first and fastest approaches will be to conduct a literature evaluation based on the stressors of interest to determine if aquatic toxicity or water quality criterion data are available. If data are available, the assessment may be done without further

testing. However, if no data are available for the contaminants or end points of interest, then testing may be necessary (described in the following sections).

Once risk exposure and effects analysis are completed, a predicted environmental concentration (PEC) and a predicted no-effect concentration (PNEC) for the stressors will be available. The PEC/PNEC ratio will determine the risk associated with the stressors. If the PEC/PNEC ratio is ≥ 1, then a risk exists to the environment. A ratio that is < 1 suggests that the potential risk to the environment is low. If adequate data are available to calculate a species sensitivity distribution, a more extensive probabilistic environmental risk assessment approach may be used to estimate the likelihood and the extent of adverse effects occurring (Verdonck et al., 2003).

Single Chemical Risk Assessment

The environmental safety of chemicals is most often assessed on an individual basis, irrespective of the fact that in the aqueous environment there is a mixture of chemicals. In a single chemical assessment scheme, a no-observed effect concentration (NOEC), the lowest-observed effect concentration (LOEC), and/or an effective concentration (EC)[2] will be derived in a series of laboratory studies with fish, algae, and invertebrates. Typically, these studies focus on higher level end points such as survival, growth, and reproduction of the test organisms. Both the EPA and the Organisation for Economic Co-operation and Development have defined methodological protocols to conduct these studies (EPA, 2010d; OECD, 2011). In the case of fish and invertebrates, the first studies that are conducted are acute or short-term assays, which are ≤ 96 hours and focus on the concentration that causes 50 percent mortality in the test organisms (lethal concentration 50 percent; LC_{50}). Following these initial mortality studies, chronic reproduction and growth studies (≥ 21 days) are often conducted with fish and invertebrates. Once

---

[2] Effective concentration (EC_x) is the concentration of a toxicant that produces X percent of the maximum physiological response. For example, an EC_{50} reflects the concentration that produces half of the maximum physiological response after a specified exposure time. NOEC and LOEC are the ecological risk parallels of the lowest observed adverse effect level (LOAELs) and no observed adverse effect level (NOAEL) for human health risk assessment as defined in Box 6-5.

a NOEC is obtained, a safety/uncertainty factor is applied, accounting for species and exposure differences, to derive the PNEC. These data can be useful in the assessment of reclaimed water because one can compare the PNEC values to the concentrations measured in the water (see Table 8-1).

Note that in this single-chemical assessment scheme, the data are usually obtained in controlled laboratory settings and do not focus on community and ecosystem attributes (e.g., nutrient cycling). Once the chemical is released into the environment, there may be interactions with other substances, such as dissolved organic carbon, that may modulate its toxicity, in addition to potential interaction with other chemical contaminants. Laboratory studies do not account for mixture interactions, where these interactions may lead to additive or greater-than-additive toxicity. Although laboratory studies can be conducted to evaluate mixtures, it is unreasonable to assume that every realistic mixture component can be studied. These sources of uncertainty with respect to potential toxicity need to be recognized. Safety or uncertainty factors can be applied with the risk assessment process to account for mixture scenarios.

A single chemical risk assessment approach is used for most trace organic chemicals, including pharmaceuticals and personal care products (see Box 8-4 for an example application of this method). In this single-chemical approach, molecular, biochemical, and physiological end points are not utilized because they are often difficult to link to higher level effects (e.g., survival, growth, and reproduction).

In the case where a PEC/PNEC ratio is >1 and the body of information suggests that a chemical may adversely affect the environment, controlled outdoor pond or stream mesocosm experiments will help to better predict its impact on populations, communities, and ecosystems. This approach was used by Kidd et al. (2007) to demonstrate that 17α-ethinyl estradiol can cause population- and community-level impacts at environmentally relevant concentrations. The benefit of these studies is that one can measure end points (e.g., species density, species richness, nutrient cycling) in a controlled exposure scenario. In addition, mesocosm experiments have been conducted where a single contaminant has been introduced into a complex effluent to evaluate potential mixture interactions (Brooks et al., 2004).

BOX 8-4
Assessing the Ecological Risk of Carbamazepine: Two Approaches

Carbamazepine is an antiepileptic drug marketed in North America and Europe. Approximately 17 percent of an ingested dose is eliminated from humans as nonmetabolized carbamazepine. Using the traditional ecological risk assessment approach, the potential ecological risk can be estimated. The predicted environmental concentration (PEC) for carbamazepine based on modeling approaches is estimated to be ≤ 0.658 µg/L (Cunningham et al., 2010), while the 90th percentile occurrence in reclaimed water meeting California's urban irrigation requirements (Title 22) is <0.400 µg/L (CSWRCB, 2010). To determine the PNEC, a wide array of available ecotoxicity data are assessed. The 96-hr LC_{50} values for *Daphnia magna* (invertebrate) and Japanese medaka (fish) are 76 and >100 mg/L, respectively (Kim et al., 2007). The 72-hr algal effective concentration 50 percent (EC_{50}; the concentration that produces a response halfway between the baseline and the maximum response) is 74 mg/L, while the 7-day duckweed growth EC_{50} was 25 mg/L (Cleuvers, 2003). Duckweed growth appears to be the most sensitive end point and because only acute data are available, a safety factor of 1,000 is applied to the EC_{50} value. The resultant PNEC is 25 µg/L. Performing the risk quotient calculation (PEC/PNEC), the risk is <0.03, indicating that adverse environmental effects are not expected in surface waters augmented with reclaimed water.

The potential environmental risk can also be estimated using the mammalian model screening approach (Huggett et al., 2003). This approach represents a rapid screening method to estimate ecological risks based on the large quantity of mammalian effects data available for pharmaceuticals. Considering the predicted environmental concentration of carbamazepine of ≤ 0.658 µg/L and an octanol water coefficient (log K_{ow}) of 1.68 (Cunningham et al., 2010), the resultant fish plasma concentration is calculated as 2.6 µg/L. The human therapeutic plasma concentration for carbamazepine is 2,170 µg/L after a single administration (Revankar et al., 1999). The calculated fish plasma concentration at estimated environmental concentrations of carbemazepine is much less than the plasma concentration known to exhibit effects in humans. Therefore, the environmental risk to fish is estimated to be low. The mammalian model screening approach yielded the same conclusion as the traditional risk assessment approach, suggesting its utility in rapid screening of environmental risk associated with pharmaceuticals.

An immense amount of mammalian pharmaceutical data (e.g., toxicological impacts, pharmacokinetics, and metabolism, often in multiple organisms) may be helpful in screening potential environmental risks associated with pharmaceuticals (Lange and Dietrich, 2002; Huggett et al., 2003, 2004). A recent analysis indicated that many human pharmaceutical therapeutic targets are present in fish (Gunnarson et al., 2008). If the therapeutic targets are similar across species, then the internal concentrations that elicit effects across species may be similar. Knowing the PEC of a pharmaceutical and its relative hydrophobicity (or aversion to water, as measured by the octanol-water coefficient, K_{ow}), a fish plasma concentration of that pharmaceutical may be calculated. This value can then be compared to the human therapeutic plasma concentration (H_TPC), which is the concentration of that drug in plasma known to cause an effect. If the fish plasma concentration exceeds the plasma concentration known to cause biological effects in humans, then the concentration of the drug in the water should be suspected of causing an ecological effect. This model can quickly help prioritize ecological risk associated with pharmaceuticals and identify specific drugs that should undergo further testing prior to ecological reuse applications (Box 8-4) (Huggett et al., 2003, 2004; Schreiber et al., 2011).

Bioconcentration and Bioaccumulation

Since the publication of *Silent Spring* by Rachel Carson in 1962, the bioaccumulation of chemicals in the environment has received growing attention. Bioconcentration has traditionally been defined as the accumulation of chemical substances from aquatic environments through nondietary routes, whereas bioaccumulation is the accumulation from nondietary and dietary routes (Barron, 1990). EPA has established criteria where a bioconcentration factor (BCF) or a bioaccumulation factor (BAF) > 1,000 (i.e., concentration in organisms 1,000× greater than water or food) must undergo further testing. Substances with a BCF or BAF >5,000 may be banned from commerce (Moss and Boethling, 1999) (Box 8-5). Several studies have shown a relationship between BCF and K_{OW} (Barron, 1990), where a log K_{OW} > 3 requires additional consideration. Both laboratory and field studies at multiple trophic levels (e.g., fish, birds) can indicate if a chemical is potentially bioaccumulated or bioconcentrated, although field measurements may be needed to confirm laboratory findings (OECD, 1996; Weisbrod et al., 2009).

BOX 8-5
Bioconcentration and Bioaccumulation of Perfluorooctane sulfonate (PFOS)

PFOS has multiple commercial uses (e.g., stain repellant) and has been detected in wastewater and reclaimed water (Plumlee et al., 2008). The log K_{ow} for PFOS is 4.4, and laboratory fish BCF values range from 210 to 5,400, which indicate that this substance is potentially bioaccumulative (Martin et al. 2003; EA, 2004; Ankley et al., 2005). Multiple field studies have measured concentrations of PFOS in invertebrates and fish at concentrations greater than that in the surrounding environment (Kannan et al., 2005; Li et al., 2008). Concentrations of PFOS have also been measured in eagles and mink from the Great Lakes region at concentrations 5–10 times greater than in their respective prey items (Kannan et al., 2005). Given that PFOS has been measured in reclaimed water, these data indicate that PFOS has the potential to move through the food chain in areas where reclaimed water is being used for environmental enhancement. The major U.S. manufacturer of PFOS has announced a voluntary phase-out of PFOS from commerce (EA, 2004).

Effluent Toxicity Testing and Monitoring

A number of toxicity testing and biomonitoring methods are available to assess the ecological effects of reclaimed water for ecological applications. These can be divided into conventional, state-of-the-science, and blended approaches.

Conventional Approaches: Whole Effluent Toxicity Tests

The WET testing program in the United States was implemented to protect water bodies from point-source municipal and industrial discharges (Heber et al., 1996). WET programs for wastewater facilities typically consist of whole-effluent bioassays to determine whether the discharges are affecting the receiving waters (Heber et al., 1996). Typical WET laboratory bioassays include acute invertebrate and fish survival studies, subchronic fish growth studies, and chronic invertebrate reproduction studies. These tests can also be conducted to determine ecological responses to a single contaminant or specific mixtures. The typical duration for most of these studies is <7 days. Field assessments of invertebrate and/or fish population and community structure can also be part of WET programs, but these assessments are not as frequent as laboratory testing.

State of the Science

While traditional ecotoxicology has focused on survival, growth, and reproduction as the main determinants of risk (e.g., WET testing), knowledge regarding the toxic modes of action (i.e., how the chemicals manifest their toxicity) has expanded available toxicity testing alternatives, including in vivo biomarkers or in vitro bioassays. These in vivo and in vitro markers may be specific or nonspecific for a class of chemicals.

In the past several decades, researchers have discovered that chemicals in the environment may interact with the normal estrogen, androgen, and thyroid signaling pathways in aquatic organisms (i.e., endocrine disruption) (Desbrow et al., 1998; Rodgers-Gray et al., 2001; Sumpter and Johnson, 2008). Through in vitro and in vivo screening of wastewater effluents (primary, secondary, and advanced secondary), researchers discovered that chemicals can interact directly with hormone receptors (e.g., estrogen receptors) and that these chemicals can induce changes in the fish egg yolk precursor vitellogenin (Desbrow et al., 1998). Desbrow et al. (1998) were unable to identify a relationship between the various wastewater treatment effluents studied (including primary, activated sludge, percolating filters, and sand filters) and vitellogenin production. From this knowledge, the yeast estrogen (YES) and yeast androgen receptor assays were developed for screening purposes (Arnold et al., 1996). These assays investigate the binding of aqueous chemicals to the estrogen or androgen receptors in yeast cells via colorimetric measurements. Ultimately, researchers can determine the extent to which estrogenic or androgenic chemicals are present in water. For instance, Holmes et al. (2010) utilized the YES assays to demonstrate a 97 percent reduction in total estrogenic activity in a reclaimed water treatment system that utilizes stabilization lagoons followed by coagulation, dissolved air flotation/filtration, and chlorination.

Vitellogenin production is directly linked to stimulation of the estrogen receptor. Circulating 17β-estradiol in female fish stimulates the production of vitellogenin in the liver, where it is released into the blood for incorporation in eggs. Chemicals that act as estrogen mimics (e.g., nonylphenol) increase vitellogenin production in fish, especially in male fish which only produce small quantities under normal conditions. Vitellogenin production, in either whole fish or liver cells, can therefore be used to evaluate estrogen content in municipal effluent, surface waters, and reclaimed waters. Filby et al. (2010) utilized vitellogenin as the primary method to determine the extent of estrogen content reduced by various wastewater treatment technologies.

Another promising nonspecific approach is through the use of gene expression profiling. Fish or other aquatic organisms are exposed to the water of interest, and the differential regulation of genes in the liver or gonad is determined (Garcia-Reyero et al., 2008). The analysis can help determine which biological pathways and processes, if any, are being altered by the water sample. Efforts are currently under way to bridge changes in biological pathways to adverse outcomes (termed adverse outcome pathways) at higher levels of biological organization, as well as develop genomic fingerprints for individual and chemical-specific classes (Kramer et al., 2011). An understanding of pathway data may be useful in developing new in vitro screening methodologies for chemicals of interest.

Blended Approaches

Conventional testing methodologies (e.g., WET) focus on higher level biological end points (i.e., growth, survival, reproduction). Research with endocrine-disrupting chemicals demonstrates that some of these methodologies (e.g., invertebrate reproduction) may not be sensitive enough to detect subtle biological changes that may take months or years to generate, while other responses (e.g., fish reproduction) offer more sensitive end points (Länge et al. 2001). The yeast screening, vitellogenin, and gene profiling assays offer the ability to generate screening-level biological data quickly to determine the presence and/or the relative levels of biologically active compounds in the matrix of interest. However, there is a need for assay standardization and training in order to achieve reliable results. There is also potential with some of these assays (e.g., YES) for false-positive or false-negative results. Further, it should be recognized that at this time there is no direct link to higher level measurements (e.g., reproduction). Neither binding of a chemical to a receptor, induction of vitellogenin, nor changes in gene expression are conclusive of a population effect. They do, however, strongly suggest that more research is needed.

Because of the advantages and shortcomings with each conventional and state-of-the-science methodology, researchers are utilizing a blended approach incorporating both methodologies (Steinberg et al., 2008). Deng et al. (2008) utilized an online, flow-through fish exposure system with reproductive, endocrine (vitellogenin), and other end points to assess the ecological effects of shallow groundwater recharged by reclaimed water in the Santa Ana River Basin, California. The advantage of using a blended monitoring system is that one can achieve the rapid screening-level data associated with the newer assays, as well examine higher level end points.

The difference in ecological risk analysis using conventional vs. more state-of-the-science techniques is evident when one considers nonylphenol (Box 8-6). For nonylphenol, EPA developed ambient water criteria using conventional toxicity testing methods, while the European Union utilized new scientific methods (Box 8-6).

CONCLUSIONS AND RECOMMENDATIONS

Currently, few studies have documented the environmental risks associated with the purposeful use of reclaimed water for ecological enhancement. Water reuse for the purpose of ecological enhancement is a relatively new and promising area of investigation, but few projects have been completed and the committee was unable to find any published research in the peer-reviewed literature investigating potential ecological effects at these sites. As environmental enhancement projects with reclaimed water increase in number and scope, the amount of research conducted with respect to ecological risk should also increase, so that the potential benefits and any issues associated with the reuse application can be identified.

The ecological risk issues and stressors in ecological enhancement projects are not expected to

---

> ### BOX 8-6
> ### Water Quality Criterion for Nonylphenol: United States vs. European Union
>
> Nonylphenol is a frequently detected wastewater contaminant, most commonly used to produce nonionic surfactants. In 1998, 104 million kg of nonylphenol was produced in the United States (Harvilicz 1999). EPA has established ambient water quality criteria for nonylphenol in both saline and freshwater systems. The acute and chronic freshwater quality criteria are 28 and 6.6 µg/L, respectively, while the acute and chronic saltwater criteria are 7 and 1.7 µg/L, respectively. Aquatic organism survival, growth, and reproduction end points were used to establish these criteria. Although nonylphenol has been demonstrated to cause estrogenicity in aquatic organisms (e.g., causes fish to produce vitellogenin), these data do not meet the acceptability requirements for water quality criteria by the EPA (EPA, 2005b). Therefore, these data were not utilized in establishing the criteria. In contrast, the European Union has restricted marketing and use of nonylphenol based in part on the potential for nonylphenol to be an estrogenic substance. The European Union risk assessment for nonylphenol cited a PNEC of 0.33 µg/L, based on a long-term algal study. Further, the resultant nonylphenol PEC/PNEC ratio was determined to be 1.8 (European Union, 2002).

---

exceed those encountered with the normal surface water discharge of municipal wastewater. The most probable ecological stressors include nutrients and trace organic chemicals, although stressors could also include temperature and salinity under some circumstances. For some of these potential stressors (e.g., nutrients) there is quite a bit known about potential ecological impacts associated with exposure. Based on the available science, there is no reason to believe that the use of reclaimed water for environmental enhancement purposes would pose greater impacts than those already occurring in many of the nation's surface waters impacted by wastewater discharge. Further, the presence of contaminants and potential ecological impacts may be lower if additional levels of treatment (e.g., nutrient removal, ozone) are applied.

Trace organic chemicals have raised some concerns with ecological enhancement projects, because aquatic organisms can be more sensitive to trace organic chemicals than humans. Although other stressors are well understood and treatment systems can be developed to reduce their concentrations to acceptable levels, less is known about the ecological effects of trace organic chemicals, including pharmaceuticals and personal care products. Endocrine disruption has been, and will likely continue to be, a scientific research area and concern. More data are needed to link population level effects in natural aquatic systems to laboratory observations.

Sensitive ecosystems may necessitate more rigorous analysis of ecological risks before proceeding with ecological enhancement projects with reclaimed water. Although conventional methods (e.g., WET) of monitoring can be used, newer, more rapid and sensitive methods of biological screening (e.g., YES) are available. However, the limitations of these assays should be recognized, and as the science develops, these limitations will likely be reduced. Site-specific considerations (e.g., species present, habitat, geology) and a priori knowledge regarding specific contaminants of concern (e.g., endocrine disruptors) may suggest a more sophisticated testing program, involving field-based testing combined with lab-based bioassays.

9

Costs

Whether water reuse makes sense for a region depends, in part, on its cost compared with the costs of other feasible water management alternatives (e.g., new supplies, expanded conservation efforts) and the cost of not pursuing any water management changes. If a community chooses not to augment its water supply, it avoids those associated costs but also misses or postpones the benefits of doing so. Because new water supply options are likely to cost more than the existing supplies (assuming no more of the existing water supply is available), the costs of water reuse need to be compared to the cost of other new-supply options.

In this chapter, the concepts of financial and economic analysis are introduced, and the costs of reuse are categorized. As described in Chapters 4 and 5, a wide variety of treatment processes can be incorporated into a reuse system to meet specific water quality goals for intended uses and to address local site-specific constraints. Thus, it is difficult to make general statements about the cost of water reuse. The committee, instead, presents example costs from potable and nonpotable reuse facilities that responded to a committee questionnaire, and where feasible, compares the costs of water reuse against other alternative water supplies.

FINANCIAL AND ECONOMIC COSTS

When assessing the economic viability of a water supply project, it is important to understand the difference between economic costs and benefits and financial accounting of costs and benefits, which are rarely, if ever, the same (NRC, 2008b). *Financial costs*

involve how much the utility has to pay to construct and operate the project, including interest costs. *Economic costs* account for all of the costs to whomever they may accrue. These include the financial costs of carrying out the project, as well as costs that take the form of impositions on or losses to anyone who is affected by the project. Examples of broadly experienced costs are odors, loss of open space, or additional greenhouse gas emissions. Examples of broadly experienced benefits are reduced nutrient discharge to surface waters and economic benefits provided by a reliable water supply.

The concept of economic cost has been captured in the idea of the "triple bottom line," which encompasses the financial, social, and environmental impacts of a project. With a triple-bottom-line approach, the project sponsor is considered to have an obligation to examine environmental and social impacts, not just profitability. The analyses undertaken in environmental impact reviews are consistent with triple-bottom-line thinking, although environmental review as an obligation ends with project certification. In contrast, triple-bottom-line approaches call for ongoing review and analyses of financial, social, and environmental costs of a project, which are often summarized in annual reports. Triple-bottom-line accounting runs into the same challenges faced by economic valuation: the difficulty of valuing environmental and social impacts (Norman and MacDonald, 2004). This difficulty means that triple-bottom-line processes offer more guidance than quantitative comparative analysis, although the concept does alert business and public agency leaders that the public is aware of difficult-to-monetize impacts of their practices and the importance of striving for full accountability for one's impacts on society and the environment.

Both financial and economic perspectives are needed when assessing water supply. If a region's water authority cannot afford a project, even one with net benefits to society, it will not get built. Subsidies are sometimes provided by local, state, or federal agencies to offset the financial costs for demonstration of new technologies or for projects with broad economic benefits that cannot be captured in an individual utility's rate structure. For example, the Metropolitan Water District of Southern California has offered a $250 per acre-foot subsidy ($767 per million gallons; $200 per thousand m^3) for up to 25 years for local water development to reduce the region's dependence on imported Colorado River water. The Bureau of Reclamation's Title XVI has also been a source of subsidies for water reuse projects since 1992 (see Box 9-1). Traditional water supplies may also receive subsidies.

FACTORS AFFECTING THE FINANCIAL COSTS OF WATER REUSE PROJECTS

Whether reclaimed water is used for nonpotable or potable uses, there are several factors that affect the costs of a water reuse program. These include the location of a reclaimed water source (i.e., the wastewater treatment facility), treatment infrastructure, plant influent water quality, customer use requirements, transmission and pumping, timing and storage needs, energy requirements, concentrate disposal, permitting, and financing costs.

Size and Location

In most cases, reclaimed water systems originate at a municipal wastewater treatment plant. Wastewater treatment plants are typically constructed at lower elevations and within close proximity to a point of discharge such as a river, lake, or ocean. As a result, there are pumping costs to bring reclaimed water to the customers or to the water treatment plant, which is typically sited at higher elevations. In U.S. cities, wastewater treatment plants have evolved into large-scale facilities serving extensive areas. This has provided economies of scale and equitable service, minimized

---

BOX 9-1
Federal Subsidies for Water Reuse Through the Title XVI Program

The Title XVI program was originally launched in 1992 in the Reclamation Projects Authorization and Adjustment Act (Public Law 102-575). The act directed the Secretary of Interior "to undertake a program to investigate and identify opportunities for reclamation and reuse of municipal, industrial, domestic, and agricultural wastewater, and naturally impaired ground and surface waters" and to support, "the design and construction of demonstration and permanent facilities to reclaim and reuse wastewater." The act also directed the Secretary "to conduct research, including desalting, for the reclamation of wastewater and naturally impaired ground and surface waters." The original act authorized cost sharing for three feasibility studies and for the construction of five reuse projects, including three in Southern California, and the act has since been amended to authorize additional projects. Title XVI has been administered through the Bureau of Reclamation.

As of November 2010, approximately $531 million has been appropriated for Title XVI projects, mostly in California, including $135 million from the American Recovery and Reinvestment Act of 2009. Unless specified by Congress, federal funding support is limited to projects in the 17 western continental states. The program has generally provided cost sharing for up to 25 percent of the total project costs, with a project maximum of $20 million. These funds historically have helped reuse projects move forward more quickly than they might have otherwise. Of the 53 authorized projects, 42 have received some funding and 16 have either been completed or have reached the maximum cost-shared funding limit. Three additional projects have received at least 80 percent of their authorized funding. As of the end of 2010, the program had a $630 million backlog for projects that have been authorized and are awaiting appropriations, a significant increase from the $354 million backlog in 2006 (Cody and Carter, 2010). Considering this growing backlog, the recent Congressional Research Service report by Cody and Carter (2010) examined program priorities and the federal role in supporting reuse.

---

impacts on nearby land uses, and centralized technical management.

Centralized treatment facilities have been preferred throughout history, but the analysis of benefits changes when one thinks of a wastewater treatment system as a source of water instead of a location for disposal of water. Multiple smaller, decentralized plants could provide several advantages as reuse systems because the location of water treatment is closer to the customers, reducing the cost of transmission and distribution infrastructure. Multiple treatment facilities could also improve system redundancy, and therefore reliability, through the interconnection of more than one source of reclaimed water. Several smaller plants may also be able to accommodate fluctuations in demand more effectively than one large centralized plant. Retrofitting centralized treatment facilities to provide redundancy can be costly if new infrastructure (e.g., transmission pipelines, pumping stations, and storage facilities) is required for the sole purpose of interconnecting more than one system or service area (Gikas and Tchobanoglous, 2009).

Treatment Infrastructure

In most cases, nonpotable uses of reclaimed water (e.g., irrigation, industrial) require a quality of water that is not much different than what a typical secondary or advanced wastewater treatment plant would produce. For the most part, turbidity, biochemical oxygen demand, and coliform standards are similar between nonpotable reuse applications and secondary treatment permit requirements, although there may be some variations in effluent quality requirements. Thus, the startup of a nonpotable reclaimed water program typically does not require a large investment in additional treatment facilities. Some facilities may need to incorporate improvements to existing infrastructure, such as improved filtration, additional chlorination for maintaining a residual, and more efficient technologies to meet regulatory requirements.

Some customers, however, may have specific water quality requirements that will necessitate a higher level of treatment. Irrigation customers, golf courses in particular, and industrial customers may impose quality restrictions that may considerably increase the capital and operating costs of a reuse program. Water reclamation treatment processes can be designed to treat or remove constituents that negatively affect the quality of the effluent or that are limited by contractual commitments with the users. In arid states, total dissolved solids (TDS) of the reclaimed water can be a concern. For example, at El Paso Water Utilities, potable water must sometimes be used to dilute the reclaimed water produced to reduce TDS to acceptable levels. This dilution step becomes costly to the utility,

considering that reclaimed water rates are typically less than potable rates. In Scottsdale, Arizona, additional treatment to lower the TDS in product water has been incorporated with use of reverse osmosis systems on a portion of the effluent prior to distribution. The cost of operation of a reverse osmosis facility depends on many factors, including quality of the source water (inflow), quality of the effluent, the cost of energy, and the cost of concentrate disposal (see also Chapter 4). As an alternative, individual industrial reclaimed water users that have specific pollutants of concern (e.g., silica for industrial cooling water) can implement point-of-use treatment systems to address these constituents, rather than requiring treatment at the water reclamation plant, thereby reducing a facility's treatment costs.

Potable reuse projects require substantially more treatment and barriers within the treatment train, and therefore require larger investments in treatment infrastructure than nonpotable projects, although the costs can vary with the treatment components selected (see Figure 4-1). Enhanced treatments steps, such as those used at the Orange County Water District (see Box 2-11), have been key to gaining public acceptance of major potable reuse projects. However, such extensive treatment is also costly and energy intensive and may not be viable in all potable reuse applications.

Influent Water Quality

Incoming water quality is a crucial factor in the production costs of reclaimed water. Typically, the source of water to a reclamation facility is the effluent of a wastewater treatment plant. Several factors can affect its quality, affecting overall treatment costs.

- **Consumer water softening.** The increased use of self-regenerating water softeners by customers has posed water quality challenges on wastewater treatment plants producing reclaimed water. High levels of salts in reclaimed water may impair its use unless additional pre- and/or post-treatment is implemented, which increases the cost of producing reclaimed water. Flow diversion programs have been developed in cities such as Las Vegas, where conductivity meters (used to measure TDS) trigger automatic valves to divert high-conductivity wastewater effluent around satellite water reclamation facilities (Crook, 2007).

- **Water conservation.** As indoor water conservation programs become more effective, the volume of wastewater discharges diminish, but the pollutant mass often remains unaffected. As a result, the concentration of constituents in wastewater increases, requiring additional treatment and therefore additional costs at the wastewater or reclaimed water facility on a volume basis.

- **Industrial pretreatment.** Implementation of a pretreatment program can limit the discharge of constituents that would negatively affect the treatment process and/or the quality of the effluent. In nearly all U.S. states, pretreatment programs are required, and certainly for those plants with a capacity greater than 5 million gallons per day (MGD; or 19,000 m^3/d). The intent of these programs is to detect and address the existence of constituents that would affect the quality of the product, compliance with regulatory entities, or contractual requirements with users, which thereby reduces reclaimed water production costs (see also Box 10-1 for a discussion of the National Pretreatment Program).

Transmission and Pumping

Delivery of reclaimed water to consumers may add a substantial capital cost to a water reuse project based on the location of the treatment facility and the distance to the service area(s). Extensive piping costs can be required when separate transmission and distribution lines need to be installed for nonpotable reclaimed water. Operating costs could also vary substantially for a system in a varied topography, where the source (the wastewater treatment plant) is typically located at lower elevations and the customers are in the higher elevations, requiring the delineation of multilayered pressure (service) zones for delivery of adequate system pressures. Additional costs include service connections to the customers and an integrated billing system.

The delivery of reclaimed water to individual customers through a dedicated network of pipes, reservoirs, and pumping stations adds considerable economic burden. Construction of piping (transmission and distribution systems), pumping, and storage facilities is comparable to the cost of the same infrastructure for a drinking water system, although specific design requirements must be observed. In the United States,

purple color coding is standardized for all reclaimed water pipes. In some states, reclaimed water pipelines must be constructed with a minimum separation from the potable water systems. For example, in Texas (30 TAC § 210), the regulatory agency for reclaimed water requires a minimum separation distance from a newly installed reclaimed water pipeline to a potable water line of 9 ft (2.7 m) horizontally and 2 ft (0.6 m) vertically (Texas Commission on Environmental Quality, 1997).

The Southwest Florida Water Management District (SWFWMD, 2006) estimated that transmission and distribution costs for reuse ranged from $5 per inch diameter/linear foot in rural areas to $9 per inch diameter/linear foot in urban areas (in 2006 dollars). In 2008, the SWFWMD estimated per lot residential distribution capital costs from $1,090 to $1,440 including the meter and related appurtenances, based on recent reuse project data. The SWFWMD estimated that these costs could be reduced by 50 percent in new subdivisions (SWFWMD, 2008). By treating water to drinking water standards, potable reuse projects alleviate the need for costly separate water transmission, distribution, and storage systems.

Existing stream channels can also be used to convey reclaimed water from a wastewater treatment plant to a downstream water treatment plant intake, assuming water rights laws allow for such conveyance. The El Paso Water Utility and the Trinity River Authority discharge treated wastewater into streams while maintaining rights to withdraw that water downstream for reuse under the Texas "Bed and Banks" statute (Texas Water Code § 11.042). This statute allows reclaimed water to be transferred substantial distances without the associated infrastructure costs required by Texas' legal definition of "direct reuse," where all reclaimed water must be transferred by constructed water infrastructure. Reuse of this water allows the utilities to get the most out of their existing water rights. See also Chapter 10 for more detailed discussions of water rights and water reuse.

In some cases, regional collaborative initiatives have been developed to enhance reuse while taking advantage of natural conveyance systems. For example, the Upper Trinity Regional Water District (See Box 2-3) discharges reclaimed water to the Trinity River which is then used as a water source for downstream municipal customers. The quantity of water available to municipal customers is based in part on those utilities' returned wastewater flows. Numerous agreements involving state and regional water agencies were needed in this collaborative initiative. Similarly, the City of Las Vegas earns gallon-for-gallon return-flow credits for advanced-treated water returned to Lake Mead.

Timing and Storage Needs

In a typical drinking water system, the distribution and storage system is designed to convey water to the customer to meet peak customer demand, which reflects an aggregate of residential, industrial, and irrigation uses. In nonpotable reclaimed water systems, the distribution and storage system is typically designed to meet a more specific customer demand, which can create challenges for the system design. For example, facilities that primarily produce reclaimed water for irrigation purposes face the dilemma of extra production during winter months when irrigation is at its lowest (Figure 9-1). Alternatives to mitigate this problem include increased discharge into surface waterways or subsurface injection to reduce seawater intrusion. At Laguna de Santa Rosa, California, low irrigation demands are offset by additional supply for industrial purposes at the Geysers Project, a geothermal power station (Crook, 2007). Agencies also take steps to limit peak demand for reclaimed water. Dunedin, Florida, imposes a fee on customers that use more than the allotted summer demand. This is an incentive to keep peak demands as low as possible and reduce the need to provide additional storage to meet these demands. Widely variable seasonal demand can add to the overall costs of the water reuse project; thus, advanced planning to minimize the unused capacity in nonpotable reuse systems is essential to optimizing the cost-effectiveness of a nonpotable reuse project.

Decreases in reclaimed water demand create another challenge: lower water quality due to primary productivity (e.g., algal growth) and the release of taste and odor compounds during the longer storage time. Some storage facilities incorporate a recirculation system to allow for continuous mixing of the water and in some cases have provisions for addition of chemicals such as sodium hypochlorite to prevent growth of organisms. Some systems include equipment that can allow pipelines to drain any water that does not meet the required quality controls back to the plant for treatment via sanitary sewer systems. These extra treatment costs are part of the overall cost of reclaimed water.

Nonpotable reuse customers also have different diurnal demand patterns. Industrial customers may also impose specific time-of-day requirements on the supply. Diurnal peak demands are typically met through a series of storage reservoirs throughout the system, which adds to a system's overall costs. However, by moving irrigation needs out of potable water systems to a separate nonpotable reuse systems, peak demands on the potable system will be reduced. Industrial customers may also impose specific time-of-day requirements on the potable supply.

Energy Requirements

Energy is needed in many phases of the reclaimed water production cycle, including wastewater treatment, transmission to the water reclamation plant, advanced treatment, distribution, and possible subsurface injection and removal costs. Many of the wastewater treatment costs would be incurred anyway to meet wastewater discharge requirements. Therefore, this section focuses on only the additional energy costs incurred by water reuse projects beyond that required for wastewater discharge.

The energy costs in reuse projects are widely variable and site specific. Variables that affect energy

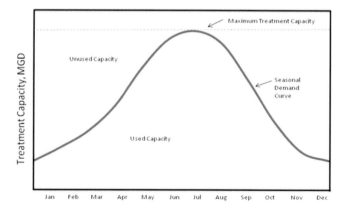

FIGURE 9-1 Seasonal demand curve for a hypothetical non-potable reuse system, showing large unused supplies in winter months.
SOURCE: CSDWD (2006).

costs include the distance of the reclamation facility from the wastewater treatment plant, the treatment technologies applied, the size of the facility (see Figure 9-2), the product water quality objectives, the extent of dual distribution systems, the topography of the service area (related to the energy required for pumping), and pumping requirements for reclaimed water injection and withdrawal in any underground storage components. Overall energy costs are also influenced by the price of energy. Understanding water reuse's energy-use profile therefore requires a comparative approach: How much energy does water reuse require in a given location compared with the feasible water supply alternatives? Generalizations on the energy cost of water supply are less useful than individual analyses of specific regions.

The amount of energy needed for water supply matters because it is a surprisingly large portion of energy use in some regions. In California, water-related energy uses consume roughly 19 percent of all electricity used in the state and 32 percent of natural gas (CEC, 2005; GEI Consutants/Navigant Consulting, 2010). Large proportions of this consumption go to conveyance costs and summer groundwater pumping. California has one of the most extensive water conveyance systems in the world, linking high-precipitation regions in the north and east with high-population re-

gions in the south and west, and mid-state agriculture. According to CEC (2005), wastewater treatment uses 1 percent of the state's electricity. Energy requirements of reclaimed water treatment and conveyance beyond that required for wastewater discharge ranged from 0.4 to 1.2 kWh/ kilogallon (kgal) (or 0.38 to 1.1 megajoule [MJ]/m^3), compared to as low as 0.1 kWh/ kgal (0.095 MJ/m^3) for traditional raw water treatment.[1] GEI Consultants/Navigant Consulting (2010) estimated the energy requirements of seawater desalination at 12.2 kWh/kgal and inland brackish water desalination at 4.0-5.5 kWh/kgal. See Table 9-1 for estimates of water-reuse–related energy consumption for several Southern California utilities (Table 9-1).

Several local comparisons of energy requirements have been published for water reuse scenarios in California. The Equinox Center (2010) estimates that potable and nonpotable reuse in San Diego requires substantially less energy than seawater desalination and water importation, and nonpotable reuse has energy requirements similar those for local surface and groundwater use (Figure 9-3). Some reuse applications also require the installation of a unique distribution system dedicated to reclaimed water, as is the case for West Basin Municipal Water District in Southern California, which supplies highly treated reclaimed water to chemical refineries. There is also a one-time energy cost incurred with the building of the needed infrastructure. Stokes and Horvath (2009) calculated comparative total energy use, considering life-cycle costs, for a hypothetical Southern California facility, and found that reclaimed water was comparable to water importation, but significantly lower than desalination (see Box 9-2).

From a policy perspective, this level of consumption of energy for water supply is insignificant from a residential consumer's point of view, because the energy cost of delivered water to a home is only a few cents per month. But in the aggregate, it influences important regional and national energy policy questions, including whether and how to expand power grids, build new power generation facilities, and meet greenhouse gas reduction targets.

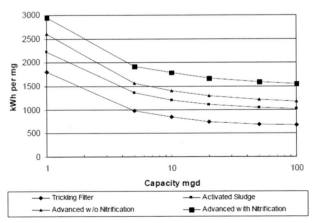

FIGURE 9-2 Variations in electricity consumption with size and wastewater treatment processes.
NOTE For this analysis, advanced treatment "is similar to the activated sludge process, but includes additional treatment in the form of filtration prior to discharge."
SOURCE: EPRI (2002).

---

[1] Adding the energy required for wastewater treatment increases the total energy use for wastewater reclamation to a range of 1.5 to 5.8 kWh/kgal (1.4 to 5.5 MJ/m^3).

TABLE 9-1 Estimates of Energy Intensity of Water Reclamation and Reuse at Three Southern California Utilities Compared with Seawater Desalination

	Project Description	Energy Intensity of Water Reuse Project	Estimated Energy Cost (assuming $0.25/kWh)
Inland Empire Utilities Agency	Nonpotable reuse; distribution of advanced-treated (Title 22) wastewater	1.02 kWh/kgal (0.97 MJ/m^3) distribution only	$0.25/kgal ($0.07/m^3)
San Diego	Nonpotable reuse; additional treatment necessary above current primary and/or secondary discharge standards, and distribution	3.53 kWh/kgal (3.36 MJ/m^3) treatment and distribution	$0.88/kgal ($0.23/m^3)
Los Angeles	Nonpotable reuse; additional treatment necessary above current secondary discharge standards, and distribution	1.84 kWh/kgal (1.75 MJ/m^3) treatment and distribution	$0.46/kgal ($0.12/m^3)
Seawater desalination	Conservative estimate for seawater desalination and distribution	12 kWh/kgal (11.4 MJ/m^3) treatment and distribution	$3.10/kgal ($0.82/m^3)

NOTES: Energy requirements associated with wastewater treatment required for discharge are not included in these totals. Thus, the entry for Inland Empire Utilities Agency, which is required to treat all wastewater to California's Title 22 standards, only includes energy costs associated with distribution.
SOURCE: California Sustainability Alliance (2008).

Concentrate Disposal Costs

Some reuse projects need to remove TDS to meet end-use requirements, and membrane treatment is the most commonly used method to accomplish this goal. Membrane treatment, such as reverse osmosis, requires that facilities manage the resultant concentrate, which represents between 15 and 50 percent of the feedwater (Asano et al., 2007). Because the salinity of membrane concentrate from wastewater reclamation is much lower than the salinity of concentrate from seawater desalination, little concern is associated with its coastal discharge (see NRC [2008b] for detailed discussions of the environmental impacts of brackish and seawater desalination concentrate disposal alternatives). Currently, inland brackish desalination facilities dispose of concentrate through deep-well injection, discharge to a wastewater treatment facility via sanitary sewer systems, discharge to surface water bodies, or evaporation ponds with burial in place or disposal via landfilling (TWDB, 2009). With water reuse systems, the most common and lowest cost alternative for inland concentrate disposal—blending and diluting the concentrate with wastewater effluent prior to surface water discharge so that it meets local water quality standards—may not be available because the wastewater effluent is being reused. Costs of concentrate disposal operations vary widely based on local factors, such as land costs, hydrogeological conditions, energy

cost, and concentrate quality. For inland desalination systems, concentrate disposal costs have been reported as high as twice that of the desalination process cost (NRC, 2008b).

Technologies are being studied that reduce the volume of concentrate produced during desalination activities. Use of pretreatment additives to decrease concentrate production (i.e., increase water recovery) may reduce the concentrate volume destined for disposal. Increasing feedwater temperatures to lower water viscosity and increase flux may also increase water recovery, although sometimes at the expense of water quality (i.e., allowing more salt to pass through the membrane). However, the energy required to increase inflow temperature may be costly and typically will exceed the savings unless a lower-cost energy source is proved to offset capital investment (Tarquin, 2009).

Permitting and Environmental Review

Nearly all water supply augmentation projects require permitting and environmental review. A reuse project differs from ocean and brackish desalination in that it also requires public health review. The permitting and review process poses direct costs to the utility, but another cost frequently noted by water utility representatives is the cost of delay due to public opposition to a proposed project. Costs of delay include additional months or years of not enjoying the full benefit of the

BOX 9-2
Life-Cycle Assessments of Energy and Environmental Effects

The results of full "life-cycle" cost analysis of water reuse will be highly site specific, but there are a few case studies in the literature that assess some of the life-cycle environmental and energy impacts of utility operations and expansion plans, including water reuse (Lundie et al., 2004; Stokes and Horvath, 2006, 2009). These illustrate the importance of taking a holistic approach to understanding how water supply investments affect economic, financial, and environmental outcomes. A systems or life-cycle approach emphasizes two especially attractive features of water reuse alternatives. First, water reuse typically reduces the quantities of bulk water supply that a utility must obtain from external raw water sources (e.g., rivers, groundwater). Second, the amount of treated wastewater discharged to aquatic ecosystems is reduced. These environmental benefits of lower raw water abstractions and reduced wastewater discharges are highly site specific, but in a particular location can be quite important.

Lundie et al. (2004) used a life-cycle assessment approach to model water supply planning options for the water and wastewater utility in Sydney, Australia ("Sydney Water"). One investment option they examined was increasing the level of treatment at wastewater treatment plants along the coast from primary to advanced. Lundie et al. (2004) concluded that this would increase total energy use and greenhouse gas emissions without any significant environmental benefits in terms of improved quality of the receiving water body. Their life-cycle assessment showed that this option of moving toward increased wastewater treatment would not be justified unless "additional environmental benefits can be generated by offsetting the demand for potable water through water recycling."

Stokes and Horvath (2006) used a hybrid life-cycle cost assessment approach to evaluate the energy use of three different water supply alternatives for two utilities—importation, nonpotable reuse, and desalination (seawater desalination for the Marin Municipal Water District [MMWD] in Northern California and brackish groundwater desalination for the Oceanside Water Department [OWD] in San Diego County, California). Their analyses showed that the "global warming potential" of nonpotable reuse was substantially less than desalination, but larger than water importation, largely due to the distribution system pumping requirements (Figure below).

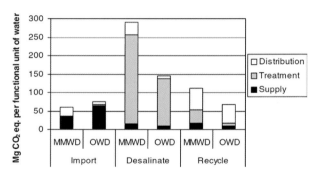

Carbon dioxide (in megagrams) produced per unit of water supplied for three water supply alternatives in Northern California (Marin Municipal Water District [MMWD]) and Southern California (Oceanside Water Department [OWD]). The analysis considered seawater desalination at MMWD, brackish groundwater desalination at OWD, and nonpotable reuse for both locations. SOURCE: Stokes and Horvath (2006).

Stokes and Horvath (2009) conducted a similar analysis focused on the energy use, air emissions, and greenhouse gas effects from different water supply alternatives in a hypothetical Southern California case study. No other environmental effects or nonmonetized benefits were included in the analysis. The authors concluded that nonpotable reuse was comparable, if slightly lower, than the imported water case scenario in energy use and greenhouse gas emissions, and was much lower in these factors than brackish water or seawater desalination (see table below).

Life-Cycle Assessment Results Comparing Air Emissions from Five Water Supply Alternatives

Water Source	Energy (MJ/m^3)	GHG (g CO_2 equiv/m^3)	NO$_x$ (g/m^3)	PM (g/m^3)	SO$_x$ (g/m^3)
Imported water	18	1093	1.9	0.40	2.9
Desalinated ocean water, conventional pretreatment	42	2465	3.4	0.77	6.9
Desalinated ocean water, membrane pretreatment	41	2395	2.9	0.71	9.4
Desalination brackish groundwater	27	1628	2.0	0.41	4.2
Reclaimed water	17	1023	1.0	0.48	2.9

SOURCE: Stokes and Horvath (2009).

completed project and possible cost increases over time in construction, as well as possible additional interest expenses. The cost of personnel, consultants, and legal counsel may significantly add to the cost of a project, especially when the permitting process and environmental review are prolonged. Typically, assuming a project is not categorically excluded from the National Environmental Policy Act process, it takes a minimum of 1 year to complete an environmental assessment and may take significantly longer if there is strong opposition. Public review of proposed projects is a right that the committee does not dispute, but it is important to evaluate the efficiency of the review process.

Reclamation System Financing

A water agency will use its existing financial resources (i.e., savings and revenue flows), its preferred bond status (if such status exists), and its access to state and federal grants and loans to finance water reuse projects. Medium- to large-sized water customers committing to long-term agreements helps secure the bonds by securing the revenue sources. Reclamation facilities typically cannot cover their costs in their early years while expanding their customer base. Bond financing and other agency revenues cover the cost difference during this period. Provision of state and federal subsidies shortens this time period.

The choice to invest in water reclamation draws down an agency's financial ability to make other capital investments. The processes of planning, financing, and building a facility are themselves costly. Launching a water reuse program requires a review of the agency's overall investment priorities to confirm that reuse is the top investment priority at the time (Asano and Mills, 1998). An otherwise desirable reuse project may be beyond the means of a water agency if certain cost categories, such as separate piping for nonpotable use, are too high. In addition to reviewing investment priorities, an agency should realistically assess the market for nonpotable reclaimed water and what it can expect in terms of revenues from water sales (Asano and Mills, 1998).

Forms of financing themselves impose differential costs on an agency. The lowest cost financing is a "pay as you go" approach, because no interest fees or investment placement and management fees are required.

This approach is beyond the reach of most agencies given the high capital costs of water treatment systems. Agencies can draw down existing investment pools, identify and pursue interest and capital subsidies (e.g., state revolving funds), raise water rates, and enter the short- and long-term capital markets in an effort to minimize the cost of a system without exposing the agency to excessive financial risk.

NONMONETIZED COSTS AND BENEFITS OF REUSE

The impacts of water reuse projects are both positive and negative, with amounts varying project by project, but many of the benefits and some of the costs are difficult to monetize. Some of the economic, environmental, and social considerations that are frequently not monetized, which may or may not apply to a particular reuse project, are listed in Table 9-2. Although factors such as improved reliability are frequently not monetized, methods exist to develop estimates of its value (e.g., see Kidson et al., 2009). Also, scientists have used life-cycle assessment approaches to evaluate the relative environmental impacts, including greenhouse gas emissions, from various water supply alternatives (see Box 9-2).

Greenhouse Gas Emissions

An environmental impact of growing interest is the carbon footprint, or greenhouse-gas emissions, resulting from water reuse. The impacts of greenhouse gases are largely not monetized in the United States, although several other countries have established or are developing carbon taxes (e.g., India, Australia) or emissions trading schemes (e.g., the European Union, China). In the absence of a system to monetize greenhouse gas emissions, the energy requirements of various water supply alternatives, discussed earlier in this chapter, can serve as an analog for comparing the carbon footprint of water supply alternatives, assuming that all facilities are powered by traditional sources of electricity. Like energy costs, greenhouse gas emissions from the complete life cycle of water reuse projects will be widely variable and site specific, based on factors such as the level of treatment (see Figure 9-2), pumping requirements, and new pipeline required. Thus, no

TABLE 9-2 Possible Nonmonetized Costs and Benefits of Reuse

Nonmonetized Benefits and Costs of Reuse	Description
Nonmonetized Benefits	
Improved reliability	Wastewater reuse provides a reliable, local supply of water during regional shortages. By diversifying a utility's water supply portfolio, a community is better able to meet the needs of its water users and the environment in both wet and dry periods and under other stresses.
Enhanced self-sufficiency	By reducing dependence on water imports and providing a local water supply, water reuse can increase a community's self-sufficiency (see Rygaard et al., 2011).
Enhanced reputation for environmental stewardship	By embracing water reuse, communities can gain positive recognition for their environmental stewardship.
Enhanced regional economic vitality	By meeting increased water demands with new sources, communities may enhance local economic growth.
Increased water for the environment	If some existing surface or groundwater supplies are replaced by water reuse, more water can be made available to meet environmental needs (e.g., instream flows for environmental restoration, reducing withdrawals of overtapped aquifers).
Improved surface water quality	By diverting discharge of nutrient-laden waters from sensitive surface waters or estuaries to landscape or agricultural irrigation, the net discharge of nutrients to surface water can be reduced. Irrigation with reclaimed water may also reduce the need for additional fertilizers.
Nonmonetized Costs	
Effects on the overall carbon footprint of water supplies	Unless offset by low-carbon energy sources, some water reuse approaches may increase the overall carbon footprint of a water supply compared to existing supplies.
Public health effects	Poor cross-connection control (see Box 6-4) or inadequate protections against equipment failures (see Chapter 5) could expose the public to pathogens causing acute gastrointestinal illness or low levels of hazardous chemicals.
Public perception of reduced quality	Public concern over the perceived lower quality of the drinking water supply could lead to increased stress among some individuals and increased expenditures on bottled water. See also Chapter 10.
Effects on downstream flows	If reclaimed water is used for irrigation or other consumptive uses, water reuse will reduce downstream flows, with potential adverse ecological effects (such as in surface water or estuarine ecosystems) and reduced supply to downstream water users. Where "return flow credits" are offered, as in the Colorado River, water reuse can reduce these credits.
Water quality impacts	If reclaimed water irrigation rates exceed the capacity for the plants to take up the nutrients, groundwater and surface water can become nutrient-enriched, which can lead to human health effects and environmental impacts, such as eutrophication and algal blooms. See also Chapter 3. Multiple cycles of nonconsumptive water reuse can increase the salinity and contaminant load in the water unless treatment is designed to remove it.
Effects on soils and plants	Excess salinity can be detrimental to plant growth and high levels of sodium can adversely impact soil structure.

SOURCES: Asano et al. (2007); EPA (2008b).

universal conclusion can be made about the relative greenhouse gas emissions of water reuse versus other water supply sources, although some generalizations are possible. From the comparative energy analyses noted in this chapter (see Tables 9-1. Figures 9-3, and the figure and table in Box 9-2), the energy use and resulting greenhouse gas emissions from potable and nonpotable water reuse can be significantly less than from desalination. In studies of Southern California, greenhouse gas emissions for nonpotable reuse were comparable or greater than for water importation when considering life-cycle costs (see figure and table in Box 9-2; Stokes and Horvath, 2006, 2009).

Understanding greenhouse gas emissions also requires an examination of the energy sources used in the region (e.g., fossil fuels, nuclear) and the costs and availability of low-carbon energy supplies. Some water utilities, such as Santa Cruz, California, are building solar energy systems in advance of expansion of water treatment facilities to offset or mitigate increases in carbon emissions. In Perth, Australia, a major seawater desalination facility is powered by wind energy to address concerns about the greenhouse gas implications of this energy-intensive water supply.

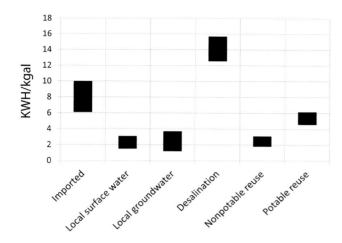

FIGURE 9-3 Power consumption for water supply alternatives for San Diego County.
SOURCE: Data from Equinox Center (2010).

REPORTED REUSE COSTS

Because of the dearth of information in the literature on the costs of water reuse facilities, the committee chose to address its task question (see Box S-1) on reuse costs by requesting this information from utilities directly. National Research Council (NRC) staff sent a questionnaire (see Appendix C) to 20 water utilities known to supply reclaimed water, reflecting both large and small utilities and potable or nonpotable applications (or both). This questionnaire was not developed to achieve a statistically defensible estimate of reuse costs but to identify an approximate range of cost across a variety of different treatment processes. Fourteen utilities responded and cost data for nine utilities were complete enough for general comparison purposes, representing seven nonpotable reuse operations and six potable reuse operations (see Tables 9-3 and 9-4).

Among those who responded to the questionnaire, projects dated back as far as 1962, although most reclaimed water projects described were implemented after the year 2000. Reported capital costs were converted to 2009 dollars based on the Consumer Price Index. These inflation adjustments were based on the midpoint of the construction period provided for a particular phase or project. The committee recognizes that this is an assumption that may introduce some

error into the final capital cost data. It should also be noted that the committee was not able to audit the data reported by the individual utilities, although Tables 9-3 and 9-4 were sent to each of the utilities for fact checking.

Wastewater treatment is required before effluent can be discharged, and the discharge requirements can vary widely depending on the sensitivity of local surface water ecosystems and state and local regulations. Therefore, the committee designed the cost questionnaire to separate the capital and operating costs associated with (or required for) effluent discharge into the environment from the costs of additional treatment or distribution lines associated with nonpotable or potable reuse projects. Treatment costs required for wastewater discharge into the environment are not included in the costs reported here because these costs would be incurred regardless of whether reuse projects were implemented.

Capital Costs

Reported capital costs for potable and nonpotable facilities include the design and construction of treatment plants, distribution pipelines, well fields, and engineered natural systems as well as related administrative costs. All costs are reported as dollars per kilogram capacity per year in 2009 dollars (Tables 9-3 and 9-4). Hypothetical annual costs amortized at 6 percent interest over 20 years are also presented to allow comparison with O&M costs.

Nonpotable Reuse

Reported capital costs for nonpotable reuse vary widely, from $1.14 to $18.75/kgal capacity per year. Despite this wide variability, several conclusions about cost can be made. For example, the specific nonpotable applications affect the degree of additional treatment costs. Of the six facilities listed in Table 9-3 that provided detailed capital costs, two reported capital costs associated with additional treatment beyond that required for wastewater discharge. For example, Denver provides additional treatment for cooling applications (see Box 2-5), and West Basin provides a range of treatment levels to meet several end uses, including irrigation and industrial cooling. Four facilities reported

TABLE 9-3 Financial Costs from Nonpotable Reuse Facilities

	Durango Hills Las Vegas, NV	Desert Breeze Las Vegas, NV	Trinity River Authority, TX	Denver, CO	West Basin, CA	Tucson, AZ	Inland Empire, CA
Capacity (MGD)	10	5	16.4	30	40	30	40
Average output (MGD)	3.0	2.9	1	6	18	15.2	15.2
Reclaimed water uses	Landscape irrigation	Landscape irrigation	Landscape irrigation, amenity reservoirs	Landscape irrigation, industrial cooling, zoo	Irrigation; cooling and boilers with additional treatment	Landscape irrigation, toilet flushing	Irrigation, industrial cooling, laundry, paper processing
Treatment	Activated sludge secondary treatment, automatic backwash filters, ultraviolet disinfection	Activated sludge secondary treatment, automatic backwash filters, ultraviolet disinfection	Advanced activated sludge treatment.	Biologically aerated filters, flocculation, sedimentation, mono-media filtration, disinfection	Coagulation, flocculation, sedimentation, mono-media filtration, disinfection	Filtration or activated sludge treatment via membrane bioreactor, chlorine disinfection	Activated sludge secondary treatment with biological nutrient removal, filtration, chlorine disinfection
Year constructed	1999–2004	2001–2004	1987	2000–present	1995–2006	1982+	2001–2010
O&M costs ($/kgal) in 2009 dollars							
Personnel	0.07	0.05	0.01	0.54	0.20	0.13	1.00
Energy	0.36	0.21	0.01	0.19	0.22	0.25	0.18
Other	0.25	0.09	0.03	0.33	0.60	0.12	0.00
Total O&M ($/kgal)	0.68	0.35	0.05	1.06	1.02	0.50	1.18
Capital Costs ($/kgal capacity/yr) in 2009 dollars							
Treatment facility	0.00	0.00	0.00	8.12	9.62	*Not reported*	0.00
Pipelines	4.23	5.73	1.14	3.58	9.14	*Not reported*	9.77
Other				1.88			
Total capital costs in 2009 dollars ($/kgal per year)	4.23	5.73	1.14	13.57	18.75	*Unable to correct for inflation*	9.77
Annualized capital cost in $/kgal[a]	0.37[a]	0.50[a]	0.10[a]	1.18[a]	1.63[a]	*Unable to calculate*	0.85[a]
Total Annual Costs (Annualized Capital + O&M) in $/kgal in 2009 dollars (also shown in $/m³)							
Total Annual cost in $/kgal[a] ($/m³)	1.05 (0.28)	0.85 (0.22)	0.15 (0.04)	2.24 (0.59)	2.65 (0.70)	*Unable to calculate*	2.35 (0.62)

NOTE: The capital costs are reported prior to any subsidies received.

[a]Assumes amortization at 6 percent over 20 years. Facilities each have different interest rates, but for the sake of comparison, a common interest rate was applied.

TABLE 9-4 Financial Costs from Potable Reuse Facilities

	Orange Co. GWRS, California	El Paso, Texas	Casey WRF/ Huie Wetlands Clayton Co, GA	Shoal Creek/Panhandle Clayton Co, GA	West Basin, CA	Inland Empire, CA
Capacity (MGD)	70	10	24	4.4	12.5	20
Average output (MGD)	Not reported	5.5	17.4	Not reported	9	7.1
Treatment	Enhanced primary treatment, activated sludge and trickling filter secondary treatment, microfiltration, reverse osmosis, advanced oxidation (ultraviolet light and hydrogen peroxide)	Activated sludge secondary treatment with denitrification, anaerobic digestion, lime treatment, sand filtration, ozonation, biologically active granular activated carbon filtration, final disinfection	Activated sludge secondary treatment with biological nutrient removal, sodium hypochlorite disinfection; treatment wetlands	Activated sludge secondary treatment with biological nutrient removal ultraviolet disinfection; treatment wetlands	Microfiltration, reverse osmosis, advanced oxidation (ultraviolet light and hydrogen peroxide), corrosion control	Activated sludge secondary treatment with biological nutrient removal, filtration, chlorine disinfection, soil aquifer treatment
Year(s) constructed	2004-2008	1984	2004-2010	2002-2003	1995-2006	2001-2010
O&M Costs ($/kgal) in 2009 Dollars						
Personnel	0.14	0.13	0.16	0.14	0.70	1.00
Energy	0.57	0.06	0.08	0.08	0.41	0.18
Other	0.45	0.14	0.11	0.09	1.27	0.00
Total O&M ($/kgal)	1.16	0.33	0.35	0.31	2.38	1.18
Capital Costs ($/kgal capacity/yr) in 2009 Dollars						
Treatment	12.42	Not reported	3.92[a]	5.53[a]	28.98	1.49
Pipelines	2.63	Not reported	Not reported	Not reported	1.74	9.77
Other costs	4.95					
Total Capital costs in 2009 dollars ($/kgal/yr)	20.00	23.46	3.92[a]	5.53[a]	30.72	11.26
Annualized capital cost ($/kgal)[b]	1.74[b]	2.05[b]	0.34[b]	0.48[b]	2.68[b]	0.98[b]
Total Annual Costs (Annualized Capital + O&M) in $/kgal in 2009 dollars ($/m³)						
Total annual cost in $/kgal[b] ($/m³)	2.90 (0.77)	2.38 (0.63)	0.69 (0.18)	0.79 (0.21)	5.06 (1.34)	2.16 (0.57)

NOTE: The capital costs are reported prior to any subsidies received.
[a] Includes engineered wetlands, and cost per thousand gallons for UV disinfection at drinking water plant.
[b] Assumes amortization at 6 percent over 20 years. Facilities each have different interest rates, but for the sake of comparison, a common interest rate was applied.

that they incur no additional treatment costs for their nonpotable applications beyond that required for effluent discharge. Distribution lines make up a sizeable extent of the capital costs of nonpotable reuse facilities, making up between 26 and 100 percent of the capital costs for the seven facilities. Projects where the effluent is used at or near the treatment plant are much less costly than systems with many miles of pipeline.

Potable Reuse

Capital costs for potable reuse projects are also widely ranging, from $3.90 to $31/kgal capacity per year in 2009 dollars (Table 9-4). The dataset demonstrates the variability in capacity and technologies that characterize water reuse today. Water reuse is a rapidly growing and technologically changing endeavor, and the evolution is reflected in the widely varying capital costs. The varying cost data suggest that future projects also will vary widely in cost, depending on the many factors raised in this chapter.

O&M Costs

Reported operation and maintenance costs also contain substantial variability. Total O&M costs for nonpotable reuse facilities range from $0.05/kgal to $1.18/kgal (Table 9-3), with an average of $0.69/kgal. Reported O&M costs for potable reuse facilities ranged from $0.31/kgal to $2.38/kgal (Table 9-4), with an average of $0.95/kgal. For nonpotable facilities, personnel costs account for about 40 percent, energy for about 30 percent, and all other costs at 30 percent of the total O&M budget. These percentages are quite similar to the percentages for reported potable reuse O&M costs (40 percent personnel, 24 percent energy, 36 percent other). Energy costs are affected by the extent of treatment required and the degree of pumping required to transmit the reclaimed water to the end user. Facilities using reverse osmosis reported much higher O&M costs than the other potable reuse facilities, although it should be noted that the dataset is too small to draw firm conclusions.

Subsidies

Six of the nine utilities reported capital subsidies in the form of grants from federal, state, and local

---

BOX 9-3
West Basin Municipal Water District Reuse Costs

West Basin Water Recycling Program provides reclaimed water for nonpotable and potable reuse applications. The program was developed in three phases. The first phase was completed in 1995, the second in 1997, and the last major phase was completed in 2006. West Basin's recycled water estimated annual production capacity is 27 MGD (100,000 m^3/d), of which 18 MGD (68,000 m^3/d) are for nonpotable uses that include irrigation and industrial applications and 9 MGD (34,000 m^3/d) for potable water uses, such as groundwater recharge.

Treatment processes for nonpotable uses include coagulation, flocculation, sedimentation, monomedia filtration, and disinfection. The potable reuse component of the program includes treatment of secondary effluent plus additional treatments that include microfiltration, reverse osmosis, disinfection with ultraviolet radiation and hydrogen peroxide, and corrosion control.

West Basin has received subsidies to support its reuse program from the Bureau of Reclamation (Title XVI; $50M), California Department of Water Resources ($9.4M), U.S. Army Corps of Engineers ($23.5M), Los Angeles Department of Water and Power ($2.7M), and the Metropolitan Water District ($91M), totaling approximately $177 million. In addition, it received over $168 million from the Uniform Standby Charge, a tax on undeveloped land parcels.

Capital cost and operating costs are shown in Tables 9-3 and 9-4. Approximately 5 percent of the total cost is attributed to concentrate management. Brine is disposed of through an existing 5-mile outfall that is owned and operated by the City of Los Angeles.

Reclaimed water is billed at $1.34/kgal for irrigation customers inside the West Basin Service Area. This represents the highest tier of a declining tiered rate structure that encourages users to purchase more reclaimed water. Potable reclaimed water for the barrier project is billed at $1.41/kgal. These are approximately two-third the cost of traditional potable water, which is billed at $2.11/kgal.

SOURCE: Mary-Ann Rexroad, Budget and Finance Officer, West Basin Municipal District.

---

entities. These subsidies ranged from $7.5 million to $344.6 million. The Bureau of Reclamation's Title XVI program (see Box 9-1) contributed grant funding to the six projects, ranging from $7.5 million to $50 million, but those facilities with large subsidies relied on multiple sources of funding to help offset the project costs, including state and local funds (see Box 9-3). The three Southern California utilities receive annual subsidies from the Metropolitan Water District of Southern California based on the volume of water produced. The costs reported in Tables 9-3 and 9-4 do not consider subsidies received by the utilities.

COMPARATIVE COSTS OF
SUPPLY ALTERNATIVES

Because site conditions vary significantly, the costs of reuse can best be assessed by comparing these projects against the costs of local water supply and conservation alternatives. Most of the utilities who responded to the committee's questionnaire, however, did not provide costs of alternative water supplies considered. Cost was cited by approximately one-third of the responding utilities as an advantage, but it was rarely the deciding factor in these reuse projects. Other factors reported by utilities as key factors that led to the decision to implement reuse included

- providing a means to diversify water supplies,
- creating a drought-resistant water supply,
- public support,
- quality of the water, and
- limited alternative sources.

Among those who did provide comparative costs, El Paso Water Utility reported that the costs of reclaimed water were slightly higher than inland desalination. Reclaimed water was much more expensive than traditional (but limited) groundwater and surface water sources but less expensive than imported water (see text and figure in Box 9-4). The extent of the distribution and concentrate disposal costs had a major impact on the overall cost of reclaimed water relative to desalination. Denver Water provided comparative costs (see Box 9-5; Table 9-5), which costs show that nonpotable reuse costs in the Denver region would be more expensive than potable reuse, considering the need to expand the service area with costly dual distribution systems, either to residential areas or major industries.

Orange County Water District also provided comparative costs. They reported that the total cost of reclaimed water to the utility ($1.80/kgal after subsidies

and contributions from Orange County Sanitation District were applied; $3.16/kgal not counting these offsets) was similar to that of imported water—$1.84/kgal. This cost was substantially lower than the cost of seawater desalination ($3.68/kgal in 2010 dollars) (Shivaji Deshmukh, Orange County Water District, personal communication, 2010).

Given the limited comparative cost data obtained from the committee's questionnaire, the committee also researched other comparative cost information available. California's Legislative Analyst's Office (CA LAO, 2008) published a comparison of water supply alternatives for the state of California. Among the eight options considered, water reuse had the second-lowest median costs, above urban water use efficiency (Figure 9-4). A similar analysis by the Los Angeles County Economic Development Corporation (Freeman et al., 2008) to assess Southern California's water strategies reported that potable water reuse (based on OCWD GWRS data) is less costly than seawater desalination, comparable to brackish groundwater desalination and surface storage, and more costly than urban water conservation, groundwater storage (Freeman et al., 2008; see Table 9-6). Comparative costs for the City of San Diego are shown in Box 9-6.

RECLAIMED WATER RATES

In this chapter, the many factors affecting the total cost of producing and delivering reclaimed water have been described. Reclaimed water rates can offset these costs, but because the cost of treatment and distribution is generally higher for reclaimed water than for conventional water sources, reclaimed water rates are frequently set at a level that does not cover the full cost of treatment. Nonpotable reclaimed water rates are frequently set lower than conventional drinking water rates to encourage its use, even though drinking water rates in many cases do not cover the full cost of conventional water treatment, delivery, and infrastructure maintenance (EPA, 2002). According to a 2007 American Water Works Association survey of approximately 30 reuse facilities, more than one-third of reuse facilities stated that they recovered less than one-quarter of their operating costs from reclaimed water rates, while approximately 25 percent of utilities reported that they recovered 100 percent of their op-

---

BOX 9-4
El Paso Water Utilities' Fred Hervey Water Reclamation Facility

El Paso's Fred Hervey Water Reclamation plant was built in 1984, along with a series of 10 injection wells for recharge in the Hueco Bolson. The 10-MGD (38,000-m³/d) capacity plant provides water for four main uses: maintenance of wetlands of ecological interest, irrigation of parks and a golf course, aquifer recharge (infiltration basins and injection wells), and industrial uses (e.g., cooling tower makeup water). Treatment processes for wastewater treatment include primary clarification, flow equalization, two-stage activated sludge with denitrification, anaerobic digestion, and biosolids dewatering/disposal. In addition, wastewater is treated to achieve potable water standards through lime treatment, sand filtration, ozonation, biologically active GAC filtration, and final disinfection. The final effluent (potable water quality) is made available for irrigation and industrial uses through the transmission system that also recharges the aquifer.

Capital and O&M costs are provided in Tables 9-3 and 9-4. All reclaimed water, regardless of intended use, distance from source, or quality of water, is billed at $1.24/kgal. This is substantially lower than the potable water tiered rate that ranges from $1.93 to $6.49/kgal.

El Paso currently reclaims a combined 10 percent of all treated wastewater from its four wastewater facilities with a goal to increase reclaimed water supply to 15 percent of all wastewater treated. The reclamation plant is undergoing a major expansion to incorporate a third treatment train that would provide redundancy to the treatment process and increase the plant's capacity by approximately 2.5 MGD (9,500 m³/d). Other water supply alternatives were considered; however, the decisive factor for implementation of this program was based on cost and need to conserve the water. Comparative costs of water supply alternative are shown in the figure below.

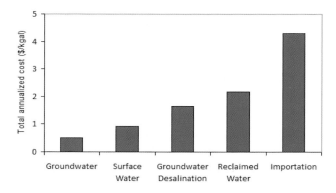

Comparative costs for alternative water supplies for El Paso Water Utilities, from 2010. This figure includes relatively low costs for desalination concentrate disposal (via deep-well injection) for the brackish groundwater desalination system.

SOURCE: Irazema Solis Rojas, P.E., EPWU Water Reclamation Engineer.

---

erating costs (Figure 9-5). However, annualized capital costs may be equal to or greater than operating costs. The state of Florida reports that 72 of its 176 utilities (41 percent) provide reclaimed water to users free of charge (FDEP, 2010).

Of the nine utilities who provided data to the committee on their nonpotable reuse rates, on average, the reclaimed water rates represented 39 percent of the rates for traditional potable sources (with ratios ranging from 11 to 75 percent).[2] While most of the potable reuse facilities combined their water supplies such that no separate charge was applied, two utilities charged separate rates to potable reclaimed water customers. Like the nonpotable reuse rates, these potable reclaimed water rates represented only a fraction (17 and 67 percent) of the traditional potable supply rates. Given the small size of this dataset, these data are not presumed to be representative of reuse rates across the United States. Because the driving motivation for water reuse is shifting from environmentally sound wastewater disposal to water supply for water-limited regions, reclaimed water rates are likely to climb so that reclaimed water resources are used as efficiently as the potable water supplies they are designed to augment.

---

[2] When utilities reported tiered water rates, the committee considered the third tiered potable rate for comparison, considering that most nonpotable reuse customers are large volume irrigators.

TABLE 9-5 Example Range of Unit Costs for Water Supply and Conservation Options in Denver, Colorado

Water Supply Alternative	Net Present Value ($/kgal/yr)
Reuse	
Expand existing nonpotable system	$250 to 300
Indirect potable	$90 to 150
Direct potable	$90 to 150
Greywater	$30 to 150
Conservation	
Advanced metering	$90 to 900
Plumbing fixture changes	$6 to 60
Landscape changes	$90 to 770
New supply	
Storage projects	$9 to 300
Pumping projects	$90 to 600

NOTE: Estimated net present value of capital, operations, and maintenance costs over 40 years divided by the annual water yield of project. Customer costs are included in conservation costs. These data are preliminary.
SOURCE: Marc Waage, Denver Water, August 2011.

though a price may be significantly lower than potable water supplies, it still may not be attractive enough if upfront costs such as installation fees, backflow prevention, and thermal expansion units are more than

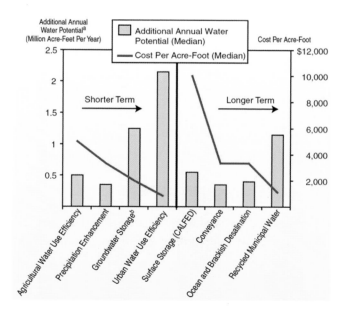

FIGURE 9-4 Costs of various water supply alternatives in the state of California. Cost estimates calculated by the California Department of Water Resources.
[a]Reflects the midrange of estimates of water supply development potential of particular solutions identified in the California Water Plan 2005.
[b]Includes integrated management of groundwater and surface water.
SOURCE: CA LAO (2008).

Other revenue options can be considered when establishing reclaimed water rates, including standby fees, property taxes, monthly minimum fees, and utility subsidies from water and wastewater fees. Organizations that provide both water and sewer services have the ability to spread some of the cost of the reuse program to wastewater treatment and/or drinking water programs, which sometimes have associated decreases in treatment and distribution costs with increased water reuse. By sharing the costs, utilities can set a reclaimed water rate that is competitive with potable water and attractive enough to prospective customers to encourage them to invest in the infrastructure to connect to the nonpotable distribution system. In some instances, even

TABLE 9-6 Estimates of Costs of Southern California Water Supply Alternatives

Water Supply Alternatives	Initial Capital Costs (million $)	Annual O&M Costs (million $)	Annualized Costs Over 30 Years ($/kgal)
Urban water conservation	0	0.5	0.64
Local stormwater capture	40-63	1-3.5	1.10
Potable reuse	480	30	3.10
Ocean desalination	300	37	3.10+
Brackish groundwater desalination	24	0.7	2.30-3.68
Transfers: agriculture to urban	na	na	2.10+
Groundwater storage	68-135	13	1.80
Surface storage	2,500	7.5-15.5	2.30-4.30

SOURCE: Freeman et al. (2008).

customers are willing to spend. In these cases, utilities must balance the need to attract customers with the costs of further subsidizing reclaimed water.

Special negotiated rates may also be considered for large customers who provide a guaranteed steady demand over an extended period of time (e.g., large industries). These customers offer an advantage of constant demand throughout the year and practically guaranteed demand for reclaimed water from one year to the next. However, customers that require a reliable supply of reclaimed water at all times may lead to increased costs for the utility if additional infrastructure must be installed to provide uninterrupted service (e.g., a redundant distribution system or provision of an alternate water supply) (Holliman, 2009).

CONCLUSIONS AND RECOMMENDATIONS

Financial costs of water reuse are widely variable because they are dependent on site-specific factors. Financial costs are influenced by size, location, incoming water quality, expectations, and/or regulatory requirements for product water quality, treatment train, method of concentrate disposal, extent of transmission lines and pumping requirements, timing and storage requirements, costs of energy, interest rates, subsidies, and the complexity of the permitting and approval process. Capital costs in particular are site specific and can vary markedly from one community to another. The lowest cost water reuse systems supply nonpotable reclaimed water to irrigation or industrial cooling operations located in close proximity to the wastewater treatment plant. Data on reuse costs are limited in the published literature, although the chapter provides reported capital and O&M costs for nine utilities (representing 13 facilities) that responded to a committee questionnaire.

Distribution system costs can be the most significant component of costs for nonpotable reuse systems. Projects that minimize those costs and use effluent from existing wastewater treatment plants are frequently cost-effective because of the minimal additional treatment needed for most nonpotable applications beyond typical wastewater disposal requirements. When large nonpotable reuse customers are located far from the water reclamation plant, the total costs of nonpotable projects can be significantly greater than potable reuse projects, which do not require separate distribution lines.

Although each project's costs are site specific, comparative cost analyses suggest that reuse projects tend to be more expensive than most water conservation options and less expensive than seawater desalination. The costs of reuse can be higher or lower than brackish water desalination, depending on concentrate disposal and distribution costs. Water reuse costs are typically much higher than those for existing water sources. The comparative costs of new water storage alternatives, including groundwater storage, are widely variable but can be less than those for reuse.

To determine the most socially, environmentally, and economically feasible alternative, water managers and planners should consider nonmonetized costs and benefits of reuse projects in their comparative cost analyses of water supply alternatives. Water reuse projects offer numerous benefits that are frequently not monetized in the assessment of project costs. For example, water reuse systems used in conjunction with a water conservation program can be effective in reducing seasonal peak demands on the potable system, which reduces capital and operating costs and prolongs existing drinking water resources. Water reuse projects can also offer improved reliability, especially in drought, and can reduce dependence on imported water supplies. Depending on the specific designs and pumping requirements, reuse projects may have a larger or smaller carbon footprint than existing supply alternatives. They

BOX 9-6
San Diego Reclaimed Water Project

The City of San Diego's recycling water program dates back to the 1980s when three small pilot plants (0.025 to 1 MGD [95 to 3,800 m³/d]) were built for irrigation and research purposes. Two larger wastewater reclamation plants (WRPs) were built in 1997 and 2000 (North City WRP and South Bay WRP respectively) committed to delivering 30 MGD (110,000 m³/d) total of nonpotable reclaimed water to large customers. The construction of these facilities was primarily driven by wastewater management issues and later to fulfill a Settlement Agreement with environmental stakeholders. In 1993, the city and the San Diego Water Authority proposed an 18 MGD (68,000 m³/d) potable reuse project with advance treatment and blending with imported water in a local surface water reservoir. The project was cancelled 7 years later because of public opposition. After the potable reuse project was canceled, the City of San Diego restructured its efforts to maximize the use of reclaimed water through nonpotable use. By 2006, its customer base included over 360 connections to the reclaimed water system using 11.6 MGD (44,000 m³/d) of the 24 MGD (91,000 m³/d) North City WRP's production capacity and 1.25 of the South Bay WRP's 13.5 MGD (4,700 of 51,000 m³/d) capacity.

With an anticipated 50 percent population increase from 2005 to 2030, the city of San Diego estimated the water supply would need to be increased about 25 percent (approximately 50 MGD [190,000 m³/d]) combined with aggressive conservation efforts. As of 2005, about 90 percent of the city's water needs were met through water importation from the Colorado River and California State Water Project. Thus, San Diego needed to expand its water supply portfolio. In 2004, San Diego City Council issued a directive for the evaluation of options to increase the beneficial use of the city's reclaimed water program to meet current and future water demands. The city released a study documenting various reuse alternatives (CSDWD, 2006) and is currently conducting a demonstration project to determine if potable reuse with reservoir augmentation is a feasible alternative for San Diego. The demonstration project is estimated to be completed by 2013.[a]

Comparative cost data considering O&M costs and annualized capital costs for San Diego's water supply alternatives show that nonpotable reclaimed water is comparable to the cost of seawater desalination, largely due to the high cost of the distribution system. Estimated potable reuse costs are lower than nonpotable reuse and desalination but substantially larger than conservation and the current costs of imported water. However, the cost of importing water is anticipated to rise faster than the other supplies, such that by 2030, the cost of potable reuse is anticipated to be comparable to imported water (Equinox Center, 2010).

[a]http://www.sandiego.gov/water/waterreuse/demo.

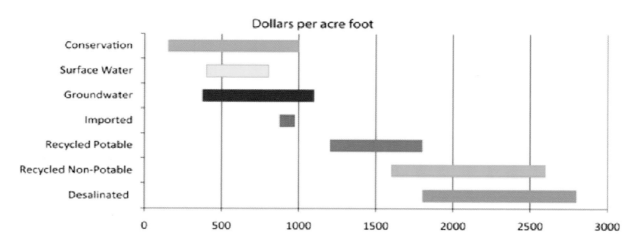

Estimated marginal costs for water in 2010 (in dollars per acre-feet) in the County of San Diego.
SOURCE: Equinox Center (2010).

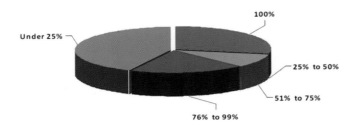

FIGURE 9-5 Percentage of annual operating costs recovered from reclaimed water rates.
SOURCE: AWWA (2008).

can also reduce water flows to downstream users and ecosystems.

Current reclaimed water rates do not typically return the full cost of treating and delivering reclaimed water to customers. Nonpotable water reuse customers are often required to pay for the connection to the reclaimed water lines; therefore, some cost incentive is needed to attract customers for a product that is perceived to be of lower quality based on its origin. Frequently, other revenue streams, including fees, drinking water programs, and subsidies, are used to offset the low rates. As the need for new water supplies in water-limited regions becomes the driving motivation for water reuse, reclaimed water rates are likely to climb so that reclaimed water resources are used as efficiently as the potable water supplies they are designed to augment.

10

Social, Legal, and Regulatory Issues and Opportunities

Water reuse projects, like any large-scale water project, affect numerous stakeholders and are affected by a complex legal and regulatory framework that spans many sectors. Water reuse, once an exceptional and little-regulated practice, is now recognized as an important component of water resources management. Our growing need and expectation of reliable water supplies have driven technological innovation in water treatment, storage, and conveyance that has created new opportunities to integrate reclaimed water into our water systems. As one might expect in any field evolving as dramatically as wastewater treatment and reuse, the regulatory, legal, economic, public understanding, and public policy aspects of water reuse are not well aligned.

In this chapter, the committee reviews the legal and regulatory framework, including water rights and regulation of water quality, that influences the application and design of water reuse projects at the local level. The chapter then describes existing state water reuse regulations, U.S. Environmental Protection Agency (EPA) guidelines, and relevant international guidelines. U.S. wastewater and drinking water regulations are also discussed as they relate to reuse. The chapter also includes an analysis of factors that contribute to positive or negative public attitudes toward reuse.

WATER RIGHTS

If one's experience with water reuse is in a water-scarce coastal city, one might assume that it is desirable for water to be treated and reused before it is released to the ocean. However, in an inland environment, water reuse may affect downstream users of the effluent. Thus, the right to use wastewater needs to be examined. The law of water rights in the United States has evolved under two distinct systems: (1) prior appropriation doctrine in the West and (2) riparian rights in the East. Broadly speaking, the prior appropriation doctrine evolved in regions where water has always been scarce, and it provides a means of allocating water in times of shortage according to the date that a right was perfected. In contrast, riparian rights evolved in more humid regions and give rights to landowners who border rivers. Within this broad construct, each state's rules have evolved within their respective borders; thus the doctrines are just a general indication of how water rights may be attributed. Finally, legislation in some states has specifically addressed water reuse and clarifies legal questions surrounding the right to reuse water.

Water Reuse Under Prior Appropriation

In accordance with each state's legal structure, treatment facilities planning to reuse water must consider the effect on downstream users. Traditionally, wastewater has been considered a liability, and municipalities have used the least expensive means to bring the water into compliance with water quality requirements so the effluent could be discharged. As communities expand and treatment and monitoring technologies improve, wastewater in some arid regions is changing from being regarded as a liability to an asset. This evolution raises important legal questions of who has rights to the treated effluent and when and how the owner can use the resource. Another perspective is to ask whether the use of wastewater constitutes a "new" water supply; it might in a region where flows otherwise are released to the ocean, but not in a region where a downstream user relies on them.

Approaches to Water Reuse Under the Prior Appropriation Doctrine

The primary conflict with respect to water rights stems from downstream water rights holders and the potential for reuse activities to impair their use of the water. Some states give water treatment facilities greater rights to treated water, whereas other states may protect downstream senior rights holders. If the water reuse proponent must purchase a separate water right to the wastewater (i.e., the locality does not have the right to retain its treated wastewater), the costs of reuse will increase substantially.

In general, the owner of a wastewater facility has the ability to reuse the water without purchasing it from another. However, this is not always the case. In Utah the right to reuse must be specified in the operator's water permit, and in New Mexico the operator's right to wastewater may be dependent on its consumptive rights (which can be less than the water it discharges). In the following paragraphs, a brief survey of how states have approached the reuse of wastewater is presented.

In Colorado, wastewater can be used by the municipal wastewater treatment plant owner when the water is "developed" water. The term is used to describe water that is not natural to a stream, such as water imported from another basin or pumped from groundwater. These wastewaters would be available for use by the city that operates the wastewater treatment works. This concept provided the ability for Denver to reuse waters that had been piped from the Colorado River basin into the Platte basin (Tarlock, 2009).[1] Further, the courts have held that there is no right in downstream entities to appropriate wastewater of another if that water has been "developed."[2]

California's reuse statute provides that "The owner of a waste water treatment plant operated for the purpose of treating wastes from a sanitary sewer system shall hold the exclusive right to the treated waste water as against anyone who has supplied the water discharged into the waste water collection and treatment system" (California Water Code § 1210).

In Utah, the right to reuse water must be specified in the original water right where wastewater reuse is included as a beneficial use (Schempp and Austin, 2007). A public agency that owns or operates a wastewater treatment facility may use, contract for the use, or reuse such water obtained under a water right under certain conditions.[3] Water rights do not automatically attach upon treatment. Most basins in Utah are fully appropriated, and therefore a significant part of the reuse program is dependent on contractual arrangements that provide wastewater treatment facility owners with rights to the treated wastewater (Schempp and Austin, 2007).

In Arizona, the State Supreme Court held that the entity that treats the wastewater is entitled to put it to any reasonable use.[4] This essentially provides wastewater reuse facilities the rights to all the water they treat. The court explained that the rule "will allow municipalities to maximize their use of appropriated water and dispose of sewage effluent in an economically feasible manner." The court added that "the spirit and purpose of Arizona water law . . . is to promote the beneficial use of water and to eliminate waste of

---

[1] See *City of Thornton v. Bijou Irrigation Co.*, 926 P.2nd 1, 65-78 (1996).

[2] The issue of water rights and water reuse was determined by the Colorado Supreme Court beginning with *Burkhar v. Meiberg*, where the Colorado Supreme Court determined there was no vested right to the captured irrigation wastewater of another (86 P. 98 (1906)).

In 1972, the court in *Metro Denver Sewage v. Farmers Reservoir* recognized that this "wastewater rule" was also applicable to municipal wastewater effluent (499 P.2d 1190 (1972)). Subsequently, the court clarified the wastewater rule distinguishing that wastewater, as opposed to return flow and seepage, was not subject to appropriation by downstream entities (*City of Boulder v. Boulder & Left Hand Ditch Co.*, 557 P.2d 1182 (1976)).

[3] Such restraints include that the water right is administered as a municipal water right, the reuse is consistent with the underlying water right, and the reuse is approved by both the Utah Water Quality Board and the State Engineers Office (Utah Code Ann. § 73-3c-201(1) and 73-3c-202(1)a-c.

[4] Senior water rights holders downstream from a municipal wastewater treatment plant alleged impairment as a result of the treatment plant's sale of its treated effluent to other parties, which significantly decreased discharges to the stream. The court held that "the 'producer' of the effluent is a senior appropriator, those who have appropriated the effluent gain no right to compel continued discharge." *Ariz. Pub. Serv. Co. v. Long*, 773 P.2d 988, 991-97 (Ariz. 1989).

this precious resource."[5] However, this reasoning has been criticized because "one equally could argue that in a highly appropriated state, the water is not wasted if it is returned to the watercourse and subsequently appropriated downstream—as was the situation in this case" (Schempp and Austin, 2007).

In New Mexico, the State Supreme Court ruled in *Reynolds v. City of Roswell* that the city's "sewage effluent is private water which the City may use or dispose of as it wishes."[6] Neither downstream users of the discharged wastewater effluent nor the state engineer can compel the continued supply of treated effluent without a contract, grant, dedication, or condemnation.[7] The Supreme Court ruled that permit conditions are allowed only to protect existing water rights.

It is important to note that the principles of water rights are not the only ones under which water flows can be protected downstream. Environmentally based standards, such as instream flow rights, also could affect the ability to reuse wastewater flows.

In summary, municipal wastewater treatment plant operators in many states have the right to reuse wastewater effluent, but in others it may be necessary to procure water rights to do so. The application process, described below, can affect these rights.

Water Rights Application Process Under the Prior Appropriation Doctrine

As would be expected, states' application processes for reuse projects range from simple to complex. Key aspects of the application process for water rights to reclaimed wastewater by state are listed in Table 10-1. A common feature is that downstream water users are protected from impairment by upstream users. Generally, impairment is used in water law to indicate that a given user's water right has been reduced or in some way negatively impacted by another user. If reuse represents a change of use, generally the applicant must demonstrate "no injury" to other users (Tarlock, 2009). States tend to acknowledge downstream uses that have become established in reliance on wastewater discharges (e.g., California). In some states environmental protection of the stream is addressed in the application

stages. Finally the burden of proving whether impairment will occur is significant, and it matters where the burden is imposed. Schemp and Austin (2007) note that when the burden is placed on the water utility, the costs of the reuse project can increase. When the burden is placed on a state agency, the utility burden is reduced but the approval time may be lengthened while the state calculates the expected consequences to the hydrological system. When the burden of proving impairment is left to the downstream user, upfront project costs are reduced but the chance of subsequent litigation is increased, with less long-term confidence in a utility's water rights.

Water Reuse Under the Riparian Doctrine

The riparian doctrine is used in the more humid Eastern states and essentially bases the right to use rivers on proximity to the waterway. Hence, the water right resides in the "riparian" land owner, in contrast to the prior appropriation doctrine where land owners who are not adjacent to the water source can acquire water rights. The doctrine has evolved with changing circumstances, and modern practice involves administrative requirements and the ability to transfer water rights. Generally the wastewater operator would be able to reuse wastewater unless it would likely cause harm to downstream riparian rights holders.

Approaches to Water Reuse Under the Riparian Doctrine

In general, water rights are less contentious in riparian states. In the eastern United States, Florida is at the forefront of water reuse and recycling activities. Water reuse is statutorily encouraged and the state recognizes that the "promotion of water conservation and reuse of reclaimed water, as defined by the department, are state objectives and considered to be in the public interest" (Fl. Stat. § 373.250[1]). All five of Florida's Water Management Districts have reuse programs and, generally, reuse is regulated under consumptive use permits. In New Jersey, the state has directed the Department of Environmental Protection (NJDEP) to encourage and promote water reuse along with water conservation (N.J. Admin. Code § 7:14A-2.1). Examples of key aspects of the water rights permitting scheme in Florida and New Jersey are provided in Table 10-1.

[5] *Id.* at 997.

[6] *Reynolds*, 654 P.2d at 539 (1982).

[7] *Reynolds v. City of Roswell*, 654 P.2d 537 (1982).

TABLE 10-1   Key Aspects of Application Process for Water Rights to Reused Wastewater for Selected States

State	Examples of Key Aspects of Water Rights Application Process in Selected States
Prior Appropriation Doctrine	
California	"Prior to making any change in the point of discharge, place of use, or purpose of use of treated wastewater, the owner of any wastewater treatment plant shall obtain approval of the board [California Water Resources Control Board (CWRCB)] for that change "(Cal. Water Code § 1211(a)).
	These provisions apply to water reuse activities unless "changes in the discharge or use of treated wastewater . . . do not result in decreasing the flow in any portion of a watercourse" (Cal. Water Code § 1211(b)).
Nevada	Can include two applications: a primary application quantifying the total discharge of the wastewater treatment facility, and a secondary application quantifying how, and what amount of, the discharge will be beneficially reused (Nev. Rev. Stat. § 533.440).
	The Nevada Division of Environmental Protection (NDEP) must confirm that proposed water reclamation projects will meet water quality standards.
	The Nevada Department of Water Resources reviews applicants proposing to reuse effluent that historically has been discharged into a water body, to determine whether the project is likely to impair the rights of downstream users.
Oregon	Water reuse projects are exempt from obtaining water appropriation permits if there are not negative impacts on fish and wildlife. Statutory focus on water quality rather than quantity (Or. Rev. Stat. § 537.131, .132(1)).
	Applications must include the traditional water right elements of source, use, amount of the use, and description and location of the conveyance mechanism to be used to transport the reuse water (Or. Rev. Stat. § 537.132[2]).
Utah	Reuse is approved under two separate applications: one to the Utah Water Quality Board and another to the State Engineer's Office for streamflow appropriation (Utah Code Ann. § 73-3c-302(2)a-c).
	Applicants must describe their water right including the diversion, depletion, and return flow requirements, in addition to the proposed water to be reused. In regard to reused water, the application must include the place, purpose, and extent of the proposed water reuse, and an evaluation of the depletion to the hydrological system caused by the reuse (Utah Code Ann. § 73-3c-302(2)g).
Washington	The distribution of water by agricultural production plants and industrial plants are exempt from traditional permit requirements (Wash. Rev. Code §§ 90.46.150, .160), easing water reuse, where water rights for the use of the reclaimed water are obtained in a single permit with associated water quality and Department of Health provisions (Wash. Rev. Code § 90.46.030).
	Statutes protect downstream users from impairment by assuring that "facilities that reclaim water under this chapter shall not impair any existing water right downstream from any freshwater discharge points of such facilities unless compensation or mitigation for such impairment is agreed to by the holder of the affected water right" (Wash. Rev. Code § 90.46.130(1)). However, the statute does not specify what constitutes "impairment" or how and by whom impairment is determined (Schempp and Austin, 2007).
Riparian Doctrine	
Florida	Reuse is generally regulated under consumptive use permits for which domestic wastewater treatment facilities must identify such factors as: the level of treatment, the volume of reclaimed water available, and the volume of reclaimed water provided to reuse customers. All wastewater facilities must reuse water of the "lowest acceptable quality" and if reclaimed water satisfies this mandate and is determined feasible, the applicant is required to implement and maximize its use.[a]
	Each Water Management District is designated as being inside or outside of a water resource caution area (FL OPPAGA, 1999), which dictates water use permitting requirements. Permittees within water resource caution areas are "required to use reclaimed water within five years and total use of reclaimed water within 20 years unless it is determined to be economically, environmentally or technically infeasible" (Fla. Admin. Code Ann. r. 40A-2.802(1)c(3)).
New Jersey	Application process requires the wastewater treatment facility to provide (1) the National Pollutant Discharge Elimination System permit associated with the reused water, (2) an operations protocol, (3) an engineer's report if application is not within the confined area, and (4) a reuse supplier and user agreement. The operations protocol section requires an applicant to provide a narrative of the project that includes the proposed procedures to be followed in applying reuse water, how the water will be transported and where the water will be applied (NJDEP, 2011).

[a]See http://www.dep.state.fl.us/water/reuse/wmdprog.htm.

Rights to Aquifer Storage

A water reuse project may rely on a reservoir to store remediated water prior to its distribution. The rights to reservoir storage are well understood: the project may own the land and the reservoir, or may buy storage rights in a reservoir owned by another. If, however, the project relies on groundwater storage, a different legal problem is presented.

The right to use of an aquifer to store water may be addressed through a statutory framework, in which case rights are likely to be defined. In some states, such as Arizona, Idaho, Oregon, and New Mexico, statutory schemes address when water may be stored and how rights to its withdrawal are governed.

Rights to store water in the subsurface are generally not controlled by the ownership of the overlying property. A recent case in Colorado[8] explained why ownership of the overlying property did not create a property right in an aquifer below the property. The proposal would have used an aquifer that covered 115 square miles of land in South Park, Colorado. The overlying landowners contended that the use of the aquifer would constitute trespass, in the absence of a contract giving permission for the use of the aquifer. The state Supreme Court rejected this argument, stating that "When parties have use rights to water they have captured, possessed, and controlled, they may place that water into an aquifer by artificial recharge and enjoy the benefit of that water as part of their decreed water use rights, if the aquifer can accommodate the recharged water without injury to decreed senior water rights."[9]

* * *

In summary, the ability to utilize wastewater for reuse is controlled by state water law. As water becomes scarcer, states will need to address the differing interests in wastewater. Generally, in regions where the wastewater generator has unambiguous ownership of the water, reuse is more easily facilitated. However, in arid states, reuse may be affected by the interests of downstream water users.

THE FEDERAL WATER QUALITY REGULATORY FRAMEWORK

As discussed in earlier chapters, effectively managing water quality concerns is central to the protection of public health and the environment in water reuse projects. Although there are no federal regulations specific to water reuse, several federal regulations have a bearing on water reuse operations. Regulations addressing the quality of discharges to surface waters (e.g., the Clean Water Act) or discharges to municipal wastewater treatment plants (e.g., the National Pretreatment

Program) affect the quality of water used for reuse, including de facto reuse scenarios. Regulations also affect the treatment level and quality of wastewater, which can affect the extent of treatment required for water reuse applications. Water quality regulations involving groundwater affect water reuse operations that use the subsurface for additional engineered natural treatment and storage. Drinking water regulations also affect the degree of reclaimed water treatment required. In summary, while many aspects of water reuse are addressed by different federal regulatory programs, there is no integrated regulatory approach to this process. The following sections outline the various federal regulatory programs that affect water reuse operations.

The Clean Water Act and Wastewater Discharge

The Clean Water Act was developed to protect the health of the nation's surface waters with the states (or tribes) given authority to determine the uses to be protected. The Act establishes the basic structure for regulating discharges of pollutants into the waters of the United States and for regulating quality standards for surface waters. Water quality standards are adopted by states and include water quality criteria, designated uses of water bodies, and antidegradation provisions. These waters may be protected to very high standards, such as the protection of a cold-water fishery, or given lesser protection. Although the use of surface waters for water supply can affect stream designation, very few rivers in the United States are classified solely on their use as a drinking water source (i.e., "drinkable"). States can take drinking water use into consideration in standard setting under the Act, and there are a few who do so.

Discharges from municipal wastewater treatment plants were regulated in the earliest days of the Clean Water Act. These facilities are subject to National Pollutant Discharge Elimination System (NPDES) permits, which reflect national standards, and state (or tribal) requirements. The Act does not protect against all sources of pollution (e.g., non-point-source pollution and certain types of agricultural return flows) so that treatment is required for almost all waters drawn from surface sources.

Clean Water Act requirements also frequently limit the discharge of saline brines (or concentrate) from

---

[8] Board of County Commissioners of the County of Park v. Park County Sportsmen's Ranch, LLP, 45 P.3d 693 (Colo, 2002))

[9] *Id.* at 703–04.

membrane treatment processes (e.g., reverse osmosis) to freshwater lakes and streams. Thus, the costs of reclaimed water treatment options for inland communities are affected by these water quality standards, which can vary across the states and even stream by stream.

One particular type of pollution—"indirect" industrial discharges to wastewater treatment plants—is regulated under the National Pretreatment Program, which was developed to reduce the discharge of industrial pollutants at their source. This program is administered locally, and reuse facilities can impose more stringent regulation for chemicals that are not sufficiently removed by conventional wastewater treatment (Box 10-1).

Future pretreatment program reviews conducted as part of requirements of the Clean Water Act (CWA § 301(d)) should be conducted with serious consideration of the increasingly intimate connection between domestic wastewater discharge and domestic water supply. Capturing contaminants at their industrial source can be an efficient method of keeping these constituents out of drinking water supplies from potable reuse projects and de facto reuse scenarios. The present list of 129 priority pollutants regulated by the National

---

BOX 10-1
The National Pretreatment Program and Expanding Source Control

The Clean Water Act (CWA), passed in 1972, was designed to eliminate the discharge of pollutants into the nation's waters and to achieve fishable and swimmable water quality levels. EPA's National Pollutant Discharge Elimination System (NPDES), one of the CWA's key components, requires that all direct discharges to the nation's waters comply with an NPDES permit, but many industries discharge through municipal wastewater treatment plants. Consequently, EPA established the National Pretreatment Program, which requires industrial and commercial dischargers to treat or control pollutants in their wastewater prior to discharge to municipal wastewater treatment plants.

Generally, wastewater treatment plants are designed to treat domestic wastewater only. Under the Pretreatment Program, local governments must implement pretreatment standards requiring that pollutants be removed from any industrial or commercial discharge to a wastewater collection system. The current objectives of the program are to

- prevent the discharge of pollutants that may pass through the municipal wastewater treatment plant untreated;
- protect wastewater treatment plants from hazards posed by untreated industrial wastewater; and
- improve the quality of effluents and biosolids so that they can be used for beneficial purposes (Alan Plummer Associates, 2010).

Under this program, wastewater authorities must adopt ordinances, issue permits, monitor compliance, and take enforcement action when violations occur. EPA has established numeric effluent guidelines for 56 categories of industry, and the Clean Water Act requires that EPA annually review its effluent guidelines and pretreatment standards to identify new categories for standards.

A summary of the Pretreatment Program's achievements (EPA, 2003b) demonstrates that it has resulted in significant reductions in the discharge of toxic chemicals to the environment. Most standards have been based on the 129 priority pollutants, which were included in the 1977 Amendments to the Clean Water Act as a result of the Toxics Consent Decree (NRDC v. Train, 421 U.S. 60 (1976)). Recently, an update has been proposed to the Universal Wastes Rule to incorporate pharmaceuticals and thereby streamline disposal of hazardous pharmaceutical wastes and reducing the amount of these chemicals in wastewater (73 Fed. Reg. 73520, Dec. 2, 2008), although no subsequent action has been taken.

In *Issues in Potable Reuse* (NRC, 1998), the committee recommended that EPA develop a priority list of contaminants of public health significance that are known or anticipated to occur in wastewater and that individual communities institute stringent industrial pretreatment and pollutant source control programs, based on this guidance. EPA has not developed such a list, although some utilities have taken actions on their own. For example, the Orange County Sanitation District, which supplies reclaimed water for the Orange County Water District's Groundwater Replenishment System (see Box 2-11), has expanded the agency's source control program to include pollutant prioritization, enhanced outreach to industry and the public, and a geographic information-system-based toxics inventory. Through its source control program the Orange County Sanitation District was able to reduce the industrial discharge of 1,4-dioxane and *N*-nitrosodimethylamine (NDMA) into the wastewater collection system. Oregon is developing rules that that will require municipal wastewater treatment plants to develop plans for reducing listed priority persistent pollutants. The Oregon list includes well-studied pollutants as well as some for which little information exists (Alan Plummer Associates, 2010). The Other programs have been developed to reduce the introduction of pharmaceutical products into the wastewater systems.[a]

[a]See http://www.nodrugsdownthedrain.org/

Pretreatment Program was established more than three decades ago as a result of the Toxics Consent Decree (Natural Resources Defense Council v. Train, 421 U.S. 60 (1976)) and the 1977 amendments to the Clean Water Act. The nation's inactive inventory of manufactured chemicals has expanded considerably since that time, as has our understanding of their significance. Updates to the National Pretreatment Program's list of priority pollutants would ensure that water reuse facilities and de facto reuse operations are protected from trace contaminants of concern. These updates can be accomplished through the existing rulemaking process. In the interim until such updates can be made, EPA should develop guidance on additional priority chemicals to include in enhanced local pretreatment programs in localities implementing potable reuse.

Consideration should also be given to expanding source control to residential releases of constituents of concern. Regional, statewide or national regulations could drive the development of less troublesome substitutes for constituents that are difficult to remove in wastewater systems. Moreover, if a pollutant source is a consumer product, regional, statewide, or national regulations may be required.

Federal Regulation for Injection or Infiltration of Reclaimed Water

As discussed in Chapters 2 and 4, numerous water reuse projects use subsurface injection or infiltration as part of the wastewater treatment and storage process. In some instances, aquifer recharge has additional purposes such as preventing subsidence or reducing saltwater intrusion into freshwater supplies. When water is stored through infiltration, rather than injection, state and local regulations rather than federal regulations, address the quality of the recharge water.

Aquifer recharge by direct injection and aquifer storage and recovery wells are regulated under the Safe Drinking Water Act (SDWA) as Class V wells under the Underground Injection Control (UIC) program (42 USC § 300h to 300h-4). The UIC program regulates the construction, operation, and permitting of wells where fluids are injected underground for storage or disposal to prevent contamination of underground drinking water resources. Reclaimed water injected into these wells is typically treated to meet both primary and secondary drinking water standards.

Under the existing federal regulations, Class V injection wells do not require a federal permit if they do not endanger underground sources of drinking water and comply with other UIC program requirements (49 CFR § 144.82). However, states may include additional requirements with regard to treatment, well construction, and water quality monitoring standards prior to permitting any injection of reclaimed water into aquifers that are currently being, or could be, used for potable supply.

U.S. Drinking Water Regulations: The Safe Drinking Water Act

The U.S drinking water regulations set standards that all drinking water treatment plants are required to meet, whether they use pristine water supply sources, supply water from potable reuse projects, or practice de facto reuse (see Box 10-2). This section provides a review of the regulatory framework and an evaluation of its adequacy for potable reuse.

BOX 10-2
Consideration of De Facto Water Reuse in U.S. Drinking Water Standards

The U.S. Public Health Service published drinking water standards in 1962 (U.S. Public Health Service, 1962) which provide some insight into concerns regarding de facto (or unplanned) water reuse. Although the standards specifically state that "The water supply should be obtained from the most desirable source which is feasible," the document goes on to state: "If the source is not adequately protected by natural means, the supply shall be adequately protected by treatment." The 1962 standards included alkylbenzene sulfonate (ABS), an anionic surfactant that was commonly used in detergent. The statement is made that "waters containing ABS are likely to be at least 10 percent of sewage origin for each mg ABS/liter present." Also of pertinent interest was the use of carbon chloroform extract (CCE) in the 1962 standards as an indicator of anthropogenic organic compounds in water. A standard of 200 µg/L CCE was established to "represent an exceptional and unwarranted dosage of the water consumer with ill-defined chemicals," whether from wastewater or other sources. The ABS and CCE standards promulgated in 1962 demonstrate that the federal government understood that de facto water reuse was occurring and that the contamination of drinking water from a diversity of synthetic organic contaminants was possible.

In 1974, Congress authorized the SDWA, which provides authority to EPA to establish and enforce national standards for drinking water to protect public health. For priority contaminants, EPA determines a maximum contaminant level goal (MCLG), the level below which there is no known or expected risk to human health. A maximum contaminant level (MCL) is the highest concentration of a contaminant that is allowed in drinking water through an enforceable primary standard. MCLs are set as close to MCLGs as possible, considering best available treatment technology and costs versus benefits. Regular testing and reporting is required to ensure that contaminants do not exceed the MCL. For some contaminants, including microorganisms, EPA instead requires specific treatment techniques (TT) be used in the drinking water treatment process in lieu of an MCL. Individual states are allowed to adopt more stringent standards, if desired. In 2009, the EPA National Primary Drinking Water Regulations included three MCLs for disinfectants, four MCLs for radionuclides, five MCLs or TTs for microorganisms, 16 MCLs or TTs for inorganic chemicals, and 53 MCLs or TTs for organic chemicals (EPA, 2009b).

To assess the occurrence of unregulated contaminants that are suspected to affect drinking water, EPA established the Unregulated Contaminant Monitoring Regulation (UCMR) program under the SDWA. Under this program and a prior related program, the presence of unregulated contaminants in drinking water has been purposefully monitored across the country since 1988. The list of contaminants to be monitored is updated in the UCMR every 5 years.

EPA's Contaminant Candidate List (CCL) process, introduced in the 1996 SDWA Amendments (Public Law 104-182), addresses unregulated contaminants that are known, or anticipated, to occur in U.S. drinking waters and that may require future regulation. The list specifically includes contaminants that (1) are not currently regulated under the SDWA, (2) may cause adverse health effects, (3) have been detected or are anticipated to occur in public water systems, and (4) may require regulation under the SDWA. The SDWA Amendments of 1996 require EPA to revise the CCL every 5 years, make regulatory determinations for at least five of the CCL contaminants, and identify up to 30 contaminants for monitoring under the UCMR.

Every 6 years, EPA also must review existing regulations to determine if modifications are required. An overview of the CCL process and its development is provided in Box 10-3.

To move a contaminant from the CCL and into regulation, EPA must show that regulation would "provide a meaningful opportunity to reduce health risk." This process can be extremely arduous, time-consuming, and controversial. The promulgation of a regulation is preceded by numerous opportunities for public comment.

New Approaches in Consideration for Contaminant Regulation

In March 2010, EPA announced a new drinking water strategy that outlines the principles to expand public health protection for drinking water (EPA, 2010a). The new strategy comprises four major points:

- Address contaminants as groups rather than one at a time so that enhancement of drinking water protection can be achieved cost-effectively.
- Foster development of new drinking water technologies to address health risks posed by a broad array of contaminants.
- Use the authority of multiple statutes to protect drinking water.
- Partner with states to share more complete data from monitoring at public water systems.

The grouping of contaminants is one of the key issues still remaining to be addressed. Addressing contaminants as groups is expected to lead to efficiencies in implementing effective treatment, provide efficiencies in developing and administering regulations based on coherent scientific and policy rationale, and foster development of new drinking water treatment technologies. Regulating groups of contaminants has been done in the past for specific contaminants (e.g., total trihalomethanes, a group of five haloacetic acids disinfection byproducts, radioactive substances).

In the new drinking water strategy, EPA continues to identify protection of source water as a key priority. Multiple statutes can be applied to control contaminants prior to their entering the water supply. This may include the use of "regulatory authority under the

---

BOX 10-3
The Contaminant Candidate List (CCL) Process

The EPA released the first CCL (CCL1) containing 60 contaminants (50 chemical and 10 biological) in March 1998. After the release of CCL1, EPA asked the National Research Council (NRC) for guidance in establishing a system to prioritize contaminants listed on the CCL (NRC, 1999b). EPA also asked the NRC to provide advice regarding the development of subsequent CCLs by identifying and prioritizing emerging contaminants. NRC (1999b) recommended that within 1 year of a CCL release, EPA should use a three-part assessment for each contaminant listed. The suggested process would review (1) existing data on health effects, (2) existing data on exposure, and (3) existing information on treatment methods and analytical procedures. Using these data, the NRC recommended that EPA conduct a preliminary risk assessment followed by separate decision documents for any contaminant to be dropped from the list, slated for additional research, or considered for regulation. NRC (1999b) further advised EPA to publish health advisories for all compounds that remain on the CCL within 3 months after completion of initial decision documents.

In a subsequent report based on a workshop on emerging drinking water contaminants, NRC (1999a) suggested that ideal CCLs should

- meet the statutory requirements of the 1996 SDWA amendments,
- identify the "entire universe of drinking water contaminants" before ranking,
- consider all routes of exposure, including dermal, inhalation, and ingestion,
- use the same identification and selection process for chemical and microbial contaminants, and
- include mechanisms to identify similarities among contaminants and contaminant classes that can be used for evaluation of individual chemicals.

The committee recommended a two-step process that would prioritize chemicals from a broad universe to a preliminary CCL (PCCL) through screening criteria and expert judgment followed by use of a prioritization tool and expert judgment to develop the final CCL. To generate the CCL, chemical attribute scores for health effects (severity and potency) and occurrence (prevalence and magnitude) were assigned to each chemical. Using both classification models and expert judgment, a draft CCL is generated and published for public comment. The NRC committee estimated that the number of contaminants in the "universe" could be close to 100,000, considering that the Toxic Substances Control Act inventory alone includes approximately 72,000 substances produced or imported at greater than 10,000 pounds/year.

In 2001, the NRC published a report that provided more detailed information regarding the suggested approaches for moving contaminants from the universe to the PCCL and eventually to the CCL (NRC, 2001). The 2001 NRC report suggested the use of selected attributes to evaluate the likelihood of a particular contaminant occurring at a concentration that could pose risk to public health through drinking water. In relationship to water reuse, NRC (2001) specifically recommended the inclusion of "any constituent of wastewater treatment or septage" within the contaminant universe. The committee also recommended the use of virulence-factor activity relationships, within which microorganisms that have the "ability to survive wastewater treatment and to re-enter drinking water" are specifically addressed.

The suggestions within NRC (2001) were not available in time to be incorporated in the second CCL (CCL2). CCL2 was published in February 2005 and contained 51 of the original 60 contaminants from CCL1. EPA determined that regulations were not required for the additional nine compounds that were then removed from the CCL.

The third CCL (CCL3) was published in 2009, largely using the processes suggested by the NRC as modified by the National Drinking Water Advisory Council (NDWAC, 2004). The EPA established a contaminant universe that contained more than 6,000 potential drinking water contaminants. The CCL3 universe includes compounds known or anticipated to occur in water supplies, considering releases to the environment, production volume, and fate characteristics. Additionally, the CCL3 universe includes contaminants with demonstrated or adverse health effects, regardless of occurrence data. EPA followed the two-step process suggested by the NRC by establishing a PCCL followed by a draft CCL. The final CCL3 contains 116 chemical and biological contaminants, including nine steroid hormones and one antibiotic, which were not included on the draft CCL3. The inclusion of these compounds suggests that wastewater-derived compounds are currently being considered in assessments of drinking water safety, although a direct responsibility to regulate potable reuse would probably cause greater scrutiny of compounds likely to be in municipal wastewater.

Federal Insecticide, Fungicide, and Rodenticide Act (FIFRA) and Toxic Substances Control Act (TSCA) to ensure that decisions made for new and existing industrial chemicals are protective of drinking water" (EPA, 2010a). Together, the recent actions by EPA suggest that the regulation of discrete chemicals along with new treatment strategies may evolve into a more holistic approach that considers mixtures and groups of contaminants according to both treatment efficacy and health risk.

Evaluation of the Sufficiency of the Federal Regulatory Framework When Applied to Reuse

The overarching question in relationship to potable water reuse is whether the CWA and the SDWA offer sufficient protection for water supplies that are derived from sources that include significant municipal wastewater effluent. As described in Chapter 2, there are many communities in the United States where municipal wastewater treatment plant discharges make significant contributions to the drinking water source. In some cases wastewater discharges are a principal source; thus, it can be argued that the SDWA has already been given this assignment. The SDWA and the CWA are the federal laws in place to protect the public from contaminants of wastewater origin. The SDWA alone applies to groundwater resources where septic systems or other sources of pollution contribute to the overall groundwater replenishment. Potable reuse projects may also be required to meet local or state regulations, above the requirements of the SDWA (state reuse regulations are discussed later in the chapter). However, de facto reuse scenarios are not subject to additional regulations.

As outlined earlier, the SDWA does provide limits (MCLs) for many chemical and biological contaminants, and a great deal of research, careful thought, and public dialogue underlies each of these limits. For contaminants regulated through MCLs, it is logical that the same limits would apply regardless of the source of the water. Where potable reuse is concerned, unregulated organic contaminants are an issue of special interest. The question remains as to the adequacy of existing drinking water regulations to protect public health where unregulated trace organic contaminants are concerned. In the following section, the committee examines the adequacy of CCL datasets for evaluating contaminants relevant to water reuse, the challenge of unknown contaminants, and the concern of greater microbial risks when raw water supplies contain significant amounts of municipal wastewater effluent.

Adequacy of CCL Data for Prioritizing Chemicals Relevant to Water Reuse

The CCL process (Box 10-3) is the primary mechanism for considering trace organic contaminants for regulation under the SDWA. Therefore, the committee first evaluated whether the CCL process

adequately targeted contaminants for water reuse applications. From a review of the history of the CCL (see Box 10-3), it is evident that the process used to gather data for the CCL is evolving to become increasingly comprehensive in character. This becomes particularly clear in the third CCL (CCL3). Nevertheless, expanding the water quality monitoring datasets that inform the CCL process, particularly targeting contaminants encountered in municipal effluents, could improve the effectiveness of the CCL for reuse applications.

The CCL3 universe encompasses a wide array of potential water contaminants, both chemical and microbial. To generate the CCL3 universe, EPA relies primarily on databases that are electronically accessible at no charge. Although some databases include data on contaminants in municipal effluents, much of the data published in peer-reviewed literature is not included. The UCMR program under SDWA monitors unregulated contaminants in drinking water, but this program does not directly target contaminants in water reuse systems or municipal wastewater. At present, the data on unregulated contaminants in wastewater discharges primarily originate from research efforts conducted by utilities and academic research funded by water industry research foundations. The program would benefit from an effort to include these data in the CCL as well. Also, a federal monitoring program for unregulated contaminants directed toward wastewater effluents, mirroring the UCMR program for drinking water, would be highly beneficial in characterizing the occurrence of emerging contaminants in reuse (and de facto reuse) applications.

The Challenge of Unknown Contaminants

Although the SDWA provides protection to public health from priority chemicals and microbial contaminants, unknown chemical compounds (i.e., those that have not yet been identified through chemical analysis or whose occurrence has not been characterized) represent a primary concern in potable reuse projects that is not currently addressed by the SDWA. This concern also applies to conventional supplies to the extent that they are influenced by wastewater sources or exposed to independent sources of contamination. The current paradigm for discrete chemical monitoring of a pre-identified suite of contaminants will not be capable of addressing the large number of potential but currently unknown contaminants within wastewater effluents. Although the inclusion of production volume and fate characteristics in the CCL3 is a reasonable start, truly identifying unknown chemicals will likely require advanced instrumental techniques and biological assays to provide more holistic and comprehensive screening tools to assess overall biological potency. Addressing contaminants by groups, in addition to individually, as employed by EPA in the original trihalomethane regulation (EPA, 1979), in subsequent regulations on disinfection byproducts (EPA, 1998b, 2003c) and as recently proposed by EPA for addressing contemporary issues (EPA, 2010a) could provide a useful strategy to address the challenge of unknowns.

An example of the emergence of one previously unknown chemical is *N*-nitrosodimethylamine (NDMA), which is commonly detected in potable reuse practices using combined chlorine for disinfection (see Box 3-2). Prior to widespread awareness of the chemical, NDMA was likely present in reclaimed and potable waters for quite some time at concentrations far greater than 0.7 ng/L, an EPA-established groundwater cleanup level (EPA, 2010b). Although nitrosamines were known to occur in potable water systems as early as the 1970s, NDMA did not gain widespread attention until the 1990s when it was discovered in elevated levels in California reuse systems (Najm and Trussell, 2001). NDMA was added to the CCL in 2009 and was included in the UCMR2.

Protection Against Greater Microbial Risks

As previously discussed, under the SDWA, viruses and protozoa are regulated by treatment techniques rather than MCL. Under the original Surface Water Treatment Rule (SWTR [42 USCA 300g-1(b)(2)(c)), all surface water treatment plants (unless exempt by waiver) had to have treatment sufficient to achieve 99.9 percent reduction in Giardia and 99.99 percent reduction in viruses, and the operational characteristics of treatment steps needed to achieve this were defined in guidance manuals. Bacterial pathogens are also presumed to be reduced. Under the Long Term 2 SWTR (LT2SWTR), utilities have been required to take measurements of the source water concentrations of Cryptosporidium to determine if further reductions of Cryptosporidium are required. This additional reduction (either by additional processes or by more intensive

application of existing processes) would also result in increased reduction of bacteria, viruses, and Giardia. It is uncertain whether this regulatory framework is sufficient when source waters contain a high proportion of wastewater.

Failure of any of the treatment processes used to control pathogens would carry a risk of sporadic "breakthrough" of pathogens. To the degree that high levels of pathogen reduction are achieved by engineered processes, rather than use of a protected watershed (with lower levels of pathogens), it becomes more critical to maintain multiple barriers designed to improve reliability (see Chapter 5), whether in a planned reuse situation or in a conventional water system treating impaired surface waters.

Assessment of the Existing Federal Regulatory Framework for Potable Reuse

Reclaimed water used for potable reuse ultimately is required to meet all physical, chemical, radiological, and microbiological standards for drinking water. The SDWA will provide a measure of human health protection in terms of discrete chemicals based upon standards established and enforced by EPA (whether in the form of a numerical MCL or a treatment technique). However, as established earlier in this section, the SDWA does not yet establish standards for all potentially harmful constituents that may be present in wastewater. At present, the rules promulgated under the CWA and SDWA do not sufficiently address the public health concerns associated with reclaimed water for potable reuse. Also, the datasets used to develop the universe of contaminants considered for regulation are not yet sufficient to capture the range of contaminants that may be present in reclaimed water for potable reuse applications. More detailed reuse regulations exist in some states to address some, but not all, of these concerns (discussed in the next section). A discussion of potential advantages and disadvantages of federal reuse regulations follows the discussion of state reuse regulations. However, it is critical to understand that many drinking water systems in the United States utilize source waters with significant contributions from treated wastewater. Therefore, a revised regulatory paradigm that provides greater protection for potable reuse applications would need to consider the extent of de facto reuse to provide equivalent protection for all consumers.

WATER REUSE REGULATIONS AND GUIDELINES

There are no federal regulations specifically governing water reclamation and reuse in the United States; hence, the regulation of water reuse rests with the individual states. However, the federal government does provide guidance to states via EPA's *Guidelines for Water Reuse*, which "presents and summarizes recommended water reuse guidelines for the benefit of the water and wastewater utilities and regulatory agencies" (EPA, 2004). Regulations differ from guidelines in that regulations are legally adopted, enforceable, and mandatory, whereas guidelines are advisory and compliance is voluntary. Guidelines sometimes become enforceable requirements if they are incorporated into state regulations or water reuse permits.

Water reuse regulations and guidelines can be based on a variety of considerations but are directed principally at public health protection. For nonpotable reclaimed water applications, criteria generally address only microbiological and environmental concerns; however, existing regulations/guidelines for nonpotable reuse generally are not risk based. For potable reuse applications, health risks associated with pathogenic microorganisms and chemical constituents are both addressed. Reuse guidelines also generally address proper controls and safety precautions implemented at areas where nonpotable reclaimed water is used (e.g., warning signs, color-coded pipes, cross-connection control provisions). Additionally, guidelines may include water quality considerations that are unrelated to public health or environmental protection but are important to the success of specific nonpotable reuse applications (e.g., irrigation, industrial cooling).

The following sections summarize the federal reuse guidelines and state guidance and/or regulations for nonpotable and potable reuse.

EPA Guidelines for Water Reuse

EPA's *Guidelines for Water Reuse* (EPA, 2004), which cover both potable and nonpotable reuse, are intended to provide reasonable guidance, with supporting information, for utilities and regulatory agencies in the United States. The guidelines are particularly useful for states that have not developed their own water reuse regulations or are revising or expanding existing regulations. The guidelines contain a plethora of information on various aspects of water reuse, including treatment technology, public health concerns, legal and institutional issues, public involvement programs, and suggested water quality treatment and quality requirements for different reuse applications. The remainder of this section focuses on the suggested water treatment and quality requirements included in the guidelines.

Table 10-2 summarizes the treatment processes and water quality limits in the guidelines for a variety of nonpotable and potable reclaimed water applications. Also included are monitoring frequencies, setback distances, and other controls for each water reuse application. The suggested guidelines pertaining to treatment and water quality are based primarily on wastewater reclamation and reuse data from the United States. The guidelines apply to the reclamation of domestic wastewater from treatment plants with limited industrial waste inputs and "are not intended to be used as definitive water reclamation and reuse criteria" (EPA, 2004).

Nonpotable Reuse

The EPA (2004) guidelines recommend two different levels of disinfection for nonpotable uses of reclaimed water. For applications where direct or indirect reclaimed water contact is probable or expected, and for dual-water systems where cross-connections are always possible, disinfection to a level of no detectable fecal coliform organisms/100 mL is advised (based on the median value of the last 7 days for which analyses have been completed). In any given sample, EPA (2004) also recommends that fecal coliforms not exceed 14/100 mL. For applications where no direct public or worker contact with reclaimed water occurs, the guidelines recommend disinfection to achieve a fecal coliform concentration not exceeding 200/100 mL (based on the median value of the last 7 days of analyses). It is noteworthy that the EPA guidelines for nonpotable reuse applications are not based on rigorous health risk assessment methodology. The World Health Organization and Australia do have nonpotable water reuse guidelines based on risk assessment, as described in Box 10-4.

TABLE 10-2   U.S. EPA Suggested Guidelines for Water Reuse Applications

Type of Use	Treatment	Reclaimed Water Quality
Urban uses,[a] food crops eaten raw, recreational impoundments[b]	• Secondary[c] • Filtration • Disinfection	• pH = 6–9 • ≤10 mg/L BOD • ≤2 NTU[d] • No detectable fecal coli/100 mL[e] • ≥1 mg/L Cl$_2$ residual[f]
Restricted access area irrigation,[g] surface irrigation of orchards and vineyards, processed food crops,[h] nonfood crops,[i] aesthetic impoundments,[j] construction uses,[k] industrial cooling,[l] environmental reuse[m]	• Secondary[c] • Disinfection	• pH = 6–9 • ≤30 mg/L BOD • ≤30 mg/L TSS • ≤200 fecal coli/100 mL[e] • ≥1 mg/L Cl$_2$ residual[f] (except for environmental reuse)
Groundwater recharge of nonpotable aquifers by spreading	• Site specific and use dependent • Primary (minimum)	• Site specific and use dependent
Groundwater recharge of nonpotable aquifers by injection	• Site specific and use dependent • Secondary (minimum)	• Site specific and use dependent
Groundwater recharge of potable aquifers by spreading	• Site specific • Secondary[c] and disinfection (minimum) • May also need filtration and/or advanced wastewater treatment	• Site specific • Meet drinking water standards after percolation through vadose zone
Groundwater recharge of potable aquifers by injection	Includes the following: • Secondary[c] • Filtration • Disinfection • Advanced wastewater treatment	Includes, but not limited to, the following: • pH = 6.5–8.5 • ≤2 NTU[d] • No detectable total coli/100 mL[e] • ≥1 mg/L Cl$_2$ residual[f] • ≤3 mg/L TOC • ≤0.2 mg/L TOX • Meet drinking water standards
Groundwater recharge of potable aquifers by augmentation of surface supplies	Includes the following: • Secondary[c] • Filtration • Disinfection • Advanced wastewater treatment	Includes, but not limited to, the following: • pH = 6.5–8.5 • ≤ 2 NTU[d] • No detectable total coli/100 mL[e] • ≥1 mg/L Cl$_2$ residual[f] • ≤3 mg/L TOC • Meet drinking water standards

[a]All types of landscape irrigation, toilet and urinal flushing, vehicle washing, use in fire protection systems and commercial air conditioner systems, and other uses with similar access or exposure to the water.

[b]Fishing boating, and full body contact allowed.

[c]Secondary treatment should produce effluent in which both the BOD and TSS do not exceed 30 mg/L.

[d]Should be met prior to disinfection. Average based on a 24-hour time period. Turbidity should not exceed 5 NTU at any time. If TSS is used in lieu of turbidity, the TSS should not exceed 5 mg/L.

[e]Based on the median value of the last 7 days for which analyses have been completed.

[f]After a minimum contact time of 30 minutes.

[g]Sod farms, silviculture sites, and other areas where public access is prohibited, restricted, or infrequent.

[h]Undergo chemical or physical processing sufficient to destroy pathogens prior to sale to the public or others.

[i]Pasture for milking animals; fodder, fiber, and seed crops.

[j]Pubic contact with reclaimed water is not allowed.

[k]Includes soil compaction, dust control, aggregate washing, making concrete.

[l]Once-through cooling. Reclaimed water for recirculating cooling towers may need additional treatment.

[m]Wetlands, marshes, wildlife habitat, stream augmentation.

SOURCE: Adapted from EPA (2004).

Additional recommendations for nonpotable reuse applications not listed in Table 10-2 include

- clear, colorless, odorless, and nontoxic water;
- a setback distance of 50 feet between areas irrigated with reclaimed water and potable water supply wells;
- maintenance of a chlorine residual of greater than or equal to 0.5 mg/L in the distribution system;
- reliable treatment and emergency storage or disposal alternatives for inadequately treated water;

BOX 10-4
Risk-Based Water Reuse Guidelines Using DALYs

The World Health Organization (WHO, 2006a,b,c) published risk-based guidelines for the use of wastewater and greywater in agriculture and aquaculture in 2006. The guidelines are directed principally at microbial health risks but also include recommended maximum tolerable soil concentrations for various organic and inorganic pollutants based on human health protection. They were based on the quantitative microbial risk assessment, complemented by epidemiological evidence.

The WHO guidelines use disability adjusted life years (DALYs), a common summary measure of population health, to compare disease outcome from one exposure pathway to another. DALYs represent a measure of time lost due to disability or death from a specific disease compared to an ideal long life, free of disease and disability. DALYs are calculated as the sum of the probable years of life lost to premature mortality and the years of productive life lost due to disability associated with a particular disease. Thus, DALYs account for both acute and chronic health effects, including morbidity and mortality. DALYs have been useful in elucidating the choices of water disinfection technologies (balancing the risks of microorganisms and disinfection byproducts) in the Netherlands (Havelaar et al., 2000), although DALYs have been also subject to some criticism (Anand and Hanson, 1997; Govind et al., 2009).

WHO determined that a waterborne disease burden of 10^{-6} DALYs per person per year is a tolerable risk (WHO, 2004). This disease burden is approximately equivalent one mild diarrheal illness (assuming a low fatality rate) per 1,000 people per year, or 1 in 10 risk of mild illness over a lifetime (WHO, 2008). Health-based targets based on DALYs can be achieved through a combination of health protection measures, such as wastewater treatment, crop restriction, wastewater application techniques that minimize contamination, chemotherapy and immunization, and washing, disinfecting, and cooking produce.

Australia has also embraced the use of DALYs to set health-based targets related to the use of reclaimed water in its *Australian Guidelines for Water Recycling: Managing Health and Environmental Risks (Phase 1)* (NRMMC/EPHC/NHMC, 2006), which deals with the reuse of wastewater, stormwater, and greywater for nonpotable purposes. Although the guidelines are not mandatory and have no formal legal status, their adoption provides a shared national objective and allows states and/or local jurisdictions to independently adopt them or to use their own legislative and regulatory tools to refine them into their own guidelines. The Australian guidelines address both human health (mainly microbial pathogen risks) and environmental risks (mainly chemical risks) using a risk management approach. In managing risks to human health, the guidelines use DALYs to convert the risk of illness into burdens of disease, and—as with the WHO guidelines—the Australian guidelines establish the tolerable risk as 10^{-6} DALYs per person per year, which is then used to develop health-based targets. In managing risks to the environment from reclaimed water, environmental guidelines related to impacts on specific endpoints or receptors within the environment are used in place of DALYs and health-based targets.

The Phase 2 report of the Australian guidelines, which focuses on potable reuse (NRMMC/EPHC/NHMRC, 2008; see also Box 5-2) uses DALYs, performance targets, and reference pathogens for the evaluation of microbial risk, based on the approach described in the World Health Organization Guidelines for Drinking-Water Quality (WHO, 2008). As with nonpotable applications of reclaimed water, the tolerable microbial risk adopted in the Australian potable reuse guidelines is 10^{-6} DALYs per person per year.

- cross-connection control devices; and
- color-coded nonpotable water lines and appurtenances.

The guidelines include similar design and operational recommendations for the other reclaimed water applications.

Potable Reuse

EPA's guidelines provide some specific wastewater treatment and reclaimed water quality recommendations for potable reuse via groundwater recharge and surface water augmentation, as indicated in Table 10-2.

The guidelines outline the extensive treatment, water quality, and monitoring requirements that are likely to be imposed for potable reuse projects and are based principally on California's draft groundwater recharge regulations and Florida's potable reuse regulations in place at the time the guidelines were written. The guidelines recommend that potable reuse projects meet drinking water standards and also monitor for hazardous compounds (or classes of compounds) that are not included in the drinking water standards (EPA, 2004). The EPA guidelines' focus on end-point water quality differs significantly from the risk management strategies of the Australian potable reuse guidelines, described in Box 5-2.

State Water Reuse Regulations and Guidelines

States generally develop water reuse regulations or guidelines in response to a need to regulate water reuse activities that are occurring or expected to occur in the near future. Water reuse criteria vary among the states that have developed regulations, and some states have no regulations or guidelines. Some states have regulations or guidelines directed at land treatment of wastewater or land application as a means of wastewater disposal rather than regulations oriented to the intentional beneficial use of reclaimed water. Water reuse regulations typically include wastewater treatment process requirements, treatment reliability requirements, reclaimed water quality criteria, reclaimed water conveyance and distribution system requirements, and area use controls. No state's regulations cover all potential applications of reclaimed water, and few states have regulations that address potable reuse. When state regulations do not address specific reuse applications, they are not necessarily prohibited; instead, these applications may be evaluated and permitted on a case-by-case basis. The following sections provide an overview of state approaches to nonpotable and potable reuse regulations.

State Guidelines and Regulations for Nonpotable Reuse

Examples of state regulations for various nonpotable applications are summarized in Table 10-3. The table includes water quality limits and, where imposed, treatment process requirements. Water quality requirements usually include maximum limits based on averages or geometric means over a specific time period or median values for a specific number of consecutively collected samples. They also usually include maximum values (particularly for microbial indicator organisms) that cannot be exceeded at any time, although these limits are not included in Table 10-3.

Table 10-3 shows clear variations in the treatment and quality requirements among the states for the types of uses listed. Key areas of significant variation are discussed below.

Microbial Indicator Organisms. Some states use total coliforms as the indicator organism, whereas others use fecal coliforms, *Escherichia coli*, or enterococci. Total coliforms represent a more conservative measure of the microbial water quality and include fecal coliforms and some nonfecal bacteria, such as soil bacteria. Some states have based their requirements on the EPA guidelines (EPA, 2004), which suggest using fecal coliforms as the indicator organism. Regulatory decisions regarding the selection of which indicator organism to use is somewhat subjective, as is the acceptable limit. The rationale regarding the selection of which indicator organism to use and the methods used to determine whether acceptable microbial limits have been met are not consistent in all states. For example, in California the total coliform reporting limit is based on a running median of the last 7 days for which analyses have been completed, whereas in Florida the fecal coliform limit must be met in at least 75 percent of the samples over a 30-day period. Daily sampling is required in both states.

Turbidity vs. Total Suspended Solids (TSS). For uses where human contact with the reclaimed water is expected or likely, some states specify turbidity limits whereas others specify TSS limits. The removal of suspended matter is related to health protection. Particulate matter can reduce the effectiveness of disinfection processes, such as chlorine and UV radiation (see Chapter 4). To ensure that pathogens are inactivated during disinfection, state water reuse regulations and guidelines generally recommend that particulate matter in reclaimed water be reduced to low levels (e.g., 2 nephelometric turbidity units [NTU] or 5 mg/L TSS). Low turbidity or suspended solids values by themselves do not indicate that reclaimed water is devoid of microorganisms. As such, turbidity and suspended solids measurements are not used as an indicator of microbiological quality but rather as a quality criterion for wastewater prior to disinfection.

Treatment Requirements. Most states adhere to the premise that water quality requirements for indicator organisms alone do not adequately characterize the microbial quality of the water. Thus, most states prescribe specific treatment processes (e.g., secondary treatment followed by filtration and disinfection) that, in conjunction with water quality requirements for parameters such as microbial indicator organisms and turbidity, have been shown to reduce pathogenic

TABLE 10-3 Examples of State Water Reuse Criteria for Selected Nonpotable Applications

State	Fodder Crop Irrigation[a] Quality Limits	Treatment Required	Processed Food Crop Irrigation[b] Quality Limits	Treatment Required	Food Crop Irrigation[c,d] Quality Limits	Treatment Required	Restricted Recreational Impoundments[e] Quality Limits	Treatment Required
Arizona	• 1,000 fecal coli/100 mL	• Secondary (stabilization ponds)	Not covered	Not covered	• No detectable fecal coli/100 mL • 2 NTU	• Secondary • Filtration • Disinfection	• No detectable fecal coli/100 mL • 2 NTU	• Secondary • Filtration • Disinfection
California	Not specified	• Secondary	Not specified	• Secondary	• 2.2 total coli/100 mL • 2 NTU	• Secondary • Coagulation[f] • Filtration • Disinfection	• 2.2 total coli/100 mL	• Secondary • Disinfection
Colorado	Not covered	Not covered	Not covered	Not covered	Not covered	Not covered	Not covered	Not covered
Florida	• 200 fecal coli/100 mL • 20 mg/L CBOD • 20 mg/l TSS	• Secondary • Disinfection	• No detectable fecal coli/100 mL • 20 mg/L CBOD • 5 mg/L TSS	• Secondary • Filtration • Disinfection	• No detectable fecal coli/100 mL • 20 mg/L CBOD • 5 mg/L TSS	• Secondary • Filtration • Disinfection	• No detectable fecal coli/100 mL • 20 mg/L CBOD • 5 mg/L TSS	• Secondary • Filtration • Disinfection
Utah	• 200 fecal coli/100 mL • 25 mg/L BOD • 25 mg/L TSS	• Secondary • Disinfection	• No detectable fecal coli/100 mL • 10 mg/L BOD • 2 NTU	• Secondary • Filtration • Disinfection	• No detectable fecal coli/100 mL • 10 mg/L BOD • 2 NTU	• Secondary • Filtration • Disinfection	• 200 fecal coli/100 mL • 25 mg/L BOD • 25 mg/L TSS	• Secondary • Disinfection
Texas	• 200 fecal coli or E. coli/100 mL • 35 enterococci/100 mL • 20 mg/L BOD • 15 mg/L CBOD	Not specified	• 200 fecal coli or E. coli/100 mL • 35 enterococci/100 mL • 20 mg/L BOD • 15 mg/L CBOD	Not specified	• 20 fecal coli or E. coli/100 mL • 4 enterococci/100 mL • 3 NTU • 5 mg/L BOD or CBOD	Not specified	• 20 fecal coli or E. coli/100 mL • 4 enterococci/100 mL • 3 NTU • 5 mg/L BOD or CBOD	Not specified
Washington	• 240 total coli/100 mL	• Secondary • Disinfection	• 240 total coli/100 mL	• Secondary • Disinfection	• 2.2 total coli/100 mL • 2 NTU	• Secondary • Coagulation • Filtration • Disinfection	• 2.2 total coli/100 mL	• Secondary • Disinfection

[a] In some states more restrictive requirements apply where milking animals are allowed to graze on pasture irrigated with reclaimed water.

[b] Physical or chemical processing sufficient to destroy pathogenic microorganisms. Less restrictive requirements may apply where there is no direct contact between reclaimed water and the edible portion of the crop.

[c] Food crops eaten raw where there is direct contact between reclaimed water and the edible portion of the crop.

[d] In Florida and Texas, "irrigation of edible crops that will be peeled, skinned, cooked, or thermally processed before consumption is allowed. Direct contact of the reclaimed water with such edible crops is allowed." "Irrigation of edible crops that will not be peeled, skinned, cooked, or thermally processed before consumption is allowed if an indirect application method is used which will preclude direct contact with the reclaimed water (such as ridge and furrow irrigation, drip irrigation, or a subsurface distribution system) is used" (30 Texas Administrative Code § 210.24).

[e] Recreation is limited to fishing, boating, and other nonbody contact activities.

[f] Not needed if filter effluent turbidity does not exceed 2 NTU, the turbidity of the influent to the filters is continually measured, the influent turbidity does not exceed 5 NTU for more than 15 minutes and never exceeds 10 NTU, and there is capability to automatically activate chemical addition or divert the wastewater should the filter influent turbidity exceed 5 NTU for more than 15 minutes.

TABLE 10-3 Continued

State	Restricted Access Irrigation[g] Quality Limits	Treatment Required	Unrestricted Access Irrigation[b] Quality Limits	Treatment Required	Toilet Flushing[i] Quality Limits	Treatment Required	Industrial Cooling Water[j] Quality Limits	Treatment Required
Arizona	• 200 fecal coli/100 mL	• Secondary • Disinfection	• No detectable fecal coli/100 mL • 2 NTU	• Secondary • Filtration • Disinfection	• No detectable fecal coli/100 mL • 2 NTU	• Secondary • Filtration • Disinfection	Not covered	Not covered
California	• 23 total coli/100 mL	• Secondary • Disinfection	• 2.2 total coli/100 mL • 2 NTU	• Secondary • Coagulation[k] • Filtration • Disinfection	• 2.2 total coli/100 mL • 2 NTU	• Secondary[k] • Filtration • Disinfection	• 2.2 total coli/100 mL • 2 NTU	• Secondary • Coagulation[k] • Filtration • Disinfection
Colorado	• 126 $E.\ coli$/100 mL • 30 mg/L TSS	• Secondary • Disinfection	• No detectable $E.\ coli$/100 mL • 3 NTU	• Secondary • Filtration • Disinfection	Not covered	Not covered	• 126 $E.\ coli$/100 mL • 30 mg/L TSS	• Secondary • Disinfection
Florida	• 200 fecal coli/100 mL • 20 mg/L CBOD • 20 mg/l TSS	• Secondary • Disinfection	• No detectable fecal coli/100 mL • 20 mg/L CBOD • 5 mg/L TSS	• Secondary • Filtration • Disinfection	• No detectable fecal coli/100 mL • 20 mg/L CBOD • 5 mg/L TSS	• Secondary • Filtration • Disinfection	• No detectable fecal coli/100 mL • 20 mg/L CBOD • 5 mg/L TSS	• Secondary • Filtration • Disinfection
Utah	• 200 fecal coli/100 mL • 25 mg/L BOD • 25 mg/L TSS	• Secondary • Disinfection	• No detectable fecal coli/100 mL • 10 mg/L BOD • 2 NTU	• Secondary • Filtration • Disinfection	• No detectable fecal coli/100 mL • 10 mg/L BOD • 2 NTU	• Secondary • Filtration • Disinfection	• 200 fecal coli/100 mL • 25 mg/L BOD • 25 mg/TSS	• Secondary • Disinfection
Texas	• 200 fecal coli or $E.\ coli$/100 mL • 35 enterococci/100 mL • 20 mg/L BOD • 15 mg/L CBOD	Not specified	• 20 fecal coli or $E.\ coli$/100 mL • 4 enterococci/100 mL • 3 NTU • 5 mg/L BOD or CBOD	Not specified	• 20 fecal coli or $E.\ coli$/100 mL • 4 enterococci/100 mL • 3 NTU • 5 mg/L BOD or CBOD	Not specified	• 200 fecal coli or $E.\ coli$/100 mL • 35 enterococci/100 mL • 20 mg/L BOD • 15 mg/L CBOD	Not specified
Washington	• 23 total coli/100 mL	• Secondary • Disinfection	• 2.2 total coli/100 mL • 2 NTU	• Secondary • Coagulation • Filtration • Disinfection	• 2.2 total coli/100 mL • 2 NTU	• Secondary • Coagulation • Filtration • Disinfection	• 2.2 total coli/100 mL • 2 NTU	• Secondary • Coagulation • Filtration • Disinfection

[g]Classification varies by state; generally includes irrigation of cemeteries, freeway medians, restricted-access golf courses, and similar restricted-access areas.

[b]Includes irrigation of parks, playgrounds, schoolyards, residential lawns, and similar unrestricted access areas.

[i]Not allowed in single-family residential dwelling units.

[j]Cooling towers where a mist is created that may reach populated areas.

[k]Not needed if filter effluent turbidity does not exceed 2 NTU, the turbidity of the influent to the filters is continually measured, the influent turbidity does not exceed 5 NTU for more than 15 minutes and never exceeds 10 NTU, and there is capability to automatically activate chemical addition or divert the wastewater should the filter influent turbidity exceed 5 NTU for more than 15 minutes.

SOURCE: Adapted from Washington Department of Health and Washington Department of Ecology (1997), Colorado Department of Health and Environment (2007), Florida Department of Environmental Protection (2007), CDPH (2009), Texas Commission on Environmental Quality (2011), Arizona Department of Environmental Quality (2010), Utah Department of Environmental Quality (2011)

organisms to very low or nondetectable levels in the reclaimed water. A few states rely solely on the water quality of the product water and do not specify treatment process requirements.

Reclaimed Water Uses. No state water reuse regulations include requirements for all potential nonpotable reuse applications; they generally include the most common or likely types of use. Regulations in many states allow types of use not specifically included in their regulations if they are shown to the satisfaction of the regulatory agency to provide an adequate degree of health or environmental protection. States listed in Table 10-3 that have uses that are not covered in their regulations do not necessarily prohibit such uses. Instead, those uses (and their attendant reclaimed water treatment and quality requirements) may be evaluated and accepted on a case-by-case basis.

Other Variables. Many state water reuse regulations include requirements for water quality monitoring frequency, treatment reliability, cross-connection control (see Box 10-5), emergency storage and disposal, and use area controls (e.g., setback distances, signage). As with treatment and reclaimed water quality requirements, these requirements are not uniform from state to state.

State Guidelines and Regulations for Potable Reuse

Some states (e.g., Hawaii) have guidelines that address potable reuse; in those states, regulatory agencies evaluate projects on a case-by-case basis. Many states do not have potable regulations, and several states rely on the EPA underground injection control regulations to protect potable groundwater basins. A few states, such as California (draft regulations), Florida, Washington, and Massachusetts, have developed comprehensive water reuse regulations for potable reuse (most of them for groundwater recharge), but the absence of state criteria for potable reuse does not necessarily prohibit potable reuse applications. Some states evaluate potable reuse projects on a case-by-case basis, even without guidelines or regulations. To date, no regulations have been adopted for potable reuse without the use of an environmental buffer (sometimes called direct potable reuse; see also Chapter 2) anywhere in the United States.

BOX 10-5
Cross-Connection Control

State nonpotable reuse regulations often address cross-connection control by specifying requirements that reduce the potential for cross connections, including the following:

- Identification of transmission and distribution lines and appurtenances via color coding, taping, or other means
- Separation of potable water and reclaimed water lines
- Allowable pressures
- Operation and maintenance procedures
- Monitoring and testing
- Surveillance
- Backflow protection devices to reduce the potential of contaminating the potable water system in the event of a cross connection at a use area

California has additional cross-connection control requirements where reclaimed water is used in buildings for toilet and urinal flushing or for fire protection. The requirements stated in the California *Water Recycling Criteria* (CDPH, 2009) for reclaimed water in dual-plumbed facilities include the following:

1. Internal use of reclaimed water within any individually owned residential unit, including multiplexes and condominiums, is prohibited.

2. Facilities that produce or process food products or beverages can use reclaimed water internally only for fire suppression systems.

3. Reclaimed water cannot be used within a building until a detailed description of the intended use areas, plans and specifications, and cross-connection control provisions and testing procedures is submitted and approved by the regulatory agency.

4. The dual-plumbed system within each facility or use area must be inspected for cross connections prior to the initial operation and annually thereafter. Additionally, the reclaimed water system must be tested at least once every four years for possible cross connections.

5. The California Department of Public Health must be notified of any incidence of backflow from the nonpotable reclaimed water system into the potable water system within 24 hours of the incident's discovery.

Direct connections between potable and nonpotable distribution systems are not allowed in any state (Asano et al., 2007). Detailed information on cross-connection control measures is available in manuals published by the American Water Works Association (AWWA, 2004, 2009) and the U.S. Environmental Protection Agency (EPA, 2003c).

As examples of regulations, existing and draft potable reuse regulations for groundwater recharge in California and adopted groundwater recharge and surface water augmentation regulations in Florida are summarized in Boxes 10-6 and 10-7. California published new draft regulations in November 2011 and expects to finalize them in the first half of 2012.

National Standards for Reuse?

The previous section highlights how water reuse regulations and guidelines vary considerably from state to state in terms of the reuse applications covered, treatment and water quality requirements, design or operational controls, the rationale for setting requirements, and the specific objectives of the regulations or guidelines. Although the EPA's Guidelines for Water Reuse (EPA, 2004) were developed for states that have not yet developed their own regulations or are updating their existing regulations, they have not significantly affected the lack of uniformity among state regulations. Further, they were not developed in a rigorous manner comparable to, for example, the SWDA or CWA, and thus were not subjected to the scrutiny required of formal federal regulatory processes.

The imbalance that results from different standards in each state is demonstrated by food crops grown with reclaimed water where, for example, lettuce grown in one state may have been irrigated with different quality water than lettuce grown in another state, yet both may be sold anywhere. A consumer does not know the different standards in each state, but rather assumes that the level of protection is the same regardless of where the lettuce was grown. From the industry perspective, an instance of food contamination will injure agricultural growers everywhere, so that even a grower in a state with stricter standards could be negatively affected by a product from a state with more relaxed regulations.

The typical model in environmental regulation is one in which Congress creates a regulatory program in broad outline, and EPA is entrusted by Congress with giving it more specificity, including setting standards for health and environmental protection. Most federal statutory schemes allow EPA to delegate the administration of the program to a state (or tribal) agency. Delegation is contingent upon the state creating and maintaining a program that is as stringent as the federal program. EPA sets standards for pollutants using health, technology, cost, or some combination of these elements. The standard-setting process allows for participation and allows for appeals if certain criteria are met.

There are several potential advantages of developing national regulations for water reuse. First, it would be more efficient for EPA to develop risk-based regulations than the effort that would be required if regulations were developed by each individual state. EPA could tap its internal experts with various areas of expertise that would be needed to establish scientifically supportable criteria (e.g., public health, microbiology, treatment technology, risk assessment). Further, national water reuse regulations may reduce the potential of local regulatory decisions that may not be supportable from a public health or environmental standpoint.

On the basis of a survey of stakeholders, including water reuse practitioners and state and federal regulators, Nellor and Larson (2010) identified the following advantages of national regulations for water reuse:

- Because the development of regulations is a rigorous process with public input, compliance with the regulations should provide enhanced public confidence that a water reuse project is safe.
- The regulations should establish credibility of and public confidence in water reuse.
- The regulations should create minimum uniform standards relative to the end use that are applied across the country, thereby eliminating concerns about lack of consistency among state regulations/guidelines in terms of public health protection.
- The regulations should eliminate the gap for states without rules.

There are also some disadvantages outlined by Nellor and Larson (2010) that may result from the promulgation of national regulations for reuse:

- It would be necessary to amend the CWA or SWDA, or create a new enabling federal law to provide authorization for the development of regulations for these uses. Changes to national statutes are difficult and resource intensive.
- To address national variation and uncertainty, federal regulations generally incorporate a margin of safety. The resulting standards may be very conservative.
- More conservative standards could create ob-

BOX 10-6
California Draft Regulations for Potable Water Reuse

The California Department of Public Health's (CDPH's) existing California Water Recycling Criteria, which were adopted in 2000, outline the requirements for recharging water supply aquifers with reclaimed water via surface spreading. According to the regulations, reclaimed water used to recharge water supply aquifers "shall be at all times of a quality that fully protects public health" (CDPH, 2009). Under the regulations, the CDPH can make project-specific recommendations based on factors such as treatment employed, effluent quality and quantity, soil characteristics, hydrogeology, residence time, and distance to withdrawal. CDPH embarked on drafting comprehensive groundwater recharge regulations for both surface spreading and injection projects several years ago that would replace the existing language in the Water Recycling Criteria and, although the draft regulations have gone through several iterations in the last decade, they have yet to be finalized and adopted. Until criteria are formally adopted, proposed groundwater recharge projects will be regulated on the basis of the most recent draft regulations (summarized in Table 10-4; CDPH, 2011), which are subject to substantial revision prior to adoption.

The draft groundwater recharge regulations apply to planned projects that are operated for the purpose of recharging a groundwater basin designated as a source of municipal and domestic water supply or a project determined to be a groundwater replenishment reuse project by a California Regional Water Quality Control Board based on a project's existing or projected replenishment of an affected groundwater basin.

Based on a bill passed by the California Senate and approved by the governor in 2010 (California State Senate, 2010), the California Water Code (CSWRCB, 2011) was amended in 2010 to require CDPH to (1) adopt uniform water reuse criteria for indirect potable reuse for groundwater recharge by December 13, 2013; (2) develop and adopt uniform water reuse criteria for surface water augmentation by December 31, 2016, if an expert panel convened in response to the legislation finds that the criteria would adequately protect public health; and (3) "investigate and report to the Legislature on the feasibility of developing uniform water recycling criteria for direct potable reuse" by December 31, 2016.

TABLE 10-4 Draft California Regulations for Groundwater Recharge into Potable Aquifers

Water Quality Limits for Recycled Water	Treatment Required	Other Selected Requirements
• ≥12-log virus reduction • ≥10-log *Giardia* cyst reduction • ≥10-log *Cryptosporidium* oocyst reduction • Drinking water MCLs (except for nitrogen) • Action levels for lead and copper • ≤10 mg/L total nitrogen[a] • TOC[b] ≤0.5 mg/L/RWC[c]	***Spreading*** • Oxidation[d] • Filtration[e] • Disinfection[f] • Soil aquifer treatment ***Spreading with full advanced treatment*** • Oxidation • Reverse osmosis • Advanced oxidation process • Soil aquifer treatment ***Injection*** • Oxidation • Reverse osmosis • Advanced oxidation process	• Industrial pretreatment and source control program • Initial maximum RWC ≤20% for spreading tertiary treated water • Initial maximum RWC for injection based on California Department of Public Health (CDPH) review of engineering report and other information from public hearing • ≥2-month retention (response) time underground[g] • 1-log virus reduction credit automatically given per month of subsurface retention • 10-log *Giardia* reduction and 10-log *Cryptosporidium* reduction credit given to spreading projects that have at least 6 months' retention time underground • Monitor recycled water and monitoring wells for priority toxic pollutants, chemicals with state notification levels specified by CDPH, and unregulated constituents specified by CDPH • Operations plan • Contingency plan • Spreading projects with full advanced treatment must meet the requirements for injection projects, except that after one year of operation the project sponsor may apply for a reduced monitoring frequency for any monitoring requirement

[a]The total nitrogen limit can be met in the recycled water or in the combination of recycled water and diluent water applied at the recharge site.

[b]Total organic carbon.

[c]The recycled water contribution (RWC) is the quantity of recycled water applied at a recharge site divided by the sum of the quantity of recycled water applied at the site and diluent water.

[d]Oxidized wastewater is wastewater in which the organic matter has been stabilized, contains dissolved oxygen, and is not liable to become putrid.

[e]Filtered wastewater is oxidized wastewater that (1) has been coagulated, filtered through media, does not exceed an average turbidity of 2 NTU, does not exceed 5 NTU more that 5% of the time within a 24-hour period, and does not exceed 10 NTU at any time; or (2) has received membrane treatment and does not exceed an average turbidity of 0.2 NTU more than 5% of the time within a 24-hour period and does not exceed 0.5 NTU at any time.

[f]Disinfected recycled water is water that has been disinfected by either chlorine that provides a CT (product of total chlorine residual and modal contact time) ≥450 at all times with a modal contact time of at least 90 minutes; or a disinfection process that inactivates/removes at least 5 logs of MS2 bacteriophage or polio virus. The 7-day median total coliform concentration in the disinfected water cannot exceed 2.2/100 mL.

[g]Must be verified by a tracer study.

SOURCE: Adapted from CDPH (2011).

```
┌─────────────────────────────────────────────────────────────────────────────────┐
│                                  BOX 10-7                                         │
│                       Florida Potable Reuse Regulations                           │
│                                                                                   │
│       The Florida reuse rule (Fla. Admin. Code, Chapter 62-610) includes          │
│  treatment and water quality requirements for groundwater recharge via            │
│  infiltration basins or injection and for indirect potable reuse by surface       │
│  water augmentation (Table 10-5). The rules address rapid-rate infiltration       │
│  basin systems and absorption field systems, both of which may result in          │
│  groundwater recharge. Although groundwater recharge projects located             │
│  over potable aquifers are not specifically designated as indirect potable        │
│  reuse systems, they could function as an indirect potable reuse system.          │
│  However, rapid-rate land application systems that result in the collection        │
│  and discharge of more than 50 percent of the applied reclaimed water are         │
│  considered as effluent disposal systems. Loading to these surface infiltration   │
│  systems is limited to 9 inches/d (23 cm/d). Reclaimed water from                 │
│  systems having higher loading rates or a more direct connection to an aquifer    │
│  than normally encountered must receive at least secondary treatment,             │
│  filtration, and disinfection. The treated water must meet primary and secondary  │
│  drinking water standards.                                                        │
│       The Florida regulations include requirements for planned indirect potable   │
│  reuse by injection into water supply aquifers and augmentation of                │
│  surface supplies. For injection, a minimum horizontal separation distance of     │
│  500 ft (150 m) is required between reclaimed water injection wells and           │
│  potable water supply wells. The injection regulations pertain to groundwaters    │
│  that are classified as potable aquifers. The Florida reuse regulations           │
│  identify discharges to Class I surface waters (public water supplies) as         │
│  indirect potable reuse. Wastewater discharges to watercourses that are           │
│  less than 24 hours' travel time upstream from Class I waters also fall under     │
│  the definition of indirect potable reuse. Wastewater outfalls for surface        │
│  water discharges cannot be located within 500 ft (150 m) of existing or          │
│  approved potable water intakes within Class I surface waters. Pilot testing      │
│  is required prior to implementation of injection or surface water augmentation   │
│  projects.                                                                        │
└─────────────────────────────────────────────────────────────────────────────────┘
```

TABLE 10-5 Florida Rules for Groundwater Recharge and Indirect Potable Reuse

Type of Use	Treatment	Water Quality Limits
Groundwater recharge (Rapid infiltration basins)	• Secondary • Disinfection	• ≤200 fecal coli/100 mL • ≤20 mg/L CBOD • ≤0 mg/L TSS • ≤12 mg/L NO$_3$ (as N)
Groundwater recharge (Rapid infiltration basins in unfavorable hydrogeological conditions [e.g., karst areas])	• Secondary • Disinfection • Filtration	• No detectable fecal coli/100 mL • ≤20 mg/L CBOD • ≤5.0 mg/L TSS • ≤10 mg/L total N • Primary[a] and secondary drinking water standards
Groundwater recharge (Injection to groundwaters having TDS < 3,000 mg/L)	• Secondary • Disinfection • Filtration • Multiple barriers for control of pathogens and organics • Pilot testing required	• No detectable total coli/100 mL • ≤20 mg/L CBOD • ≤5.0 mg/L TSS • ≤3.0 mg/L TOC • ≤0.2 mg/L TOX[b] • ≤10 mg/L total N • Primary[a] and secondary drinking water standards
Groundwater recharge (Injection to groundwaters having TDS 3,000–10,000 mg/L)	• Secondary • Disinfection • Filtration	• No detectable total coli/100 mL • ≤20 mg/L CBOD • ≤5.0 mg/L TSS • ≤10 mg/L total N • Primary drinking water standards[a]
Indirect potable reuse (Discharge to Class I surface waters (used for public water supply)	• Secondary • Disinfection • Filtration	• No detectable total coli/100 mL • ≤20 mg/L CBOD • ≤5.0 mg/L TSS • ≤3.0 mg/L TOC • ≤10 mg/L total N • Primary[a] and secondary drinking water standards • WQBELs[c] may apply

[a] With some exceptions, e.g., asbestos.
[b] TOX = total organic halogen.
[c] WQBELs are water quality-based effluent limitations to ensure that water quality standards in a receiving body of water will not be violated.

SOURCE: Adapted from Fla. Admin. Code, Chapter 62-610.

stacles for promoting and/or continuing to implement reuse projects in states with existing standards that are less stringent than the federal regulations.

- Almost certainly, states would retain the legal authority to prescribe more stringent regulations, thereby eliminating uniformity.
- The development and promulgation of the regulations may take a significant amount of time and resources.

There are other potential disadvantages associated with national regulations. National standards may not be sensitive to local or regional conditions and could limit flexibility at the local level. Conflicts could arise regarding compatibility with existing state wastewater discharge requirements, environmental controls, or other regulations or statutes. It may be difficult to reconcile differences or conflicts between national criteria and existing state water reuse standards, policies, or guidelines. For example, if national criteria were more restrictive than a state's criteria, the national criteria would override local criteria. In such cases, it may result in considerable cost to upgrade existing projects, call into question past practices in the state, and potentially damage the credibility of the regulatory agency. All these present challenges that a national regulatory program would need to address.

The committee concludes that there are important inconsistencies among existing water reuse regulations/guidelines. Reclaimed water is of ever-growing importance as an integral component of the nation's water resources portfolio, and action to embark on the development and implementation of risk-based national water reuse regulations would allow the nation to more efficiently and effectively maximize this resource. Regulations can be crafted that do not stifle innovation but allow for new and innovative treatment and quality assurance processes.

PUBLIC INVOLVEMENT AND ATTITUDES

Planning for water reuse projects regularly involves public involvement and evaluation, which influence the type of reuse projects pursued and whether the project will move forward (Hartley, 2006). Proposed water reuse projects (especially potable reuse projects) have numerous aspects for the public to consider, including public health, public finance, local land use, regional

environmental protection, and economic growth. Public policy processes take the form of feasibility studies, environmental review, approval of funding, and zoning and siting of facilities, nearly all of which are subject to public hearings. There are also robust dialogues in letters to the editors, blogs, public meetings, and elsewhere. The goals of these processes are to inform the public of pending decisions, seek public input, and in some cases to seek direct public approval. Another source of public review occurs when state or national funding is sought for reuse projects that have extensive nonlocal benefits.

In this section, research on public perception with respect to water reuse is discussed. Additionally, the role of communication in successful reuse projects is examined. The bulk of the research on these issues has occurred in countries outside of the United States. In this section, the committee briefly reviews research findings on public perception worldwide, but examines data from the United States in somewhat more detail.

Public perception with respect to water reuse has been studied with increasing interest in the United States and Australia since the mid-1990s (summarized in Russell and Lux, 2006), and with interest expanding globally since the early 2000s (e.g., Jeffrey, 2002; Al-Kharouf et al., 2008; Ching, 2010; Domenech and Sauri, 2010). The long and challenging drought experienced by Australia in the 2000s focused intellectual and policy attention on water reuse, with extensive research on public perception and policy processes emerging. Beliefs about the importance of public perception to the successful establishment of water reuse projects range from "crucial importance" (Marks et al., 2008) to one factor among many (Stenekes et al., 2006).

Fear of contaminated water (or anything that is perceived to be contaminated) is a common human response. Numerous factors influence risk perception with respect to water, including sensory input (odor and taste), delivery context (tap vs. bottle, visual cues from surface waters), prior experience with the water, sources of information (informal, interpersonal), level of trust in the water purveyor, and one's perceived control over the quality of the water (Doria, 2010). Water reuse projects necessarily involve the use of water that was once contaminated. The perception that something is contaminated can trigger a strong, immediate reaction of revulsion (see Box 10-8; Rozin and Royzman, 2001;

BOX 10-8
Public Discourse on Water Reuse in Pembroke Pines, Florida

A new water reuse facility has been proposed for Pembroke Pines, Florida. The city of 150,000 people plans to inject 7 MGD of wastewater into the Biscayne Aquifer, rather than piping it to an ocean outfall. The effluent would receive primary, secondary, and reverse osmosis membrane treatment prior to injection. Restoring flows into the Biscayne Aquifer, which is shared by several cities, is required by the regional water management authority.

Although this project is still in the study phase, patterns of communication surrounding the disgust response and concerns over trace organic chemicals are already emerging. A local newspaper began its review article of the project with this sentence: "The water in Pembroke Pines toilet bowls may soon show up in the drinking glasses of South Floridians from Miami to Boca Raton" (Barkhurst, 2011). The article quotes an environmental activist: "You can't remove all pharmaceuticals from the water. It can't be done. You are putting drugs into our drinking water—Tylenol, birth control medication, antipsychotics." The article later quotes a water agency official who comments positively on available water treatment technologies.

This is a common pattern in public communication over proposed water reuse facilities. The debate has been framed as disgusting water source that threatens public health vs. scientific demonstrations of water need and safety. The debate also is framed as the public (in opposition) vs. the water agency (in support), which departs from the ideal of water agencies playing the role of neutral implementer of the public's wishes. Instead, the public would be best served by informed public discourse on a wide range of topics pertaining to water reuse, including relative risks compared to other water supply alternatives and sources already used widely today (see Chapter 7).

Nemeroff and Rozin, 1994). Although technology is available to treat such water to meet or exceed drinking water standards (see Chapter 4), members of the public may remain skeptical of such claims (Haddad et al., 2010). The history of water matters to many people more than the type and concentrations of impurities remaining in the water. This can result in a public preference for lower quality water emerging from a "natural" aquifer or river over higher quality water emerging directly from an advanced wastewater reclamation facility.

The research field of judgment, risk perception, and decision making is well established (Kahneman et al., 1982; Slovic, 1987, 1993; Slovic et al., 2002, 2004). Surveys and experiments have shown that people often connect perceived benefits of an activity with their evaluation of its risk: the more they think they will benefit, the lower they consider its risk. This approach is different from a scientific evaluation of risk, which would not consider the benefits in any quantitative risk assessment. Thus, there is a predisposition among those who dislike water reuse to believe it puts them at risk.

Willingness to use reclaimed water is, in part, a function of the intended use, with willingness higher for uses that minimize human contact, including irrigation, car washing, and other cleaning (Bruvold, 1988; Hills et al., 2002; Dolnicar and Schäfer, 2009; Hurlimann and Dolcinar, 2010). In a nationwide survey of attitudes toward potable reuse, Haddad et al. (2010) reported that 38 percent said they would be willing to drink "certified safe recycled water," 49 percent were uncertain, and 13 percent said they would refuse to drink the water. This result, especially the small but not insignificant number of individuals who initially say they would refuse such water, is consistent with the reported experience of water agencies that have proposed water reuse projects. The survey showed few demographic or geographic differences in attitudes toward potable reuse. However, studies outside the United States have found weak but significant demographic differences in water-related risk perception (Po et al., 2003; Hurlimann, 2008; Doria, 2010). Hurlimann (2008), for example, found that males, people older than age 50, and people with college degrees were more willing to use reclaimed water for personal uses (including showering, clothes washing, drinking).

A general criticism of this line of research is that it does not analyze actual behavior and use of reclaimed water but instead focuses on the stated intentions of respondents. Saying one is willing to reuse water in the hypothetical is not the same as actually doing so, according to Mankad and Tapsuwan (2011), who call for more research on communities already using decentralized water reuse systems (e.g., residence-scale reuse).

Part of the challenge of public acceptance of water reuse hinges on perception of the origins of the water and whether it can be considered "natural" (see also discussion of environmental buffers in Chapter 2). Survey results showed that individuals' trust in the water as a supply for drinking improved if the reclaimed water is passed through systems perceived to be natural. Aquifer storage for 10 years was favored over aquifer storage for 1 year, and passing water down a swift-flowing river for 100 miles was preferred over passing water down a 1-mile stretch. Aquifer storage overall was preferred to passage down a river (Haddad et al., 2010).

According to Haddad et al. (2010), local independent (e.g., university) scientists are viewed by the public as the most credible sources of information on reclaimed water (see Table 10-6), because they combine topical expertise and knowledge of the local situation and have no professional stake in water management decisions. Dolnicar and Hurliman (2009), in qualitative interviews, found friends and relatives to be the most trusted sources of information on whether to drink reclaimed water. However, those negatively predisposed to potable reuse were least willing to be convinced of its efficacy by anyone, although relative rankings of trusted sources were generally consistent among all respondents regardless of their willingness to drink reclaimed water (Haddad et al., 2010).

Public Communication

The choice of words matters when describing water reuse. Menegaki et al. (2009), studying farming behaviors on the Island of Crete, identify differences in farmers' willingness to pay for reclaimed water based on whether it is called "recycled water" or "treated wastewater." Haddad et al. (2010) found that even individu-

als who were strongly opposed to indirect potable reuse could be influenced by paragraphs that cast water reuse in a positive light. Macpherson and Slovic (2011) found that the water reuse profession does not have standard definitions for commonly used technical terms, and this causes confusion among customers. They have generated a glossary of terms and advocate that the profession adopt it as standard terms and definitions.

The sophistication of communication between water agencies and the public continues to evolve (Box 10-9). There is more public outreach, including visitor centers and tours at water reclamation facilities, more Web sites, and better communications with regional political leaders and media outlets. Surveys in Australia by Dolnicar et al. (2010) and in Barcelona, Spain, by Domenech and Sauri (2010) found that knowledge of the water treatment process increased acceptance of water reuse. One often cited example of public relations success is Singapore's NEWater Facility, which invested extensively in a visitor center. Positive media coverage of water reuse in Singapore compared with Australia is also recognized as a factor influencing the success of water reuse (Ching and Yu, 2010). However, it is difficult to ascertain if the absence of domestic opposition to the NEWater program is because of the successful visitor center, positive press coverage, cultural differences, national policies that limit civic discourse, or all of these reasons. In the United States, tours of water reuse facilities are common, but to date, research has not been undertaken to link tours

TABLE 10-6 Trusted Source of Information on Reclaimed Water Safety: Overall and by Willingness to Drink "Certified Safe Recycled Water" on a Scale of 0-10

	Overall[a]	Unwilling[b]	Uncertain[b]	Willing[b]
An actor or athlete you admire hired to represent the water treatment facility	2.14	1.05	1.79	2.54
Your neighbor	3.20***	2.30	2.83	3.64
A private firm hired by the water treatment facility	4.11***	2.55	3.40	4.87
The manager of the water treatment facility	4.62***	3.00	4.07	5.27
Staff of the water treatment facility	4.67	3.32	4.00	5.36
A doctor who lives nearby	4.68	3.65	4.00	5.33
Someone who has drunk reclaimed water for years	5.06**	3.18	4.60	5.74
A board made up of engineers and other representative of the community	5.70***	3.48	5.05	6.58
Engineers/inspectors from the federal government	5.88	3.78	5.02	6.85
Engineers/inspectors from the state government	5.95	4.02	5.09	6.86
A qualified scientist from a nearby university	6.59***	5.15	6.25	7.08

[a]The items are arranged from top to bottom in terms of increasing trust for the full sample (overall). Asterisks indicate that the value is significantly different from the item immediately above it. * = $p < .05$, ** = $p < .01$, *** = $p < .001$

[b]By willingness: ANOVAs on all rows for trust as a function of membership in the three groups are significant at $p < .001$.

SOURCE: Haddad et al. (2010).

BOX 10-9
Lessons Learned on Public Communication and Involvement in Redwood City, California

Redwood City, located in the San Francisco Bay area, has 75,000 residents. By 2000, the city was exceeding its assured supply of 11 MGD (41,000 m^3/d) from the Hetch Hetchy regional system, with demand projected to increase. After a study of supply alternatives, the city in 2003 settled on water conservation and water reclamation and reuse (supplying 1.8 MGD [6,800 m^3/d]). In an otherwise politically active community, only two individuals attended a mandatory public meeting on environmental impacts held in 2002. These two individuals then formed the Safewater Coalition, which objected to use of reclaimed water for landscape irrigation in residential areas and in schoolyards, playgrounds, and parks. The Safewater Coalition focused public attention on the project, effectively using the Internet and local media. The Redwood City Recycled Water Task Force was then formed, with equal balance of membership in favor and opposed to the project, and tasked to find 1.8 MGD in water reuse and/or additional water conservation. After 5 months of deliberation, the Task Force recommended and the City Council approved a plan that addressed some of the Safewater Coalition's concerns. The Task Force plan would rely on 1.6 MGD water reuse and an additional 0.2 MGD in water conservation, including artificial turf on the playing fields.

Lessons from Redwood City focus more on tactics of public communications than on fundamental changes to project review and approval. The Redwood City experience highlights the importance of public acceptance of a project *in addition to* completion and certification of formal environmental impact reviews. In the case of Redwood City, which echoed the experience of Los Angeles and San Diego in the 1980s, opposition to a proposed reuse project did not emerge until very late in the formal review process. Additionally, the project exemplifies the capacity of a very small group of people (as few as one in the case of Redwood City) to impact a project's progress and the power of the Internet as an organizing tool and source of information (and sometimes misinformation) on a proposed project. A public vote against a proposed water reuse facility in Toowoomba, Australia, also appears to have hinged on the actions of one citizen who adamantly opposed the project (van Vuuren, 2009). Water agency personnel were not, at first, prepared to respond with trusted sources of information for the community to address the Coalition's claims. The Redwood City case also highlights the importance of extensive ongoing public communication on water issues in urban areas. Water is no longer a behind-the-scenes question of infrastructure development, implementation, and financing. It is now an issue of immediate and active public concern.

Today, the Redwood City Recycled Water Project is considered to be successful and is supported by the community. In late 2002, it was perceived to be held up by a small, determined group. It represents the transition of water agencies into the current era of savvy communication between water agencies, the public, and political leaders.

SOURCES: Ingram et al. (2006); M. Milan, Data Instincts, personal communication, 2009.

---

and other improvements in public communication with achievement of other goals (e.g., maintaining or increasing public trust in the water supply, public support for investments in water infrastructure).

There are many reasons why a major infrastructure project gets delayed or canceled. Public perception that water produced from a water reclamation facility is objectionable could be one, but public perception may not be determinative. Rather, a richer understanding of the social, technical, procedural, and policy-related aspects of a particular proposal may be the more reliable determinant of whether a project proceeds (Russell and Lux, 2009). Marks and Zadoroznyj (2005) identify institutional and knowledge factors, including the extent of social capital (e.g., homeowners associations), accountability of water managers for promised water quality, public awareness of environmental problems and the benefits of water reuse, and public trust in reclaimed water and water managers as crucial to the

success of water reuse projects. Similarly, Stenekes et al. (2006), also writing in the Australian context, propose that a more productive public engagement is needed, including a better public understanding of the cost of water, greater participation of the public in water planning, and institutional reforms that would clear the way for water agencies to pursue more sustainable water technologies and strategies. Public perception and agency–public communications matter but should be understood in a larger economic, procedural, and governance context.

CONCLUSIONS

Water rights laws, which vary by state, affect the ability of water authorities to reuse wastewater. States are continuing to refine the relationship between wastewater reuse and the interests of downstream entities. Regardless of how rights are defined or assigned, projects can proceed through the acquisition of water rights after water rights have been clarified. The right to use aquifers for storage can be clarified by states through legislation or court decision. The clarification of these legal issues can provide a clearer path for project proponents.

Scientifically supportable risk-based federal regulations for nonpotable water reuse would provide uniform nationwide minimum acceptable standards of health protection and could facilitate broader implementation of nonpotable water reuse projects. Existing state regulations for nonpotable reuse are developed at the state level and are not uniform across the country. Further, no state water reuse regulations or guidelines for nonpotable reuse are based on rigorous risk assessment methodology that can be used to determine and manage risks. EPA has published suggested guidelines for nonpotable reuse, which are based, in part, on a review and evaluation of existing state regulations and guidelines and are not based on rigorous risk assessment methodology. Federal regulations would not only provide a uniform minimum standard of protection, but would also increase public confidence that a water reuse project does not compromise public health. Scientific research, which requires resources beyond the reach of most states, should inform the development of nonpotable reuse regulations at the federal level to address the wide range of potential nonpotable reuse applications and practices. If federal regulations were developed through new enabling legislation, individual states would maintain the authority to impose more stringent criteria at their discretion. Therefore, EPA should fully consider the advantages and disadvantages of federal reuse regulations to the future application of water reuse to address the nation's water needs while appropriately protecting public health.

Modifications to the structure or implementation of the SDWA would increase public confidence in the potable water supply and ensure the presence of appropriate controls in potable reuse projects. Although there is no evidence that the current regulatory framework fails to protect public health when planned or de facto reuse occurs, federal efforts to address potential exposure to wastewater-derived contaminants will become increasingly important as planned and de facto potable reuse account for a larger share of potable supplies. The SDWA was designed to protect the health of consumers who obtain potable water from supplies subject to many different sources of contaminants but does not include specific requirements for treatment or monitoring (see Chapters 4 and 5) when source water consists mainly of municipal wastewater effluent. Presently, many potable reuse projects include additional controls (e.g., advanced treatment and increased monitoring) in response to concerns raised by state or local regulators or the recommendations of expert advisory panels. Adjustment of the SDWA to consider such requirements when planned or de facto potable reuse is practiced could serve as a mechanism for achieving a high level of reliability and public health protection and nationwide consistency in the regulation of potable reuse. In the process, public confidence in the federal regulatory process and the safety of potable reuse would be enhanced.

Application of the legislative tools afforded by the CWA and SDWA to effluent-impacted water supplies could improve the protection of public health. Increasingly, we live in a world where municipal effluents make up a significant part of the water drawn for many water supplies, but this is not always openly and transparently recognized. Recognition of this reality necessitates increased consideration of ways to apply both the CWA and the SDWA toward improved drinking water quality and public health. For example, the CWA allows states to list public water supply as a designated use of surface waters. Through this mechanism, some states have set up requirements on discharge of contaminants that could adversely affect downstream water supplies.

Updates to the National Pretreatment Program's list of priority pollutants would help ensure that water reuse facilities and de facto reuse operations are protected from potentially hazardous contaminants. The National Pretreatment Program has led to significant reductions in the concentrations of toxic chemicals in wastewater and the environment. However, the list of 129 priority pollutants presently regulated by the National Pretreatment Program has not been updated since its development more than three decades ago, even though the nation's inventory of manufactured chemicals has expanded considerably since then, as has our understanding of their significance. Updates to the National Pretreatment Program's priority pollutant list can be accomplished through existing rulemaking processes. Until this can be accomplished, EPA guidance on priority chemicals to be included in local pretreatment programs would assist utilities implementing potable reuse.

Enhanced public knowledge of water supply and treatment are important to informed decision making. The public, decision makers, and decision influencers (e.g., members of the media) need access to credible scientific and technical materials on water reuse to help them evaluate proposals and frame the issues. A general investment in water knowledge, including improved public understanding of a region's available water supplies and the full costs and benefits associated with water supply alternatives, could lead to more efficient processes that evaluate specific projects. Public debate on water reuse is evolving and maturing as more projects are implemented and records of implementation are becoming available.

11

Research Needs

This report has examined key challenges and opportunities for water reuse as an approach to meet the nation's future water needs, and research will be needed to address many of the challenges ahead. In this chapter, the committee identifies key research needs that are not currently being addressed in a major way. These research areas hold significant potential to advance the safe, reliable, and cost-effective reuse of municipal wastewater where traditional sources are inadequate. This chapter also includes a discussion of the current roles of federal agencies and nongovernmental organizations (NGOs) in supporting reuse-related research, because these same entities could play a role in supporting the committee-identified research needs.

RESEARCH PRIORITIES

In the committee's review of a wide range of issues affecting the application of nonpotable and potable reuse, the committee did not identify any technological hurdles that were holding back the application of reuse to address local water supply needs. In fact, in its review of water reclamation technologies (see Chapter 4), the committee found the state of technology to be quite advanced, with room for improvements but no major limitations to their use. However, additional research could enhance the performance and quality assurance of existing processes and help address public concerns over the safety of reuse to human health and the environment.

Overall, the committee organized the proposed 14 priority research areas within two broad categories:

1. Health, social, and environmental issues
2. Performance and quality assurance

The topics are identified in Box 11-1 and are described in more detail in this chapter. The issues are not listed in order of priority.

Human Health, Social, and Environmental Issues

1. Quantify the extent of de facto (or unplanned) potable reuse in the United States.

Although population density has increased substantially in parts of the country with limited water resources, a systematic analysis of the contribution of municipal wastewater effluent to potable water supplies has not been made in the United States for over 30 years. The lack of such data impedes efforts to identify the significance and potential health impacts of de facto water reuse. Because new water reuse projects could decrease the volume of wastewater discharged to water sources where de facto reuse is being practiced, the lack of understanding of the contribution of wastewater effluent to water supplies restricts our ability to assess the net impact of future water reuse on the nation's water resource portfolio. Available hydrological modeling and monitoring tools would enable an accurate assessment of de facto water reuse. Ideally, these efforts would take advantage of existing monitoring networks (e.g., U.S. Geological Survey [USGS] streamflow gauging stations), data on wastewater effluent discharges submitted by National Pollutant Discharge Elimination Sys-

BOX 11-1
Summary of Research Priorities

These research areas hold significant potential to advance the safe, reliable, and cost-effective reuse of municipal wastewater where traditional sources are inadequate. They are not prioritized here.

Health, Social, and Environmental Issues

1. Quantify the extent of de facto (or unplanned) potable reuse in the United States.

2. Address critical gaps in the understanding of health impacts of human exposure to constituents in reclaimed water.

3. Enhance methods for assessing the human health effects of chemical mixtures and unknowns.

4. Strengthen waterborne disease surveillance, investigation methods, governmental response infrastructure, and epidemiological research tools and capacity.

5. Assess the potential impacts of environmental applications of reclaimed water in sensitive ecological communities.

6. Quantify the nonmonetized costs and benefits of potable and nonpotable water reuse compared with other water supply sources to enhance water management decision making.

7. Examine the public acceptability of engineered multiple barriers compared with environmental buffers for potable reuse.

Treatment Efficiency and Quality Assurance

8. Develop a better understanding of contaminant attenuation in environmental buffers.

9. Develop a better understanding of the formation of hazardous transformation products during water treatment for reuse and ways to minimize or remove them.

10. Develop a better understanding of pathogen removal efficiencies and the variability of performance in various unit processes and multibarrier treatment and develop ways to optimize these processes.

11. Quantify the relationships between polymerase chain reaction (PCR) detections and viable organisms in samples at intermediate and final stages.

12. Develop improved techniques and data to consider hazardous events or system failures in risk assessment of water reuse.

13. Identify better indicators and surrogates that can be used to monitor process performance in reuse scenarios and develop online real-time or near real-time analytical monitoring techniques for their measurement.

14. Analyze the need for new reuse approaches and technology in future water management.

---

tem permit holders, and hydrological models developed to study watersheds with historical concerns about the impact of effluent discharges on water quality. These efforts could be updated periodically (e.g., every 5 to 10 years) to provide decision makers with an understanding of the role of de facto reuse in the nation's potable water supply. Furthermore, an improved understanding of de facto potable reuse could spur the development and/or application of contaminant prediction tools or lead to enhanced monitoring programs that could increase public health protection.

2. Address critical gaps in the understanding of health impacts of human exposure to constituents in reclaimed water.

Potential health impacts resulting from long-term, low-level exposure to chemicals and mixtures of chemicals present in wastewater effluent have yet to be fully elucidated. It would be expensive and time-consuming to conduct batteries of in vitro and in vivo toxicity studies on all of the different chemicals in reclaimed water. However, a carefully planned research effort would be useful to inform future decisions about potable water reuse. In particular, there is a need to fill in data gaps in existing toxicological databases with respect to contaminants that are known to occur in wastewater and persist in the environment and are refractory in water reclamation and water treatment processes. The risk exemplar (Chapter 7) highlights several of these chemicals, including nitrosamines, disinfection byproducts, hormones, certain pharmaceuticals, antimicrobials, flame retardants, and perfluorochemicals. As noted in Chapter 6, there is also a need to assess the importance of indirect pathways of exposure to constituents in reclaimed water, such as bioaccumulation of trace organic chemicals in food crops.

3. Enhance methods for assessing the human health effects of chemical mixtures and unknowns.

Concerns about the health effects of unknown chemicals and contaminant mixtures remain a major challenge in public and political acceptance of water reuse. Additional research is needed to further develop in vivo and in vitro bioassay methods that can be used to rapidly and selectively screen the product water from water reclamation facilities for possible physiological effects. Improved rapid bioassays could also help in the prioritization of those chemicals, or chemical mixtures, which may necessitate longer term in vivo testing.

4. Strengthen waterborne disease surveillance, investigation methods, governmental response infrastructure, and epidemiological research tools and capacity.

Despite the frequency of acute gastrointestinal infections (AGIs) worldwide and in the United States and public concern over chemical contamination of public and private water supplies, the ability of the public health sector and the research community to attribute disease to water consumption remains problematic. Attributing waterborne disease outbreaks to a source or treatment practice will only become more difficult with the growing complexity of drinking water sources, including reclaimed water. There is no national public health epidemiological research program dedicated to tracking endemic water-associated AGI community disease trends or comparative health impacts of differing water reuse patterns. There is little public health response capacity until disease reaches epidemic outbreak status, when generic public health outbreak investigation resources become available. As water reuse increases in scope and volume, methods and expertise to determine whether AGIs are waterborne or whether community chronic health disparities are related to water reuse will be important to maintaining public acceptance of reuse practices and should be the focus of research partnerships. Disease and exposure surveillance tools, investigation practices, and human health outcomes research need to be improved and strengthened.

5. Assess the potential impacts of environmental applications of reclaimed water in sensitive ecological communities.

Reclaimed water has many potential uses for habitat restoration, but a need exists to better understand the impact of wastewater-derived contaminants in purposeful ecological enhancement projects. Many scientific studies of surface water impacts associated with municipal effluent discharges have been undertaken,

although few have focused solely on purposeful restoration projects. The location and site-specific attributes associated with the restoration project will determine the extent of the research needs, but only through several site-specific analyses can the range of potential issues be fully understood. Conventional (e.g., whole effluent toxicity) testing and risk paradigms are available, but a need exists to further develop rapid screening methodologies. Research related to purposeful ecological enhancement with reclaimed water might lead to more successful habitat restoration projects.

6. Quantify the nonmonetized costs and benefits of potable and nonpotable water reuse compared with other water supply sources to enhance water management decision making.

When making major water management decisions and weighing various competing water supply alternatives, communities and decision makers must evaluate many factors (e.g., life-cycle costs, environmental costs and benefits, public acceptance, supply reliability, water system independence) in addition to traditional financial costs. However, a full understanding of these costs and benefits is rarely available. Quantification of environmental costs and benefits, for example, should include impacts on surface water flows and ecosystems, nutrients, and greenhouse gas emissions. Although these costs and benefits are inherently site specific, a synthesis of such analyses across a number of facilities and conditions could inform broader discussions of water management alternatives. Additionally, an evaluation of existing tools that planners and water managers could use to integrate these various costs and benefits into overall project analysis would help support and better inform water management decisions.

7. Examine the public acceptability of engineered multiple barriers compared with environmental buffers for potable reuse.

As described previously in this report, environmental buffers have been an important aspect of almost all successful potable reuse projects because of particular functions they serve toward contaminant attenuation, retention, and/or blending (see Chapter 5) and because some buffers (e.g., groundwater injection) serve to disassociate reclaimed water from its source in the minds of the public (see Chapter 10). However, from a technical perspective, the public health protection that natural systems provide is often not well defined. Recent research has shown that engineered barriers can provide equivalent or superior levels of protection compared with some environmental buffers currently in use. Research is needed to understand the public acceptability of engineered buffers compared with environmental buffers used for potable reuse.

Treatment Efficiency and Quality Assurance

8. Develop a better understanding of contaminant attenuation in environmental buffers.

Research on how well different environmental buffers function under various conditions, their potential weaknesses, and their impacts on water quality is crucial to the optimization of potable reuse systems and future decisions about their design. Some researchers have examined the performance of soil aquifer treatment systems in the southwestern United States, but the performance of such systems under other hydrogeological conditions is poorly understood. Information on contaminant attenuation in wetlands, rivers, and reservoirs is also lacking.

9. Develop a better understanding of the formation of hazardous transformation products during water treatment for reuse and ways to minimize or remove them.

As described in Chapter 3, wastewater contains a rich mixture of organic constituents, and during disinfection and other treatment processes, some hazardous transformation products are formed. Continued research is needed to understand the precursors of hazardous transformation products and how precursor chemicals can be better managed to reduce the formation of hazardous chemicals. *N*-Nitrosodimethylamine (NDMA) is a particularly challenging disinfection by-product that merits additional research, because it poses a risk for cancer at very low concentrations (0.7 ng/L) and potable reuse projects frequently require expensive and energy-intensive additional treatment to remove it. Research on transformation products is important for

enhancing the safety of water reuse scenarios, including de facto reuse.

10. Develop a better understanding of pathogen removal efficiencies and the variability of performance in various unit processes and multibarrier treatment and develop ways to optimize these processes.

Because health effects can result from a single exposure to a pathogen, the variability in pathogen occurrence and removal during wastewater reclamation and distribution processes should be better understood to capture the overall variability in exposure and risk. Data developed from careful monitoring across processes in full-scale installations and showing the variations in pathogen densities over time would serve as an important database for project design. Because low levels of pathogens remain toward the end of treatment, indicator organism monitoring may be needed to assess the variability in pathogen removal. Research is needed to better understand how changes in process design and operation affect the removal of pathogens (and indicators) to develop more efficient ways to reduce risks from microorganisms in treatment systems.

11. Quantify the relationships between polymerase chain reaction detections and infectious organisms in samples at intermediate and final stages.

With the increasing use of molecular biological methods such as quantitative *polymerase chain reaction* (qPCR) for pathogen enumeration in environmental samples, occurrence data are being obtained in terms of genome copies per unit water volume (e.g., gc/L). However, for risk assessment, dose-response relationships are generally based on number of viable pathogens (e.g., colony-forming units, plaque-forming units) in a dose. The percentage of genome copies that represent viable (or infectious) units is likely to degrade during treatment and exposure to the environment, especially during exposure to oxidizing disinfectants. Therefore, to use qPCR data with more confidence in risk assessments of pathogens and in the control of advanced treatment systems, reliable data on the ratio and variability in the ratio of genome copies/viable pathogens are needed for various types of waters (e.g., source, partially treated, completely treated). Alternatively, another means is needed for quantifying infectious pathogens that cannot be grown in conventional media.

12. Develop improved techniques and data to consider hazardous events or system failures in risk assessment of water reuse.

The committee developed its risk exemplar to compare the relative risks of conventional and a de facto water reuse scenarios (see Chapter 7), but this analysis did not consider the impacts of hazardous events (e.g., earthquakes, hurricanes, disease outbreaks) or major equipment failures. Ideally, risk assessments would address these factors and include techniques for quantitative analysis of both the likelihood and consequences of specific hazardous events in order to quantify the risks. However, the data to support such an analysis are not widely available. Improved techniques and data could also facilitate increased incorporation of quality assurance strategies into treatment plant design and operation (see Chapter 5). Additionally, the level of quality assurance necessary for public health protection needs to be better defined so that potable reuse systems can be designed to provide it, with or without environmental buffers.

13. Identify better indicators and surrogates that can be used to monitor process performance in reuse scenarios and develop online real-time or near real-time monitoring techniques for their measurement.

It remains impractical to use direct measurements of most contaminants to assess actual performance of individual processes and process sequences. Therefore, development and application of surrogate and/or indicator measurements (see Chapter 5) are needed that could be used to assess the performance of individual water reclamation processes. Indicators are individual chemicals or microorganisms that represent the characteristics of other trace organic contaminants or microorganisms of concern, particularly their removal through the specific process(es) where they are measured. A surrogate is a quantifiable change of a bulk parameter that can be continuously monitored and that correlates with contaminant removal. Development of real-time or near-real time monitoring techniques, particularly for contaminants with acute effects, such as

microorganisms, could reduce the post-treatment storage capacity needed to ensure quality in potable reuse projects and could reduce the extent of contamination and potentially the exposure duration in the event of process failures.

14. Analyze the need for new reuse approaches and technology in future water management.

A review of the history of wastewater management in the United States (see Chapter 2) reveals that water reuse began as a strategy to dispose of large volumes of liquid waste generated in densely populated areas. More recently, reuse also has evolved to address local water demands, but largely working within the framework of an existing wastewater infrastructure designed in the early to mid-20th century. These existing wastewater infrastructure designs constrain water reuse in a number of ways. The strategy of draining wastewater from urban areas by gravity and managing water quality at the point of discharge to a receiving stream has favored the establishment of large centralized wastewater treatment plants. The location of these treatment plants limits the options for water reuse because large dedicated conveyance systems are costly and difficult to implement in existing urban settings, particularly when potential users are not located close to water reclamation facilities. An additional constraint on reuse is that only one quality of effluent is typically produced from wastewater treatment plants, even though potential users may have widely ranging quality requirements. Considering existing treatment train designs and site constraints, many of these existing wastewater treatment plants are not easily adaptable to the production of high-quality reclaimed water for reuse. Meanwhile, core elements of the infrastructure that embeds both water and wastewater treatment, storage, and conveyance were developed and designed during a time of inexpensive energy, smaller urban populations, and little appreciation of the need for aquatic habitat protection and control of greenhouse gas emissions (Daigger, 2009). The interdependency of water and energy has been mostly neglected, and the existing water infrastructure is rather energy intensive (e.g., water conveyance systems, need for pumping, energy-intensive treatment processes).

Many of these water and wastewater systems are

now reaching the end of their design life, and EPA has estimated that between $320 billion and $450 billion will need to be invested in wastewater infrastructure between 2002 and 2020 in the United States. Estimates of capital needs for drinking water infrastructure range from $178 billion to $475 billion (EPA, 2002). Thus, questions arise as to whether the water and wastewater infrastructure of the future will be (or should be) vastly different from that of today, and if so, what is the role of water reuse? Although this question is beyond the scope of the committee's charge, there are several important questions based on future population scenarios and future water and wastewater infrastructure designs, whose answers will affect research priorities and the generation of future technologies.

- What are the water quality implications of expanded reuse, including de facto reuse, under future population scenarios,[1] considering that contributions of wastewater in receiving streams are likely to increase under current population projections and migration trends?
- What are the implications of increased water conservation on the potential contribution of water reuse, and how will the likely associated increase in salinity and other effects on water quality affect water reuse applications?
- What are the water budget implications of various types of reuse, considering growing urban centers?
- How can future water reclamation plants be designed (or existing plants upgraded) to better take advantage of potential opportunities for water reuse?
- What advances in technology are needed to support reuse to address future water needs?
- What is the role of distributed wastewater treatment and reuse in future water management?
- What technologies can be applied to water reclamation so that new plants can recover energy and use resources most efficiently?

Additional research is needed to address these questions so that water reuse facilities constructed during this decade can provide appropriate benefits in the

[1] It is estimated that by 2030, 86% of the U.S. population will live in urban centers (U.S. Census, 2008).

decades to come, while contributing to efficient use of water and energy resources.

FEDERAL AND NONFEDERAL ROLES

As the nation seeks to meet its water needs through new water supply approaches, such as water reuse, Congress and the executive branch are increasingly asking what the federal government role should be (Cody and Carter, 2010). At present, as discussed in Chapter 10, the federal presence is primarily directed toward regulation of wastewater discharges, injection of reclaimed water, and regulation of drinking water. Various reuse projects have benefited from federal funding, perhaps through Title XVI (see Box 9-1), which is generally limited to the 17 western continental states, or as earmarks in congressional budgets. The federal EPA has administered programs for funding municipal wastewater treatment facilities in the past, and administers a revolving loan fund for these purposes. The question of the appropriateness of federal funding for water supply projects is currently being debated in Congress and the administration of the executive branch (Cody and Carter, 2010) and is not a question that this committee is appropriately constituted to resolve. Instead, the committee reviewed the research programs supported by both federal and nonfederal entities and discusses in this section appropriate roles to address the above research needs.

Federal Agency Reuse Research

There is no single lead federal agency on water reuse–related research. Seven federal agencies provide at least some research funding for water reuse: the U.S. Bureau of Reclamation (USBR), USGS, EPA, the U.S. Department of Agriculture (USDA), the Centers for Disease Control and Prevention (CDC), the Department of Energy (DOE), and the National Science Foundation (NSF).

USBR

USBR is the only federal agency with a specific directive to address water reuse–related issues, and it provides the largest amount of funding for water reuse–related research via several programs. In particular, between 2000 and 2011, the USBR provided $17 million in research funding to the WateReuse Research Foundation, through the Title XVI program (see Box 9-1), which was used to support research projects and workshops on microbial and trace organic contaminants, treatment technologies, salinity management, and social and institutional issues such as public perception, economics, and marketing. Additional programs, such as the Secure Water Act (Public Law [P.L.] 111-11, Subtitle F, enacted in 2009), which was intended to "accelerate the adoption and use of advanced water treatment technologies to increase water supply," the Rural Water Act of 2006 (P.L. 109-451), and the Water Desalination Act of 1996 (P.L. 104-298), provide some support for reuse-related research. The USBR estimates that approximately 5 percent of the research projects funded under the Water Desalination Act were specifically targeted toward water reuse, although some of the desalination research has relevance to reuse applications (C. Brown, USBR, personal communication, 2009; Kevin Price, USBR, personal communication, 2011).

EPA

EPA has many ongoing efforts that are relevant to reuse, although like most of the federal agencies discussed in this section, the agency has no specific directive driving research in water reuse. Water reuse, however, is relevant to many of the agency's cross-cutting interests—particularly at the nexus of water availability and water quality. EPA has an extensive research program on human health effects of chemicals (using screening and laboratory studies) and pathogens (using epidemiological data). Through the Unregulated Contaminant Monitoring Rule (UCMR) program described in Chapter 10, EPA supports research on analytical methods, monitoring, and treatment efficacy and conducts extensive data analysis on the occurrence of contaminants. It supports research to understand the human health and environmental effects of endocrine-disrupting chemicals at environmentally relevant concentrations. Research is also under way on pathogen monitoring, sampling, and analysis (A. Levine, EPA, personal communication, 2010).

USGS

USGS maintains an extensive water research program although there is no specific water reuse–related directive within the agency. Three areas of ongoing research with relevance to water reuse include the Water Census, aquifer storage and recovery (ASR), and wastewater-derived chemicals in the aquatic environment. The Water Census is an updated and expanded approach to prior efforts by USGS to account for water supplies and water use in the United States, including precipitation, evaporation, groundwater recharge, storage, water withdrawals, consumptive uses, return flows, and ecological needs. ASR research under way seeks to understand changing geochemistry associated with subsurface storage of water (which may or may not include reclaimed water). USGS has also conducted extensive research on the occurrence of human-use compounds in the nation's surface waters and has the measurement capabilities to detect an extensive array of human-use compounds in water and sediment. Research is currently under way to better understand the occurrence, pathways, uptake, and effects of these human-derived contaminants (J. Bales, USGS, personal communication, 2010).

USDA

In recent years, USDA has developed a strong interest in water reuse as a means to provide reliable supplies of water for irrigation in areas where water is scarce. They have cosponsored two conferences (2007, 2008) with the WateReuse Research Foundation on Agricultural Water Reuse, and starting in 2007 began funding research on water reuse in agriculture. Through its National Institute of Food and Agriculture, USDA has distributed grants for research on minimizing food safety hazards, understanding pharmaceuticals and hormones in agricultural production, impacts of reclaimed water on plants and soils, treatment methods to prevent impacts to soils, and long-term effects of irrigating with reclaimed water. It is also collecting information on the extent of the use of reclaimed water in irrigation in an annual inventory of farms conducted by its National Agricultural Statistics Service (J. Dobrowolski, USDA, personal communication, 2010).

CDC

Although CDC has no specific directive in water reuse, the agency is interested in the issues from a number of perspectives, particularly in its National Center for Environmental Health and its Division of Emergency and Environmental Health Services. CDC has supported two research efforts on the subject: an analysis of reuse as a means to protect human health during drought conditions and a research project to enhance capacity to investigate the link between wastewater, groundwater contamination, and human health (M. Zarate-Bermudez, CDC, personal communication, 2010).

DOE

DOE's National Energy Technology Laboratory is conducting research on ways to reduce water demand associated with energy production. Specific to municipal wastewater reuse, DOE is conducting research on the technical issues associated with using reclaimed wastewater for power plant cooling, on costs and benefits of various levels of reclaimed water treatment, and analyses of ongoing use of reclaimed water for this purpose.

NSF

NSF sponsors approximately 20 percent of the water resources research in the United States (NRC, 2004), although it has no specific funding emphasis on water reuse. However, water reuse-related research may be funded under related initiatives or under a new initiative on water sustainability and climate. For example, improved technology for water reuse is a focal area for an NSF-funded center on water treatment technology (the Center of Advanced Materials for the Purification of Water with Systems [WaterCAMPWS]) (B. Hamilton, NSF, personal communication, 2010).[2] NSF has also recently funded an engineering research center on reinventing the nation's urban water infrastructure (ReNWUIt) that will bring together researchers from environmental engineering, earth sciences, hydrology,

ecology, urban studies, economics, and law. The center is funded with $18.5M over the next 5 years.

Other Federal Interests in Reuse

Several federal agencies have interests in reuse, although they are not currently sponsoring research to support it. For example, the U.S. Agency for International Development (USAID) has major interests in water management and in water and sanitation for health in developing countries. USAID has sponsored projects to implement nonpotable water reuse projects in Morocco and Jordon. It anticipates that water reuse will become an increasingly important part of water management in water-poor nations, particularly as part of efforts to enhance food security during droughts (J. Franckiewicz, USAID, personal communication, 2010). Large military installations of the Department of Defense may have their own wastewater treatment plants and may practice nonpotable reuse to maximize their available water resources.

NGO-Sponsored Research

WateReuse Research Foundation

The mission of the WateReuse Research Foundation is to conduct and promote applied research on the reuse, reclamation, recycling, and desalination of water. The foundation provides $2–$4 million per year to support research, with a significant portion coming from the USBR through the Title XVI program. Between 2000 and 2011, the WateReuse Research Foundation used the $17 million funding from USBR to leverage $41 million in research, through additional contributions from state and local agencies, the private sector, universities, and others (K. Price, USBR, personal communication, 2011). Supported research categories include policy and social sciences, microbiology and disinfection, chemistry and toxicology, and treatment technologies. They also conduct periodic analysis of research needs in the area of water reuse (W. Miller, WateReuse Research Foundation, personal communication, 2010).

---

[2] See http://www.watercampws.uiuc.edu/.

Water Research Foundation

The Water Research Foundation (formerly known as the American Water Works Association Research Foundation) is a member-supported NGO established to support applied research related to drinking water. Although reuse-specific projects represent a small fraction of their overall research portfolio, the Water Research Foundation has sponsored research on SAT in water reuse projects. The foundation has recently committed up to $1 million per year for at least 5 years to research on trace organic contaminants (e.g., pharmaceuticals, personal care products) in drinking water, including assessment of exposure, improvements in analytical methods, and improved frameworks for risk communication for utilities (S. Cline, Water Research Foundation, personal communication, 2009).

National Water Research Institute

The National Water Research Institute (NWRI) supports scientific research and outreach efforts related to ensuring clean and reliable water. NWRI has six member organizations, all based in Southern California, with strong interests and vast ongoing efforts in water reuse. Since its founding in 1991, NWRI has invested over $17 million in research. Funded research topics have included disinfection guidelines for water reuse, the fate and transport of trace organic contaminants, subsurface transport of bacteria and viruses, and use of bioassays and monitoring to assess trace contaminant removal in water reuse.[3]

Water Environment Research Foundation

The Water Environment Research Foundation (WERF) is a subscriber-based organization focused on wastewater- and stormwater-related research. In general, WERF applies only a small portion of its research funding to projects that are directly focused on the reuse of municipal wastewater, but it has funded studies on public perception of water reuse and attenuation of trace organic contaminants in landscape irrigation. The organization is also interested in research on the reuse

of stormwater and greywater (D. Woltering, WERF, personal communication, 2010).

Coordination to Support Needed Research

The research needs identified in Box 11-1 cannot be addressed by a single organization or agency, because collectively, they rely on expertise that is distributed among agencies and universities. However, the agencies and NGOs with interest in reuse could collectively work to address these research needs, with improved coordination. As described in the previous sections, at least seven federal agencies and three NGOs are conducting or supporting research related to water reuse. Of these, two federal agencies (USBR and EPA) and the NGOs represent the lead contributors to water reuse–related research. This speaks to the need for improved coordination to see that these research needs are addressed.

Under the current research funding framework, the bulk of the water reuse research is focused on near-term research priorities, largely dominated by particular agency interests or issues of concern to the NGOs' subscribers. The NGOs have limited resources with which to address long-term (~5-year) research efforts. In the past, the Joint Water Reuse and Desalination Task Force, an alliance of the USBR, Sandia National Laboratories, and research organizations with interests in desalination and water reuse, was used to pool research funding toward longer term research investments, improving coordination, and reducing redundancy, although the group is not as active as it once was. The Global Water Research Coalition (GWRC), a collaboration between 12 research organizations around the globe, including organizations from Singapore, Australia, France, and the United States (WERF and the Water Research Foundation), with partnership from EPA, serves a similar function from an international perspective. The GWRC aims to leverage funding and expertise toward water quality research of global interest. Both groups, if active, could assist with coordination and leveraging resources to accomplish the needed research.

Among federal agencies, water resources research is spread among numerous agencies, based on specific issues (e.g., quality [EPA], quantity [USBR], energy [DOE]) (NRC, 2004), but water scarcity concerns

---

[3] See http://www.nwri-usa.org/researchprogram.htm.

call for a closer coordination of federal efforts. Thus, the intergovernmental Subcommittee on Water Availability and Quality (SWAQ) was formed under the executive branch's Committee on Environment, Natural Resources, and Sustainability (CENRS).[4] SWAQ is chartered to "facilitate communication and coordination among federal agencies and representatives from nonfederal sectors on issues of science, technology, and policy related to water availability and quality." Additionally, SWAQ is charged to periodically assess "priorities for research and development of systems related to enhancement of water supplies," advise the CENRS on additional research needs, and develop coordinated plans to provide the needed research (SWAQ charter provided in NRC, 2004). Thus far, SWAQ has not been used to coordinate federal efforts on reuse research, but federal leadership will be needed if the issues and obstacles to water reuse are to be addressed.

CONCLUSIONS

The committee identified 14 water reuse research priorities (see Box 11-1) that are not currently being addressed in a major way. These research priorities in

the areas of human health, social, and environmental issues, and treatment efficiency and quality assurance hold significant potential to advance the safe, reliable, and cost-effective reuse of municipal wastewater where traditional sources are inadequate.

Improved coordination among federal and nonfederal entities is important for addressing the longterm research needs related to water reuse. Addressing the research needs identified in Box 11-1 will require the involvement of several federal agencies as well as support from nongovernmental research organizations. Several mechanisms could be used to enhance the coordination of reuse research, minimize duplication, and leverage limited resources. A past example that could be built upon is the Joint Water Reuse and Desalination Task Force. Additionally, the SWAQ, which is chartered to facilitate coordination among federal agencies, could be used to enhance coordination of federal water-reuse-related research.

If the federal government decides to develop national regulations for water reuse, a more robust research effort will be needed to support that initiative with enhanced coordination among federal and nonfederal entities. Such an effort would benefit from the leadership of a single federal agency, which could serve as the primary entity for coordination of research and for information dissemination.

[4] The Committee on Environment, Natural Resources, and Sustainability reports to the Office of Science and Technology Policy's National Science and Technology Council.

References

Abrahams M. V., M. Mangel, and K. Hedges. 2007. Predator-prey interactions and changing environments: who benefits? Philosophical Transactions of the Royal Society of London, Series B, Biological Sciences. 362: 2095-2104.

Adams, C., Y. Wang, K. Loftin, and M. Meyer. 2002. Removal of antibiotics from surface and distilled water in conventional water treatment processes. Journal of Environmental Engineering 12: 253-260.

Adham, S., R. S. Trussell, P. Gagliardo, and R. R. Trussell. 1998. Rejection of MS-2 virus by RO membranes. Journal—American Water Works Association 90(9): 130-135.

ADEQ (Arizona Department of Environmental Quality). 2011. Reclaimed Water Quality Standards. Arizona Department of Environmental Quality, Phoenix, Arizona. (http://www.azsos.gov/public_services/Title_18/18-11.htm).

Agus, E., M. H. Lim, L. Zhang, and D. L. Sedlak. 2011. Odorous Compounds in Municipal Wastewater Effluent and Potable Water Reuse Systems. Environmental Science and Technology. 45: 9347-9355.

Ahel, M., W. Giger, and M. Koch. 1994a. Behavior of alkylphenol polyethoxylate surfactants in the aquatic enviromnent. Occurrence and transformation in sewage treatment. Water Research 28: 1131-1142.

Ahel, M., W. Giger, and C. Schaffner. 1994b. Behavior of alkylphenol polyethoxylate surfactants in the aquatic environment. Occurrence and transformation in rivers. Water Research 28: 1143-1152.

Al-Kharouf, S., I. Al-Khatib, and H. Shaheen. 2008. Appraisal of social and cultural factors affecting wastewater reuse in the West Bank. International Journal of Environment and Pollution 33(1): 3-14.

Alan Plummer Associates, Inc. 2010. Final Report: State of Technology of Water Reuse. Prepared for the Texas Water Development Board, Austin, TX. Available online at http://www.twdb.state.tx.us/innovativewater/reuse/projects/reuseadvance/doc/PhaseB_final.pdf. Accessed December 19, 2011.

Albinana-Gimenez, N., P. Clemente-Casares, S. Bofill-Mas, A. Hundesa, F. Ribas, and R. Girones. 2006. Distribution of human polyomaviruses, adenoviruses, and hepatitis E virus in the environment and in a drinking-water treatment plant. Environmental Science & Technology. 40: 7416-7422.

Alley, W. M., T. E. Reilly, and O. L. Franke. 1999. Sustainability of Ground-Water Resources. Circular 1186. Denver, CO: U.S. Geological Survey.

Amy, G., and J. Drewes. 2007. Soil aquifer treatment (SAT) as a natural and sustainable wastewater reclamation/reuse technology: Fate of wastewater effluent organic matter (EfoM) and trace organic compounds. Environmental Monitoring and Assessment. 129(1-3): 19-26.

Anand, S., and K. Hanson. 1997. Disability-adjusted life years: A critical review. Journal of Health Economics 16(6): 685-702.

Anderson, P. D., V. J. D'Aco, S. C. Chapra, M. E. Buzby, V. L. Cunningham, B. M. Duplessie, E. P. Hayes, F. J. Mastrocco, N. J. Parke, J. C. Rader, J. H. Samuelian, and B. W. Schwab. 2004. Screening analysis of human pharmaceutical compounds in U.S. surface waters. Environmental Science & Technology 38(3): 838-849.

Anderson, P., N. Denslow, J. Drewes, A. Olivieri, D. Schlenk, and S. Snyder. 2010. Monitoring Strategies for Chemicals of Emerging Concern (CECs) in Recycled Water. Recommendations of a Science Advisory Panel convened by the California State Water Resources Control Board, Sacramento, CA.

Ankley, G. T., D. W. Kuehl, M. D. Kahl, K. M. Jensen, A. Linnum, R. L. Leino, and D. A. Villeneuvet. 2005. Reproductive and developmental toxicity and bioconcentration of perfluorooctanesulfonate in a partial life-cycle test with the fathead minnow (Pimephales promelas). Environmental Toxicology and Chemistry 24(9): 2316-2324.

Anonymous, 1893. The use of sewage for irrigation in the west. Engineering News and American Railway Journal XXIX: 1-15.

Aragón, T. J., S. Novotony, W. Enanoria, D. J. Vugia, A. Khalakdina, and M. Katz. 2003. Endemic cryptosporidiosis and exposure to municipal tap water in persons with acquired immunodeficiency syndrome (AIDS): A case-control study. BMC Public Health 2(3).

Arizona Department of Environmental Quality. 2011. Water Quality Division. Permits: Reclaimed Water. Available online at http://www.azdeq.gov/environ/water/permits/reclaimed.html. Accessed December 19, 2011.

Armstrong, T. W., and C. N. Haas. 2007. A quantitative microbial risk assessment model for Legionnaires' disease: Animal model selection and dose-response modeling. Risk Analysis 27(6): 1581-1596.

Arnold, S. F., D. M. Klotz, B. M. Collins, P. M. Vonier, L. J. Guillette, Jr., and J. A. McLachlan. 1996. Synergistic activation of estrogen receptor with combinations of environmental chemicals. Science 272(5267): 1489-1492.

Asami, M., K. Kosaka, and S. Kunikane. 2009. Bromate, chlorate, chlorite and perchlorate in sodium hypochlorite solution used in water supply. Journal of Water Supply: Research and Technology. AQUA 58(2): 107-115.

Asano, T., and J. A. Cotruvo. 2004. Groundwater recharge with reclaimed municipal wastewater: Health and regulatory considerations. Water Research 38: 1941-1951.

Asano, T., and R. Mills. 1998. Planning and analysis of water reuse projects. In T. Asano, ed., Wastewater Reclamation and Reuse. Lancaster, PA: Technomic.

Asano, T., F. L. Burton, H. L. Leverenz, R. Tsuchihashi, and G. Tchobanoglous. 2007. Water Reuse: Issues, Technologies, and Applications. New York: McGraw-Hill.

Atkinson, S. F., D. R. Johnson, B. J. Venables, J. L. Slye, J. R. Kennedy, S. D. Dyer, B. B. Price, M. Ciarlo, K. Stanton, H. Sanderson, and A. Nielsen. 2009. Use of watershed factors to predict consumer surfactant risk, water quality, and habitat quality in the upper Trinity River, Texas. Science of the Total Environment 407(13): 4028-4037.

AWWA (American Water Works Association). 2004. Recommended Practice for Backflow Prevention and Cross-Connection Control, 3rd Ed. Manual M14, Denver, CO: AWWA.

AWWA. 2008. Water Reuse Rates and Charges: 2000 and 2007 Survey Results. Available at http://www.ecy.wa.gov/programs/wq/reclaim/AWWAreuseratessurveyJuly2008.pdf. Accessed December 14, 2011.

AWWA. 2009. Planning for the Distribution of Reclaimed Water, 3rd Ed.: Manual M24. Denver, CO: AWWA.

AWWA Disinfection Systems Committee. 2008. Committee Report: Disinfection Survey, Part 1—Recent Changes, Current Practices and Water Quality. Journal of the American Water Works Association 100(10): 76-90.

AWWA Water Quality Division Disinfection Systems Committee. 2000. Committee report: Disinfection at large and medium-size systems. Journal of American Water Works Association 92(5): 32-43.

AWWA/WEF (American Water Works Association and Water Environment Federation). 1998. Using Reclaimed Water to Augment Potable Water Resources, 1st Ed. Denver, CO: American Water Works Association.

Azadpour-Keeley, A., and C. Ward. 2005. Transport and survival of viruses in the subsurface—processes, experiments, and simulation models. Remediation 15(3): 23-49.

Baalousha, M., and J. R. Lead. 2007. Size fractionation and characterization of natural aquatic colloids and nanoparticles. Science of the Total Environment 386(1-3): 93-102.

Baccarelli, A., and V. Bollati. 2009. Epigenetics and environmental chemicals. Current Opinion in Pediatrics 21(2): 243-251.

Baker, L. C., D. L. Ashley, F. L. Cardinali, S. M. Kieszak, and J. V. Wooten. 2000. Household exposures to drinking water disinfection by-products: Whole blood trihalomethane levels. Journal of Exposure Analysis and Environmental Epidemiology 10(4): 321-326.

Barbeau, B., R. Desjardins, C. Mysore, and M. Prevost. 2005. Impacts of water quality on chlorine and chlorine dioxide efficacy in natural waters. Water Research 39: 2024-2033.

Barkhurst, A. 2011. Pembroke Pines Plans to inject treated sewage into water supply. SunSentinel. Fort Lauderdale, Florida. Available at http://articles.sun-sentinel.com/2011-01-03/news/fl-pines-sewage-treatment-20101216_1_water-supply-drinking-water-biscayne-aquifer. Accessed December 14, 2011.

Barlow, S. 2005. Threshold of Toxicological Concern (TTC). A Tool for Assessing Substances of Unknown Toxicity Present at Low Levels in the Diet. ILSI Europe Concise Monograph Series. Washington, DC: ILSI Press.

Barnes, D. G., and Dourson, M. 1988. Reference Dose (RfD): Description and Use in Health Risk Assessments. Regulatory Toxicology and Pharmacology 8: 471-486.

Barron, M. G. 1990. Bioconcentration—Will water-borne organic chemicals accumulate in aquatic animals? Environmental Science & Technology 24(11): 1612-1618.

Bartrand, T. A., B. Farouk, and C. N. Haas. 2009. Countercurrent gas/liquid flow and mixing: Implications for water disinfection. International Journal of Multiphase Flow 35(2): 171-184.

Barzilay, J. I., W. G. Weinberg, and J. W. Eley. 1999. Drinking water and infectious disease. In The Water We Drink: Water Quality and Its Effects on Health. Piscataway, NJ: Rutgers University Press, pp. 41-59.

Basista, M., and W. Weglewski. 2009. Chemically assisted damage of concrete: A model of expansion under external sulfate attack. International Journal of Damage Mechanics 18(2): 155-175.

Beach, S. A., J. L. Newstead, K. Coady, and J. P. Giesy. 2006. Ecotoxicological evaluation of perfluorooctanesulfonate (PFOS). Reviews of Environmental Contamination and Toxicology 186: 133-174.

Begier, E. M., M. S. Oberste, M. L. Landry, T. Brennan, D. Mlynarski, P. A. Mshar, P. A. K. Frenette, T. Rabatsky-Her, K. Purviance, A. Nepaul, W. A. Nix, M. A. Pallansch, D. Ferguson, M. L. Cartter, and J. L. Hadler. 2008. An outbreak of concurrent echovirus 30 and coxsackievirus A1 infections associated with sea swimming among a group of travelers to mexico. Clinical Infectious Diseases 47(5): 616-623.

Bellamy, W. D., G. R. Finch, and C. N. Haas. 1998. Integrated Disinfection Design Framework. American Water Works Association Research Foundation.

Bellona, C., and J. E. Drewes. 2007. Viability of a low pressure nanofilter in treating recycled water for water reuse applications—A pilot-scale study. Water Research 41: 3948-3958.

Bellona, C., G. Oelker, J. Luna, G. Filteau, G. Amy, and J. E. Drewes. 2008. Comparing nanofiltration and reverse osmosis for drinking water augmentation. Journal—American Water Works Association 100(9): 102-116.

Ben Ayed, L., J. Schijven, Z. Alouini, M. Jemli, and S. Sabbahi. 2009. Presence of parasitic protozoa and helminth in sewage and efficiency of sewage treatment in Tunisia. Parasitology Research 105: 393-406.

Benkel, D. H., E. M. McClure, D. Woolard, J. V. Rullan, G. B. Miller, S. R. Jenkins, J. H. Hershey, R. F. Benson, J. M. Pruckler, E. W. Brown, M. S. Kolczak, R. L. Hackler, B. S. Rouse, and R. F. Breiman. 2000. Outbreak of Legionnaires' disease associated with a display whirlpool spa. International Journal of Epidemiology 29(6): 1092-1098.

Benotti, M. J., R. A. Trenholm, B. J. Vanderford, J. C. Holady, B. D. Stanford, and S. A. Snyder. 2009. Pharmaceuticals and endocrine disrupting compounds in U.S. drinking water. Environmental Science & Technology 43(3): 597-603.

Berg, G. 1973. Removal of viruses from sewage, effluents and waters. Bulletin of the World Health Organization 49: 451-460.

Billings, M. D. 1885. Sewage disposal in cities. Harper's Magazine 577-584.

Bitton, G. 2005. Wastewater Microbiology, 3rd Ed. Hoboken, NJ: John Wiley & Sons, Inc.

Blackburn, B. G., G. F. Craun, J. S. Yoder, V. Hill, R. L. Calderon, N. Chen, S. H. Lee, D. A. Levy, and M. J. Beach. 2004. Surveillance for waterborne-disease outbreaks associated with drinking water—United States, 2001-2002. Morbidity and Mortality Weekly Report 55(SS08): 31-58.

Blatchley, E. R., W. L. Gong, J. E. Alleman, J. B. Rose, D. E. Huffman, M. Otaki, and J. T. Lisle. 2007. Effects of wastewater disinfection on waterborne bacteria and viruses. Water Environment Research 79: 81-92.

Bloom, J. 2003. Residents Sue SAWS Over 'Recycled Water. Article in 9 News San Antonio, San Antonio, Texas, July 7.

Blumenthal, W. J., D. D. Mara, A. Peasey, G. Ruiz-Palacios, and R. Stott. 2000. Guidelines for the microbiological quality of treated wastewater used in agriculture: Recommendations for revising WHO guidelines. Bulletin of the World Health Organization 78: 1104-1116.

Blute, N., C. Russell, Z. Chowdhury, X. Wu, and S. Via. 2010. Nitrosamine occurrence in the US: Analysis and interpretation of UCMR2 data. In 2010 Water Quality Technology Conference and Exposition Proceedings, Vol. 3. Denver, CO: American Water Works Association, pp. 2516-2543.

Bofill-Mas, S., N. Albinana-Gimenez, P. Clemente-Casares, A. Hundesa, J. Rodriguez-Manzano, A. Allard, M. Calvo, and R. Girones. 2006. Quantification and stability of human adenoviruses and polyomavirus JCPyV in wastewater matrices. Applied and Environmental Microbiology 72: 7894-7896.

Boxall, A. B. A., P. Johnson, E. J. Smith, C. J. Sinclair, E. Stutt, and L. S. Levy. 2006. Uptake of veterinary medicines from soils into plants. Journal of Agricultural and Food Chemistry 54: 2288-2297.

Brauch, H.-J., F. Sacher, E. Denecke, and T. Tacke. 2000. Wirksamkeit der Uferfiltration für die Entfernung von polaren organischen Spurenstoffen. GWF Wasser Abwasser 141(4): 226-234.

Brekke, L. D., J. E. Kiang, J. R. Olsen, R. S. Pulwarty, D. A. Raff, D. P. Turnipseed, R. S. Webb, and K. D. White. 2009. Climate change and water resources management—A federal perspective: U.S. Geological Survey Circular 1331, 65 pp. Available at http://pubs.usgs.gov/circ/1331/.

Brix, K. V., J. Keithly, R. C. Santore, D. K. DeForest, and S. Tobiason. 2010. Ecological risk assessment of zinc from stormwater runoff to an aquatic ecosystem. Science of the Total Environment 408(8): 1824-1832.

Brooks, B. W., P. K. Turner, J. K. Stanley, J. J. Weston, E. A. Glidewell, C. M. Foran, M. Slattery, T. W. LaPoint, and D. B. Huggett. 2003. Waterborne and sediment toxicity of fluoxetine to select organisms. Chemosphere 52(1): 135-142.

Brooks, B. W., J. K. Stanley, J. C. White, P. K. Turner, K. B. Wu, and T. W. LaPoint. 2004. Laboratory and field responses to cadmium: An experimental study in effluent-dominated stream mesocosms. Environmental Toxicology and Chemistry 23(4): 1057-1064.

Brooks, B. W., T. M. Riley, and R. D. Taylor. 2006. Water quality of effluent-dominated ecosystems: Ecotoxicological, hydrological, and management considerations. Hydrobiologia 556: 365-379.

Brown, P. H., N. Bellaloui, M. A. Wimmer, E. S. Bassil, J. Ruiz, H. Hu, H. Pfeffer, F. Dannel, and V. Romheld. 2002. Boron in plant biology. Plant Biology 4(2): 205-223.

Bruce, G. M., R. C. Pleus, M. K. Peterson, M. H. Nellor, and J. A. Soller. 2010. Development and Application of Tools to Assess and Understand the Relative Risks of Drugs and Other Chemicals in Indirect Potable Reuse: Tools to Assess and Understand the Relative Risks of Indirect Potable Reuse and Aquifer Storage & Recovery Projects, Vol. 2. Alexandria, VA: Water Reuse Research Foundation.

Bruvold, W. 1988. Public opinion on water reuse options. Journal of the Water Pollution Control Federation 60: 45-49.

Buerge, I. J., H.-R. Buser, M. Kahle, M. D. Muller, and T. Poiger. 2009. Ubiquitous occurence of the artificial sweetner acesulfame in the aquatic environment: An ideal chenical marker of domestic wastewater in groundwater. Environmental Science & Technology 43(12): 4381-4385.

Bull, R., J. Crook, M. Whittaker, and J. Cotruvo. 2011. Therapeutic dose as the point of departure in assessing potential health hazards from drugs in recycled municipal wastewater. Regulatory Toxicology and Pharmacology 60(1): 1-19.

Burian, S. J., S. J. Nix, R. E. Pitt, and S. R. Durrans. 2000. Urban wastewater management in the United States: Past, present, and future. Journal of Urban Technology 7: 33-62.

Burmaster, D. E. 1998. Lognormal distributions for total water intake and tap water intake by pregnant and lactating women in the United States. Risk Analysis 18(2): 215-219.

Burmaster, D. E., and P. D. Anderson. 1994. Principles of good practice for the use of Monte Carlo techniques in human health and ecological risk assessment. Risk Analysis 14(4): 477-481.

Burmaster, D. E., and A. M. Wilson. 1996. An introduction to second-order random variables in human health risk assessments. Human and Ecological Risk Assessment 2(4): 892-919.

Butterfield, C., and E. Wattie. 1946. Influence of pH and temperature on the survival of coliforms and enetric pathogens when exposed to free chlorine. Public Health Reports 61(6): 157-192.

Butterfield, C., E. Wattie, S. Megregian, and C. Chambers. 1943. Influence of pH and temperature on the survival of coliforms and enetric pathogens when exposed to free chlorine. Public Health Reports 58(51): 1837-1866.

Butterwick, L., N. De Oude, and K. Raymond. 1989. Safety assessment of boron in aquatic and terrestrial environments. Ecotoxicology and Environmental Safety 17: 339-371.

CA LAO (California Legislative Analyst's Office). 2008. California's Water: An LAO Primer. Available at http://www.lao.ca.gov/2008/rsrc/water_primer/water_primer_102208.pdf. Accessed November 14, 2011.

Cahill, M. J., C. U. Kloser, N. E. Ross, and J. A. C. Archer. 2010. Read length and repeat resolution: Exploring prokaryote genomes using next-generation sequencing technologies. PLOSOne 5(7): 1-9.

Caldwell, D. J., F. Mastrocco, T. H. Hutchinson, R. Länge, D. Heijerick, C. Janssen, P. D. Anderson, and J. P. Sumpter. 2008. Derivation of an aquatic predicted no-effect concentration for the synthetic hormone, 17α-ethinyl estradiol. Environmental Science & Technology 42(19): 7046-7054.

California State Senate. 2010. Senate Bill No. 918, amended June 1, 2010. Available at http://www.leginfo.ca.gov/pub/09-10/bill/sen/sb_0901-0950/sb_918_bill_20100601_amended_sen_v95.pdf). Accessed October 24, 2011.

California Sustainability Alliance. 2008. The Role of Recycled Water in Energy Efficiency and Greenhouse Gas Reduction. Available at http://www.fypower.org/pdf/CSA_RecycledH20.pdf.

Capdevielle, M., R. Van Egmond, M. Whelan, D. Versteeg, M. Hofmann-Kamensky, J. Inauen, V. Cunningham, and D. Woltering. 2008. Consideration of exposure and species sensitivity of triclosan in the freshwater environment. Integrated Environmental Assessment and Management 4(1): 15-23.

Carey, R. O., and K. W. Migliaccio. 2009. Contribution of wastewater treatment effluents to nutrient dynamics in aquatic systems: A review. Environmental Management 44(2): 205-217.

Carvalho, L. 2007. Avoiding Costly Water Mistakes in Comined Cycle Power Plant Projects: Power. Available at http://www.waterprofessionals.com/pdfs/avoiding_costly_water_treatment_mistakes_in_CCPP_Projects.pdf.

Castro-Hermida, J. A., I. Garcia-Presedo, A. Almeida, M. Gonzalez-Warleta, J. M. C. Da Costa, and M. Mezo. 2008. Contribution of treated wastewater to the contamination of recreational river areas with Cryptosporidium spp. and Giardia duodenalis. Water Resources 42: 3528-3538.

Cath, T. 2010. Advanced Water Treatment Engineering and Reuse. Graduate Course Script. Colorado School of Mines, Golden, CO.

CDPH (California Department of Public Health). 2008. Groundwater Recharge Regulation Draft. Sacramento, CA: CDPH.

CDPH. 2009. Regulations Related to Recycled Water: January 2009. California Code of Regulations, Title 22, Division 4, Chapter 3, Water Recycling Criteria. Sacramento, CA: CDPH. Available at http://www.cdph.ca.gov/certlic/drinkingwater/Documents/Lawbook/RWregulations-01-2009.pdf.

CDPH. 2011. Groundwater Replenshment Reuse Draft Regulation.November 21. Available at http://www.cdph.ca.gov/certlic/drinkingwater/Documents/Recharge/DraftRechargeReg-2011-11-21.pdf. Accessed December 14, 2011.

CEC (California Energy Commission). 2005. California's Water–Energy Relationship: Final State Report. November. Available at http://www.energy.ca.gov/2005publications/CEC-700-2005-011/CEC-700-2005-011-SF.PDF. Accessed December 14, 2011.

CH2M Hill. 1993. Tampa Water Resource Recovery Project Pilot Studies. Tampa, Fla.: CH2M Hill.

Chalmers, R. M., G. Robinson, K. Elwin, S. J. Hadfield, E. Thomas, J. Watkins, D. Casemore, and D. Kay. 2010. Detection of Cryptosporidium species and sources of contamination with Cryptosporidium hominis during a waterborne outbreak in north west Wales. Journal of Water and Health 8: 311-325.

Cheeseman, M. A., E. J. Machuga, and A. B. Bailey. 1999. A tiered approach to threshold of regulation. Food Chemical Toxicology 37: 387-412.

Ching, L. 2010. Eliminating "yuck": A simple exposition of media and social change in water reuse policies. International Journal of Water Resources Development 26(1): 111-124.

Ching, L., and J. Yu. 2010. Turning the tide: Informal institutional change in water reuse. Water Policy 12(Suppl): 121-134.

Choi, J. H., S. E. Duirk, and R. L. Valentine. 2002. Mechanistic studies of N-nitrosodimethylamine (NDMA) formation in drinking water. Journal of Environmental Monitoring 4: 249-252.

Chong, M., B. Jin, C. Chow, and C. Saint. 2010. Recent developments in photocatalytic water treatment technology: A review. Water Research 44(10): 2997-3027.

Cleuvers, M. 2003. Aquatic ecotoxicity of pharmaceuticals including the assessment of combination effects. Toxicology Letters 142(3): 185-194.

Cody, B., and N. Carter. 2010. Water Reuse and the Title XVI Program: Legislative Issues. Congressional Research Service Report R41487, November 9. Damascus, MD: Penny Hill Press.

Colford, J. M. Jr., T. J. Wade, S. K. Sandhu, C. C. Wright, S. Lee, S. Sha, K. Fox, S. Burns, A. Benker, M. A. Brookhart, M. Van Der Laan, and D. A. Levy. 2005. A Randomized, controlled trial of in-home drinking water intervention to reduce gastrointestinal illness. American Journal of Epidemiology 161(5): 472-482.

Collins, F. S., G. M. Gray, and J. R. Bucher. 2008. Transforming Environmental Health Protection. Science. February 15. 319 (5865): 906-907. Available at http://www.ncbi.nlm.nih.gov/pmc/articles/PMC2679521/pdf/nihms103919.pdf. Accessed December 14, 2011.

Colorado Department of Health and Environment. 2007. Reclaimed Water Control Regulation. 5 CCR 1002-84, Colorado Department of Health and Environment, Denver, Colorado.

Condie, L. W., W. C. Lauer, G. W. Wolfe, E. T. Czeh, and J. M. Burn. 1994. Denver Potable Water Reuse Demontration Project: Comprehensive chronic rat study. Food Chemical Toxicology 32(11): 1021-1030.

Costanza, R., O. Perez-Magueo, M. L. Martinez, P. Sutton, S. J. Anderson, and K. Mulder. 2008. The value of coastal wetlands for hurricane protection. Ambio 37: 241-248.

Costanzo, S. D., A. J. Watkinson, E. J. Murby, D. W. Kolpin, and M. W. Sandstrom. 2007. Is there a risk associated with the insect repellent DEET (N,N-diethyl-m-toluamide) commonly detected in aquatic environments? Science and the Total Environment 384: 214-220.

Cramer, G. M., R. A. Ford, and R. L. Hall. 1978. Estimation of toxic hazard-decision tree approach. Food and Cosmetics Toxicology 16: 255-276.

Craun, G. F. 1991. Causes of waterborne outbreaks in the United States. Water Science and Technology 24: 17-20.

Craun, M., G. Craun, R. Calderon, and M. Beach. 2006. Waterborne outbreaks reported in the United States. Journal of Water and Health 4(Suppl 2): 19-30.

Crittenden, J., R. R. Trussel, D. Hand, K. Howe, and G. Tchobanoglous. 2005. Water Treatment: Principles and Design, 2nd Ed. Hoboken, NJ: John Wiley & Sons, Inc.

Crook, J. 1999. Water reuse in the United States. Journal of New England Water Environmental Association 33(2): 106-125.

Crook, J. 2004. Innovative Applications in Water Reuse: Ten Case Studies. Alexandria, VA: WateReuse Association.

Crook, J. 2005a. St. Petersburg, Florida Dual Water System: A Case Study. In: Water Conservation, Reuse, and Recycling, Proceedings of an Iranian-American Workshop. 177-186. Washington, D.C.: The National Academies Press.

Crook, J. 2005b. Irrigation of Parks, Playgrounds, and Schoolyards with Reclaimed Water: Extent and Safety. Alexandria, VA: WateReuse Association.

Crook, J. 2007. Innovative Applications in Water Reuse and Desalination 2: Ten Case Studies. Alexandria, VA: WateReuse Association.

Crook, J. 2010. Regulatory Aspects of Direct Potable Reuse in California. National Water Research Institute. Fountain Valley, California.

Crook, J., and D. A. Okun. 1993. Water reuse: Past, present, and future. In Proceedings of the 1993 AWWA Annual Conference Denver, CO: American Water Works Association, pp. 159-180.

Crook, J., J. J. Mosher, and J. M. Casteline. 2005. Status and Role of Water Reuse: An International View. London, UK: Global Water Research Coalition. Available at http://www.watereuse.org/sites/default/files/images/04-007-01.pdf.

Crump, K. S. 2003. Quantitative risk assessment since the Red Book: Where have we come and where should we be going? Human and Ecological Risk Assessment 9(5): 1105-1112.

CSDWD (City of San Diego Water Department). 2005. City of San Diego Water Reuse Study 2005 Interim Report. Available at http://www.sandiego.gov/water/waterreuse/pdf/ir05toc.pdf.

CSDWD. 2006. Water Reuse Study: Final Draft Report, March. Available online at http://www.sandiego.gov/water/waterreuse/waterreusestudy/news/fd2006.shtml. Accessed January 15, 2011.

CSWRCB. 2010. Monitoring Strategies for Chemicals of Emergening Concern (CECs) in Recycled Water. Recommendations of a Science Advisory Panel. Report dated June 25, 2010. Available at http://www.swrcb.ca.gov/water_issues/programs/water_recycling_policy/docs/cec_monitoring_rpt.pdf.

CSWRCB. 2011. California Water Code. Available at http://www.leginfo.ca.gov/cgi-bin/calawquery?codesection=wat&codebody=&hits=20

Cunningham, V. L., C. Perino, V. J. D'Aco, A. Hartmann, and R. Bechter. 2010. Human health risk assessment of carbamazepine in surface waters of North America and Europe. Regulatory Toxicology and Pharmacology 56(3): 343-351.

Daigger, G. T. 2009. Evolving urban water and residuals management paradigms: Water reclamation and reuse, decentralization, and resource recovery. Water Environment Research 81(8): 809-823.

Dasgupta, P. K., P. K. Martinelango, W. A. Jackson, T. A. Anderson, K. Tian, R. W. Tock, and S. Rajagopalan. 2005. The origin of naturally occurring perchlorate: The role of atmospheric processes. Environmental Science & Technology 39(6): 1569-1575.

Daughton, C. G., and T. A. Ternes. 1999. Pharmaceuticals and personal care products in the environment: Agents of subtle change? Environmental Health Perspectives 107(6): 907-938.

Davis, S., K. Drake, and K. Maier. 2002. Toxicity of boron to the duckweed, *Spirodella polyrrhiza*. Chemosphere 48:615-620.

Day, H. P., and K. D. Conway. 2009. Case 1: No water, no watts; Case 2: No watts, no water. In Proceedings of the 24th Wate Reuse Annual Symposium, Seattle, Washington. Alexandria, VA: WateReuse Association.

Day, J. W., J. Y. Ko, J. Rybczyk, D. Sabins, R. Bean, G. Berthelot, C. Brantley, L. Cardoch, W. Conner, J. N. Day, A. J. Englande, S. Feagley, E. Hyfield, R. R. Lane, J. Lindsey, W. J. Mitsch, E. Reyes, and R. R. Twilley. 2004. The use of wetlands in the Mississippi Delta for wastewater assimilation: A review. Ocean and Coastal Management 47(11-12): 671-691.

de Koning, J., D. Bixio, A. Karabelas, M. Salgot, and A. Schafer. 2008. Characterization and assessment of water treatment technologies for reuse. Desalination 218: 92-104.

Deborde, M., S. Rabouan, H. Gallad, and B. Legube. 2004. Aqueous chlorination kinetics of some endocrine disruptors. Environmental Science & Technology 38: 5577-5583.

Deng, X., M. Carney, D. E. Hinton, S. Lyon, C. Woodside, C. N., S. D. Kim, and D. Schlenk. 2008. Biomonitoring recycled water in the Santa Ana river basin in Southern California. Journal of Toxicology and Environmental Health A 71: 109-118.

Desbrow, C., E. J. Routledge, G. C. Brighty, J. P. Sumpter, and M. Waldock. 1998. Identification of estrogenic chemicals in STW effluent. 1. Chemical fractionation and in vitro biological screening. Environmental Science & Technology 32(11): 1549-1558.

Dethloff, G. M., W. A. Stubblefield, and C. E. Schlekat. 2009. Effects of water quality parameters on boron toxicity to *Ceriodaphnia dubia*. Archives of Environmental Contamination and Toxicology 57(1): 60-67.

Dickenson, E., J. E. Drewes, S. A. Snyder, and D. L. Sedlak. 2011. Indicator compounds: An approach for using monitoring data to quantify the occurrence and fate of wastewater-derived contaminants in surface waters. Water Research 45: 1199-1212.

Dizer, H., A. Nasser, and J. M. Lopez. 1984. Penetration of different human pathogenic viruses into sand columns percolated with distilled water, groundwater, or wastewater. Applied and Environmental Microbiology 47(2): 409-415.

Dolnicar, S., and A. Hurlimann. 2009. Drinking water from alternative water sources: Differences in beliefs, social norms and factors of perceived behavioural control across eight Australian locations Water Science and Technology 60(6): 1433-1444.

Dolnicar, S., and A. Schäfer. 2009. Desalinated versus recycled water: Public perceptions and profiles of the accepters. Journal of Environmental Management 90(2): 888-900.

Dolnicar, S., A. Hurlimann, and L. Nghiem. 2010. The effect of information on public acceptance—the case of water from alternative sources. Journal of Environmental Management 91(6): 1288-1293.

Domenech, L., and D. Sauri. 2010. Socio-technical transitions in water scarcity contexts: Public acceptance of greywater reuse technologies in the metropolitan area of Barcelona. Resources Conservation and Recycling 55(1): 53-62.

Donohue, J. M., and W. M. Miller. 2007. Summary of the development of federal drinking water regulations and health-based guidelines for chemical contaminants. In R. A. Howd and A. M. Fan, eds., Risk Assessment for Chemicals in Drinking Water. Hoboken, NJ: John Wiley & Sons, pp. 17-33.

Donohue, J. M., and J. Orme-Zavaleta. 2003. Toxicological basis for drinking water risk assessment. In F. W. Pontius, ed., Drinking Water Regulation and Health. New York: John Wiley & Sons, pp. 133-146.

Doria, M. 2010. Factors influencing public perception of drinking water quality. Water Policy 12(1): 1-19.

Drewes, J. E., and P. Fox. 2000. Effect of drinking water sources on reclaimed water quality in water reuse systems. Water Environment Research 72(3): 353-362.

Drewes, J. E., and S. Khan. 2010. Water reuse for drinking water augmentation. J. Edzwald, ed., Water Quality and Treatment, 6th Ed. New York: McGrawHill, pp. 16.1-16.48.

Drewes, J. E., T. Heberer, T. Rauch, and K. Reddersen. 2003a. Fate of pharmaceuticals during groundwater recharge. Journal of Ground Water Monitoring and Remediation 23(3): 64-72.

Drewes, J. E., M. Reinhard, and P. Fox. 2003b. Comparing microfiltration-reverse osmosis and soil-aquifer treatment for indirect potable reuse of water. Water Research 37: 3612-3621.

Drewes, J. E., C. Hoppe, and T. Jennings. 2006. Fate and transport of N-nitrosamines under conditions simulating full-scale groundwater recharge operations. Water Environment Research 78(13): 2466-2473.

Drewes, J., D. Sedlak, S. Snyder, and E. Dickenson. 2008. Development of Indicators and Surrogates for Chemical Contaminant Removal During Wastewater Treatment and Reclamation. WRF-03-014. Alexandria, VA: WateReuse Foundation.

Drewes, J. E., P. Anderson, N. Denslow, A. Olivieri, D. Schlenk, and S. Snyder. 2010. Monitoring Strategies for Chemicals of Emerging Concern (CECs) in Recycled Water: Recommendations of a Science Advisory Panel. Sacramento: California State Water Resources Board. Available at http://www.waterboards.ca.gov/water_issues/programs/water_recycling_policy/docs/cec_monitoring_rpt.pdf.

du Pisani, P. L. 2005. Direct reclamation of potable water at Windhoek's Goreangab Reclamation Plant. In S. J Khan, M. H. Muston, and A. I Schafer, eds., Integrated Concepts in Water Recycling 2005, pp. 193-202. Wollongong, Australia: University of Wollongong.

Dupont, H., C. Chappell, C. Sterling, P. Okhuysen, J. Rose, and W. Jakubowski. 1995. Infectivity of Cryptosporidium parvum in healthy volunteers. New England Journal of Medicine 332(13): 855.

Durand, R., and G. Schwebach. 1989. Gastrointestinal effects of water reuse for public park irrigation. American Journal of Public Health 79(12): 1659-1660.

Durando, M., L. Kass, P. Julio, C. Sonnenschein, A. M. Soto, E. H. Luque, and M. Muñoz-de-Toro. 2007. Prenatal bisphenol A exposure induces preneoplastic lesions in the mammary gland in Wistar rats. Environmental Health Perspectives 115(1): 80-86.

EA (United Kingdom Environment Agency). 2004. Environmental risk evaluation report: Perfluorooctanesulphonate (PFOS). Available at http://www.pops.int/documents/meetings/poprc/submissions/Comments_2006/sia/pfos.uk.risk.eal.report.2004.pdf.

EAO, Inc. 2000. Orange County Water District/Orange County Sanitation District Groundwater Replenishment System. Water Quality Evaluation. Task 7: Conduct Risk Assessment. Final Report, November.

Elovitz, M. S., U. von Gunten, and H. P. Kaiser. 2000. Hydroxyl radical/ozone ratios during ozonation processes. II. the effect of temperature, pH, alkalinity, and DOM properties. Ozone: Science & Engineering 22(2): 123-150.

Engineering-Science. 1987. Monterey Wastewater Reclamation Study for Agriculture: Final Report. Prepared for the Monterey Regional Water Pollution Control Agency, Pacific Grove, CA. Available at http://www.sanjoseca.gov/sbwr/PDFs/MontereyCountyRW_AG_1987.pdf.

EPA (Environmental Protection Agency). 1979. National Interim Primary Drinking Water Regulations; Control of Trihalomethanes in Drinking Water; Final Rule. Fed. Reg. 44(231): 68624-68707.

EPA. 1986. Environmental Regulations and Technology: The National Pretreatment Program, July. Available at http://www.epa.gov/npdes/pubs/owm0018.pdf.

EPA. 1993a. Constructed Wetlands for Wastewater Treatment and Wildlife Habitat: Arcata, California—A Natural System for Wastewater Reclamation and Resource Enhancement. Available at http://www.epa.gov/owow/wetlands/pdf/Arcata.pdf. Accessed September 16, 2010.

EPA. 1993b. Wetland Treatment Systems: A Case History—The Orlando Easterly Wetlands Reclamation Project. EPA/832/R-93/0051. NTIS No. PB95-227740. Springfield, VA: National Technical Information Service.

EPA. 1998a. Guidelines for Ecological Risk Assessment. EPA/630/R-95/002F. Available at http://www.epa.gov/raf/publications/pdfs/ECOTXTBX.PDF. Accessed December 15, 2011.

EPA. 1998b. National Drinking Water Regulations; Disinfectants and Disinfectant Byproducts: Final Rule. Federal Register 63(241): 69390.

EPA. 2000. Risk Characterization Handbook. Office of Science Policy. Office of Research and Development. EPA/100/B-00/002. Available at http://www.epa.gov/spc/pdfs/rchandbk.pdf.

EPA. 2001. Low-Pressure Membrane Filtration for Pathogen Removal: Application, Implementation, and Regulatory Issues. Available at http://www.epa.gov/ogwdw/disinfection/lt2/pdfs/report_lt2_membranefiltration.pdf.

EPA. 2002. Clean Water and Drinking Water Infrastructure Gap Analysis Report. EPA-816-R-02-020. Available at http://water.epa.gov/aboutow/ogwdw/upload/2005_02_03_gapreport.pdf.

EPA. 2003a. Cross-Connection Control Manual. EPA/816/R-03/002, Available at http://www.epa.gov/ogwdw/pdfs/cross-connection/crossconnection.pdf.

EPA. 2003b. EPA's National Pretreatment Program, 1973-2003: Thirty Years of Protecting the Environment. Available at http://www.epa.gov/region8/water/pretreatment/pdf/EPAsNationalPretreatmentProgram.pdf

EPA. 2003c. National Drinking Water Regulations; Stage 2 Disinfectants and Disinfectant Byproducts: Final Rule. Fed. Reg. 68(159): 49547.

EPA. 2004. Guidelines for Water Reuse. EPA/625/R-04/108. Available at http://www.epa.gov/ord/NRMRL/pubs/625r04108/625r04108.htm.

EPA. 2005a. Guidelines for Carcinogen Risk Assessment. EPA-630/P-03/001F. Available at http://www.epa.gov/raf/publications/pdfs/CANCER_GUIDELINES_FINAL_3-25-05.PDF.

EPA. 2005b. Ambient aquatic life water quality criteria—nonylphenol final. Office of Water, Office of Science and Technology, Washington, DC. EPA-822-R-05-005.

EPA. 2006a. National Primary Drinking Water Regulations, Long Term 2 Enhanced Surface Water Treatment Rule: Final Rule. Federal Register 71: 654-786.

EPA. 2006b. National Primary Drinking Water Regulations. Stage 2 Disinfectants and Disinfection Byproducts Rule: Final Rule. Federal Register 71: 388.

EPA. 2008a. Aquatic Life Criteria for Contaminants of Emerging Concern. Prepared by the OW/ORD Emerging Contaminants Workgroup. Available at http://www.epa.gov/waterscience/criteria/library/sab-emergingconcerns.pdf. Accessed August 27, 2010.

EPA. 2008b. Clean Watershed Needs Survey 2004 Report to Congress. Available at http://water.epa.gov/scitech/datait/databases/cwns/upload/2008_01_09_2004rtc_cwns2004rtc.pdf.

EPA. 2008c. Clean Watersheds Needs Survey 2008 Report to Congress. Available at http://water.epa.gov/scitech/datait/databases/cwns/2008reportdata.cfm. Accessed November 2010.

EPA. 2009a. Draft 2009 Update Aquatic Life Ambient Water Quality Criteria For Ammonia—Freshwater. EPA-822-D-09-001. Available at http://water.epa.gov/scitech/swguidance/standards/criteria/aqlife/pollutants/ammonia/upload/2009_12_23_criteria_ammonia_2009update.pdf. Accessed December 15, 2011.

EPA. 2009b. National Primary Drinking Water Regulations. EPA/816/F-09/04. Available at http://www.epa.gov/safewater/consumer/pdf/mcl.pdf.

EPA. 2010a. A New Approach to Protecting Drinking Water and Public Health. March 2010. Available at http://www.epa.gov/ogwdw000/sdwa/pdfs/Drinking_Water_Strategyfs.pdf.

EPA. 2010b. Emerging Contaminant—N-Nitrosodimethylamine (NDMA). EPA/505/F-10/005. Available at http//www.epa.gov/fedfac/documents/emerging_contaminant_ndma.pdf.

EPA. 2010c. Human Health Risk Assessment: Dose-Response Assessment. Available at http://www.epa.gov/risk/dose-response.htm.

EPA. 2010d. OPPTS harmonized test guidelines—master list. Available online at http://www.epa.gov/ocspp/pubs/frs/publications/Test_Guidelines/OPPTS-TestGuidelines_Master List-2010-08-04.pdf.

EPA. 2011. Working in Partnership with States to Address Phosphorus and Nitrogen Pollution through Use of a Framework for State Nutrient Reductions. Memorandum from Nancy Stoner, Acting Assistant Administrator, Office of Water: March 16. Available online at: http://water.epa.gov/scitech/swguidance/standards/criteria/nutrients/upload/memo_nitrogen_framework.pdf.

EPRI (Electric Power Research Institute). 2002. Water & Sustainability (Volume 4): U.S. Electricity Consumption for Water Supply and Treatment—The Next Half Century. Palo Alto, CA: EPRI.

Equinox Center. 2010. The Potential of Purified Recycled Water. Available at http://www.equinoxcenter.org/assets/files/pdf/EquinoxPotentialofrecycledwaterjuly2010finalrev.pdf. Accessed January 15, 2011.

Erickson, R. J., D. A. Benoit, V. R. Mattson, H. P. Nelson, Jr., and E. N. Leonard, 1996. The Effects of Water Chemistry on the Toxicity of Copper to Fathead Minnows. Environmental Toxicology and Chemistry 15(2): 181-193.

Escher, B. I., N. Bramaz, M. Maurer, M. Richter, D. Sutter, C. Von Kanel, and M. Zschokke. 2005. Screening test battery for pharmaceuticals in urine and wastewater. Environmental Toxicology and Chemistry 24(3): 750-758.

Esplugas, S., D. M. Bila, L. G. Krause, and M. Dezotti. 2007. Ozonation and advanced oxidation technologies to remove endocrine disrupting chemicals (EDCs) and pharmaceuticals and personal care products (PPCPs) in water effluents. Journal of Hazardous Material 149(3): 631-642.

European Commission, Scientific Committee on Consumer Safety. 2009. Opinion on Triclosan—Antibiotic Resistance. Available at http://ec.europa.eu/health/scientific_committees/consumer_safety/docs/sccs_o_023.pdf. Accessed November 6, 2011.

European Union. 2002. 4-Nonylphenol (branched) and nonylphenol. European Union Risk Assessment Report. EUR 20387 EN.

Everts, C. M., and A. H. Dahl. 1957. The Federal Water-Pollution Control Act of 1956. American Journal of Public Health and the Nations Health 47(3): 305-310.

Fabrega, J., S. N. Luoma, C. R. Tyler, T. S. Galloway, and J. R. Lead. 2011. Silver nanoparticles: Behaviour and effects in the aquatic environment. Environment International 37: 517-531.

Fair, G. M., J. C. Morris, S. L. Chang, I. Weil, and R. P. Burden. 1948. The behaviour of chlorine as water disinfectant. Journal of the American Water Works Association 1: 1051-1061.

Farée, M., K. Döder, L. Hearn, Y. Poussade, J. Keller, and W. Gernjak. 2011. Understanding the Operational Parameters Affecting NDMA Formation at Advanced Water Treatment Plants. Journal of Hazardous Materials 185: 1575-1581.

Faustman, E. M., and G. S. Omenn. 2008. Risk Assessment. In C. D. Klassen, ed., Cassarett and Doull's Toxicology. The Basic Science of Poisons, 7th Ed. New York: McGraw-Hill, pp. 107-128.

FDA (U.S. Food and Drug Administration). 1995. Food additives: Threshold of regulation for substances used in food-contact articles (final rule). Federal Register 60: 36582-36596.

FDA. 2011. FDA MRTD Database. Available at http://www.fda.gov/aboutfda/centersoffices/cder/ucm092199.htm.

FDEP (Florida Department of Environmental Protection). 2006. 2005 Reuse Inventory: Tallahassee: FDEP, Water Reuse Program, 21 pp.

FDEP. 2010. 2009 Reuse Inventory. Tallahassee: FDEP, Water Reuse Program.

FDEP. 2011. 2010 Reuse Inventory. Tallahassee: FDEP, Water Reuse Program. Available at http://www.dep.state.fl.us/water/reuse/inventory.htm.

Fernandez Escartin, E., J. Saldana Lozano, O. R. Garcia. and D. O. Cliver. 2002. Potential Salmonella transmission from ornamental fountains. Journal of Environmental Health 65(4): 9-12, 22.

Filby, A. L., J. A. Shears, B. E. Drage, J. H. Churchley, and C. R. Tyler. 2010. Effects of advanced treatments of wastewater effluents on estrogenic and reproductive health impacts in fish. Environmental Science & Technology 44(11): 4348-4354.

Finch, G. R., C. N. Haas, J. A. Oppenheimer, G. Gordon, and R. R. Trussell. 2001. Design criteria for inactivation of cryptosporidium by ozone in drinking water. Ozone: Science and Engineering, 23(4): 259-284.

Finkel, A. M. 1990. Confronting Uncertainty in Risk Management. Washington, DC: Resources for the Future, Center for Risk Management.

FL OPPAGA. 1999. Florida Water Policy, OPPAGA Report No. 99-06. Available at http://www.oppaga.state.fl.us/MonitorDocs/Reports/pdf/9906rpt.pdf.

Fleming, P. A. 1990. Water supply, reclamation and reuse at Grand Canyon: A case study. In Proceedings of Conserv 90, August 12-16, Phoenix, Arizona. Dublin, OH: National Water Well Association, pp. 129-133.

Florida Department of Environmental Protection. 2007. Reuse of Reclaimed Water and Land Application. Florida Administrative Code Chapter 62-610. Tallahassee: Florida Department of Environmental Protection.

Focazio, M. J., D. W. Kolpin, K. K. Barnes, E. T. Furlong, M. T. Meyer, S. D. Zaugg, L. B. Barber, and M. E. Thurman. 2008. A national reconnaissance for pharmaceuticals and other organic wastewater contaminants in the United States. II. Untreated drinking water sources. Science of the Total Environment 402(2-3): 201-216.

Fong, T. T., L. S. Mansfield, D. L. Wilson, D. J. Schwab, S. L. Molloy, and J. B. Rose. 2007. Massive microbiological groundwater contamination associated with a waterborne outbreak in Lake Erie, South Bass Island, Ohio. Environmental Health Perspectives 115(6): 856-864.

Fong, T. T., M. S. Phanikumar, I. Xagoraraki, and J. B. Rose. 2010. Quantitative detection of human adenoviruses in wastewater and combined sewer overflows influencing a Michigan river. Applied and Environmental Microbiology 76: 715-723.

Fono, L. J., E. P. Kolodziej, and D. L. Sedlak. 2006. Attenuation of wastewater-derived contaminants in an effluent-dominated river. Environmental Science & Technology 40(23): 7257-7262.

Fox, P., S. Houston, P. Westerhoff, J. E. Drewes, M. Nellor, W. Yanko, R. Baird, M. Rincon, R. Arnold, K. Lansey, R. Bassett, C. Gerba, M. Karpiscak, G. Amy, and M. Reinhard. 2001. Soil Aquifer Treatment for Sustainable Water Reuse. Denver, CO: American Water Works Association Research Foundation.

Freeman, G., M. Poghosyan, and M. Lee. 2008. Where Will We Get the Water? Assessing Southern California's Future Water Strategies. Los Angeles County Economic Development Corporation. Available at http://www.laedc.org/sclc/documents/Water_SoCalWaterStrategies.pdf Accessed January 15, 2011.

Frerichs, R. R. 1984. Epidemiologic monitoring of possible health reactions of wastewater reuse. Science of the Total Environment 32(3): 353-363.

Frerichs, R. R., E. M. Sloss, and K. P. Satin. 1982. Epidemiologic impact of water reuse in Los-Angeles County. Environmental Research 29(1): 109-122.

Frick, H. 1985. Boron tolerance and accumulation in the duckweed, *Lemnn minor*. 1. Plant Nutrition 8: 1123-1129.

Fu, C. Y., X. Xie, J. J. Huang, T. Zhang, Q. Y. Wu, J. N. Chen, and H. Y. Hu. 2010. Monitoring and evaluation of removal of pathogens at municipal wastewater treatment plants. Water Science Technology 61: 1589-1599.

Fujita, Y., W. H. Ding, and M. Reinhard. 1996. Identification of wastewater dissolved organic carbon characteristics in reclaimed wastewater and recharged groundwater. Water Environment Research 68(5): 867-876.

Gan, J., S. Bondarenko, F. Ernst, W. Yang, S. Ries, and D. Sedlak. 2006. Leaching of *N*-Nitrosodimethylamine (NDMA) in turfgrass soils during wastewater Irrigation. Journal of Environmental Quality 35(1): 277-284.

Garcia-Reyero, N., R. J. Griffitt, L. Liu, K. J. Kroll, W. G. Farmerie, D. S. Barber, and N. D. Denslow. 2008. Construction of a robust microarray from a non-model species largemouth bass, *Micropterus salmoides* (Lacepede), using pyrosequencing technology. Journal of Fish Biology 72(9): 2354-2376.

Garrison, A. W., J. D. Pope, and F. R. Allen. 1975. GC/MS analysis of organic compounds in domestic wastewaters. In L. H. Keith, ed., Identification and Analysis of Organic Pollutants in Water. Ann Arbor, MI: Ann Arbor Science, pp. 517-556.

GEI Consutants/Navigant Consulting. 2010. Embedded Energy in Water Studies—Study 1: Statewide and Regional Water-Energy Relationship. Prepared for the California Public Utilities Commission, Energy Division, San Francisco, CA, August 31.

Gennaccaro, A. L., M. R. McLaughlin, W. Quintero-Betancourt, D. E. Huffman, and J. B. Rose. 2003. Infectious *Cryptosporidium parvum* oocysts in final reclaimed effluent. Applied and Environmental Microbiology 69: 4983-4984.

Gerba, C. P., J. B. Rose, and C. N. Haas. 1994. Waterborne disease: Who is at risk? In Proceedings of the American Water Works Association's Water Quality Technology Conference, San Francisco. Denver, CO: American Water Works Association.

Gikas, P., and G. Tchobanoglous. 2009. The role of satellite and decentralized strategies in water resources management. Journal of Environmental Management 90: 144-152.

Gleick, P. 2003. Global freshwater resources: Soft-path solutions for the 21st century. Science 302: 1524.

Gold, L. S., C. B. Sawyer, R. Magaw, G. M. Backman, M. De Veciana, R. Levinson, N. K. Hooper, W. R. Havender, L. Bernstein, R. Peto, M. C. Pike, and B. N. Ames. 1984. A carcinogenic potency database of the standardized results of animal bioassays. Environmental Health Perspectives 58: 9-319.

Gold, L. S., G. M. Backman, N. K. Hooper, and R. Peto. 1986. Ranking the potential carcinogenic hazards to workers from exposures to chemicals that are tumorigenic in rodents. Environmental Health Perspectives 76: 211-219.

Gold, L. S., T. H. Slone, G. M. Backman, R. Magaw, M. Da Costa, P. Lopipero, M. Blumenthal, and B. N. Ames. 1987. Second chronological supplement to the carcinogenic potency database: Standardized results of animal bioassays published through December 1984 and by the National Toxicology Program through May 1986. Environmental Health Perspectives 74: 237-329.

Gollnitz, W. D., J. L. Clancy, B. L. Whitteberry, and J. A. Vogt. 2003. RBF as a microbial treatment process. Journal—American Water Works Association 95(12): 56-66.

Gollnitz, W. D., J. L. Clancy, B. McEwen, and S. C. Garner. 2005. Riverbank filtration for IESWTR compliance. Journal—American Water Works Association 97(12): 64-76.

Gong, Z., T. Yan, Y. Tong, S. Shalin, H. S. Lee, E. Lee, and W. E. Hawkins. 2008. Singapore's evaluation of carcinogenic and estrogenic potentials of reclaimed water in medaka fish. In Proceedings of the IWA World Water Congress, Vienna, Sept. 8-12 (on CD-ROM).

Goodman, A. M., G. G. Ganf, G. C. Dandy, H. R. Maier, and M. S. Gibbs. 2010. The response of freshwater plants to salinity pulses. Aquatic Botany 93(2): 59-67.

Gordon, G., L. C. Adam, B. P. Bubnis, C. Kuo, R. S. Cushing, and R. H. Sakaji, 1997. Predicting liquid bleach decomposition. Journal—American Water Works Association 89(4): 142-149.

Govind, P., W. Alan, and J. E. Ezekiel. 2009. Principles for allocation of scarce medical interventions. Lancet 373(9661): 423-431.

Graf, W. 1999. Dam nation: A geographic census of American dams and their large-scale hydrologic impacts. Water Resources Research 5(4): 1305-1311.

Gray, N. F. 2008. Drinking water standards and risk. In Drinking Water Quality: Problems and Solutions. Cambridge, UK: Cambridge University Press.

Gray, J. L., and D. L. Sedlak. 2005. The fate of estrogenic hormones in an engineered treatment wetland with densed macrophytes. Water Environment Research 77: 24-31.

Greyshock, A. E., and P. J. Vikesland. 2006. Triclosan reactivity in chloraminated waters. Environmental Science & Technology 40(8): 2615-2622.

Grischek, T., D. Schoenheinz, E. Worch, and K. Hiscock. 2002. Bank filtration in Europe—An overview of aquifer conditions and hydraulic controls. In P. Dillon, ed., Management of Aquifer Recharge for Sustainability: Proceedings of the 4th International Symposium of Artificial Groundwater Recharge, Adelaide, Australia. Lisse, The Netherlands: A. A. Balkema, pp. 485-488.

Gross, B., J. Montgomery-Brown, A. Naumann, and M. Reinhard. 2004. Occurrence and fate of pharmaceuticals and alkylphenol ethoxylate metabolites in an effluent-dominated river and wetland. Environmental Toxicology and Chemistry 23(9): 2074-2083.

Grünheid, S., G. Amy, and M. Jekel. 2005. Removal of bulk dissolved organic carbon and trace organic compounds by bank filtration and artificial recharge. Water Research 39: 3219-3228.

Gunnarson, E., G. Axehult, G. Baturina, S. Zelenin, M. Zelenina, and A. Aperia. 2008. Identification of a molecular target for glutamate regulation of astrocyte water permeability. Glia 56: 587-596.

Gupta, V., W. P. Johnson, P. Shafieian, H. Ryu, A. Alum, M. Abbaszadegan, S. A. Hubbs, and T. Rauch-Williams. 2009. Riverbank filtration: Comparison of pilot scale transport with theory. Environmental Science & Technology 43(3): 669-676.

Haarhof, J., and B. Van der Merwe. 1996. Twenty Five Years of Wastewater Reclamation in Windhoek, Namibia. Water Science and Technology 33(10-11): 25-35.

Haas, C. N. 1983. Estimation of risk due to low doses of microorganisms: A comparison of alternative methodologies. American Journal of Epidemiology 118(4): 573-582.

Haas, C. N. 1996. How to average microbial densities to characterize risk. Water Research 30(4): 1036-1038.

Haas, C. N. 2010. Chemical Disinfection. Water Quality and Treatment Handbook, 6th Ed. New York: McGraw-Hill.

Haas, C. N., C. Crockett, J. B. Rose, C. Gerba, and A. Fazil. 1996. Infectivity of *Cryptosporidium parvum* oocysts. Journal—American Water Works Association 88: 131-136.

Haas, C. N., and R. R. Trussell. 1998. Frameworks for assessing reliability of multiple, independent barriers in potable water reuse. Water Science and Technology 38(6): 1-8.

Haas, C., J. Rose, and C. Gerba. 1999. Quantitative Microbial Risk Assessment. New York: John Wiley & Sons.

Haas, C. N., A. Thayyar-Madabusi, J. B. Rose, and C. P. Gerba. 2000. Development of a dose-response relationship for *Escherichia coli* O157:H7. International Journal of Food Microbiology 56(2-3): 153-159.

Haddad, B., P. Rozin, P. Slovic, and C. Nemeroff. 2010. The Psychology of Water Reclamation and Reuse: Survey Findings and Research Roadmap. Arlington, VA: WateReuse Foundation.

Hamlin, C. 1980. Sewage—waste or resource. Environment 22(8): 16.

Hammes F. A., and T. Egli. 2005. New method for assimilable organic carbon determination using flow-cytometric enumeration and a natural microbial consortium as inoculum. Environmental Science & Technology 39: 3289-3294.

Hannah, R., V. J. D'Aco, P. D. Anderson, M. E. Buzby, D. J. Caldwell, V. L. Cunningham, J. F. Ericson, A. C. Johnson, N. J. Parke, J. H. Samuelian, and J. P. Sumpter. 2009. Exposure assessment of 17α-ethinyl estradiol in surface waters of the United States and Europe. Environmental Toxicology and Chemistry 28(12): 2725-2732.

Haramoto, E., H. Katayama, K. Oguma, H. Yamashita, A. Tajima, H. Nakajima, and S. Ohgaki. 2006. Seasonal profiles of human noroviruses and indicator bacteria in a wastewater treatment plant in Tokyo, Japan. Water Science and Technology 54: 301-308.

Haramoto, E., H. Katayama, K. Oguma, and S. Ohgaki. 2007. Quantitative analysis of human enteric adenoviruses in aquatic environments. Journal of Applied Microbiology 103: 2153-2159.

Hartley, T. 2006. Public perception and participation in water reuse. Desalination 187: 115-126.

Harvilicz, H. 1999. NPE demand remains strong despite environmental concerns in Europe. Chemical Market Reporter, October 18, 1999, p. 15.

Harwood, V. J., A. D. Levine, T. M. Scott, V. Chivukula, J. Lukasik, S. R. Farrah, and J. B. Rose. 2005. Validity of the indicator organism paradigm for pathogen reduction in reclaimed water and public health protection. Applied and Environmental Microbiology 71(6): 3163-3170.

Havelaar, A. H. 1994. Application of HACCP to drinking water supply. Food Control 5(3): 145-152.

Havelaar, A. H., M. van Olphen, and J. F. Schijeven. 1995. Removal and inactivation of viruses by drinking water treatment processes under full scale conditions. Water Science and Technology 31(5-6): 55-62.

Havelaar, A. H., A. E. De Hollander, P. F. Teunis, E. G. Evers, H. J. Van Kranen, J. F. Versteegh, J. E. Van Koten, and W. Slob. 2000. Balancing the risks and benefits of drinking water disinfection: Disability adjusted life-years on the scale. Environmental Health Perspectives 108(4): 315-321.

Hazen, A. 1909. Clean Water and How to Get It. New York: John Wiley and Sons.

He, J. W., and S. Jiang. 2005. Quantification of enterococci and human adenoviruses in environmental samples by real-time PCR. Applied and Environmental Microbiology 71: 2250-2255.

Heber, M. A., D. K. Reed-Judkins, and T. T. Davies. 1996. USEPA's whole effluent toxicity testing program: A national regulatory perspective. In D. R. Grothe, K. L. Dickson, and D. K. Reed-Judkins, eds., Whole Effluent Toxicity Testing: An Evaluation of Methods and Prediction of Receiving Water Impacts. Pensacola, FL: SETAC Press, pp 9-15.

Heberer, T. 2002. Occurrence, fate, and assessment of polycyclic musk residues in the aquatic environment of urban areas: A review. Acta Hydrochimica et Hydrobiologica 30(5-6): 227-243.

Heerden, J., M. M. Ehlers, and W. O. Grabow. 2005. Detection and risk assessment of adenoviruses in swimming pool water. Journal of Applied Microbiology 99(5): 1256-1264.

Hejkal, T. W., F. M. Wellings, P. A. LaRock, and A. L. Lewis. 1979. Survival of poliovirus within organic solids during chlorination. Applied and Environmental Microbiology 38: 114.

Hemmer, J., C. Hamann, and D. Pickard. 1994. Tampa Water Resource Recovery Project. San Diego Water Reuse Health Effects Study. In 1994 Water Reuse Symposium Proceedings, Feebruary 27-March 2, Dallas, TX. Denver CO: American Water Works Association.

Heringa, M. B., D. J. H. Harmsen, E. F. Beerendonk, A. A. Reus, C. A. M. Krul, D. H. Metz, and G. F. IJpelaar. 2011. Formation and removal of genotoxic activity during UV/H$_2$O$_2$–GAC treatment of drinking water. Water Research 45(1): 366-374.

Hignite, C., and D. L. Azarnoff. 1977. Drugs and drug metabolites as environmental contaminants: Chlorophenoxyisobutyrate and salicylic acid in sewage water effluent. Life Sciences 20(2): 337-341.

Hijnen, W., E. Beerendonk, and G. Medema. 2005a. Inactivation credit of UV radiation for viruses, bacteria and protozoan (oo) cysts in water: A review. Water Research 40(1): 3-22.

Hijnen, W. A., A. J. Brouwer-Hanzens, K. J. Charles, and G. J. Medema. 2005b. Transport of MS2 Phage, Escherichia coli, Cryptosporidium parvum, and Giardia intestinalis in a gravel and a sandy soil. Environmental Science & Technology 39(20): 7860-7868.

Hills, S., R. Birks, and B. McKenzie. 2002. The Millennium Dome "Watercycle" experiment to evaluate water efficiency and customer perception at a recycling scheme for 6 million visitors. Water Science and Technology 46(6-7): 233-240.

Hinckley, G. T., C. J. Johnson, K. H. Jacobson, C. Bartholomay, K. D. McMahon, D. McKenzie, J. M. Aiken, and J. A. Pedersen. 2008. Persistence of pathogenic prion protein during simulated wastewater treatment processes. Environmental Science & Technology 42(14): 5254-5259.

Hiscock, K. M., and T. Grischek. 2002. Attenuation of groundwater pollution by bank filtration. Journal of Hydrology 266(3-4): 139-144.

Hoekman, D. 2010. Turning up the Heat: Temperature influences the relative importance of top-down and bottom-up effects. Ecology 91(10): 2819-2825.

Holliman, P. T. 2009. Manual of Practice: How to Develop a Water Reuse Program. Alexandria, VA: WateReuse Association.

Hollis, L., L. Muench, and R. C. Playle. 1997. Influence of dissolved organic matter on copper binding, and calcium on cadmium binding, by gills of rainbow trout. Journal of Fish Biology 50: 703-720.

Holmes, M., A. Kumar, A. Shareef, H. Doan, R. Stuetz, and R. Kookana. 2010. Fate of indicator endocrine disrupting chemicals in sewage during treatment and polishing for non-potable reuse. Water Science and Technology 62(6): 1416-1423.

Hopkins, J., R. Williamson, and R. Simonsen. 2002. The Impact of Recycled Water on Water Quality in Coyote Creek in San Jose, California in 2001. Prepared for the City of San Jose, CA. Available at http://www.sanjoseca.gov/esd/PDFs/Recycled WaterImpact-CoyoteCreek_0702.pdf.

Hoppe-Jones, C., G. Oldham, and J. E. Drewes. 2010. Attenuation of total organic carbon and unregulated trace organic chemicals in U.S. riverbank filtration systems. Water Research 44: 4643-4659.

Howard, I., E. Espigares, P. Lardelli, J. L. Martín, and M. Espigares. 2004. Evaluation of microbiological and physicochemical indicators for wastewater treatment. Environmental Toxicology 19(3): 241-249.

Howe, P. D. 1998. A review of boron effects in the environment. Biological Trace Element Research 66: 153-166.

Hoxie, N. J., J. P. Davis, J. M. Vergeront, R. D. Nashold, and K. A. Blair. 1997. Cryptosporidiosis-associated mortality following a massive waterborne outbreak in Milwaukee, Wisconsin. American Journal of Public Health 87(12): 2032-2035.

Hua, G., and D. A. Reckhow. 2007. Comparison of disinfection byproduct formation from chlorine and alternative disinfectants. Water Research 41(8): 1667-1678.

Hua, G., and D. A. Reckhow. 2008. DBP formation during chlorination and chloramination: Effect of reaction time, pH, dosage, and temperature. Journal—American Water Works Association 100(8): 82-95.

Huang, C. H. and D. L. Sedlak. 2001. Analysis of Estrogenic Hormones in Municipal Wastewater Effluent and Surface Water Using Enzyme-Linked Immunosorbent Assays and Gas Chromatography/Tandem Mass Spectrometry. Environmental Toxicology and Chemistry 20: 133-139.

Hudson, H. E., and F. W. Gilcreas. 1976. Health and economic aspects of water hardness and corrosiveness. Journal—American Water Works Association 68(4): 201-204.

Huertas, E., M. Salgot, J. Hollender, S. Weber, W. Dott, S. Khan, A. Schafer, R. Messalem, B. Bis, A. Sharoni, and H. Chikurel. 2008. Key objectives for water reuse concepts. Desalination 218: 120-131.

Huggett, D. B., J. C. Cook, J. F. Ericson, and R. T. Williams. 2003. A theoretical model for utilizing mammalian pharmacology and safety data to prioritize potential impacts of human pharmaceuticals to fish. Human and Ecological Risk Assessment 9(7): 1789-1799.

Huggett, D. B., J. F. Ericson, J. C. Cook, and R. T. Williams. 2004. Plasma concentrations of human pharmaceuticals as predictors of pharmacological responses in fish. In K. Kummerer, ed., Pharmaceuticals in the Environment. Heidelberg, Germany: Springer, pp. 373-386.

Huitric, S., J. Kuo, C. Tang, M. Creel, R. Horvath, and J. Stahl. 2005. Fate of NDMA in tertiary water reclamation plants. In Proceedings of the Water Environment Federation Technical Exhibition and Conference 2005. Alexandria, VA: Water Environment Federation, pp. 582-595.

Huitric, S., C. Tang, J. Kuo, M. Creel, D. Snyder, P. Ackman, and R. Horvath. 2007. Sequential chlorination for reclaimed water disinfection. In Proceedings of the Water Environment Federation Technical Exhibition and Conference 2007. Alexandria, VA: Water Environment Federation, pp. 2552-2564.

Hurlimann, A. 2008. Urban versus regional—how public attitudes to recycled water differ in these contexts. Water Science and Technology 57(6): 891-899.

Hurlimann, A., and S. Dolnicar. 2010. Acceptance of water alternatives in Australia—2009. Water Science and Technology 61(8): 2137-2142.

Hutcheson, M. S., K. A. Martin, and D. M. Mangaro. 1990. Guide to the Regulation of Toxic Chemicals in Massachusetts Waters. Massachusetts Department of Environmental Protection. Boston, MA.

Hutson, S. S., N. L. Barber, J. F. Kenny, K. S. Linsey, D. S. Lumia, and M. A. Maupin. 2000. Estimated Water Use in the United States in 2000. USGS Circular 1268. Reston, VA: U.S. Geological Survey.

Icekson-Tal, N., O. Avraham, J. Sack, and H. Cikurel. 2003. Water reuse in Israel—the Dan Region Project: Evaluation of water quality and reliability of plant's operation. Water Science and Technology: Water Supply 3(4): 231-237.

Illing, H. P. A. 2006. Risk assessment. In J. H. Duffus and H. G. Worth, eds., Fundamental Toxicology. Cambridge, UK: RSC Publishing, pp. 56-71.

Ingram, P., V. Young, M. Millan, C. Chang, and T. Tabucchi. 2006. From controversy to consensus: The Redwood City recycled water experience. Desalination 187(1-3): 179-190.

Isaacson, M., and A. R. Sayed. 1988. Health aspects of the use of recycled water in Windhoek, SWA/Namibia, 1974-1983. South African Medical Journal 73: 596-599.

Isaacson, M., A. R. Sayed, and W. H. J. Hattingh. 1987. Studies on health aspects of water reclamation during 1974-1983 in Windhoek, South West Africa/ Namibia. 38 (1): 87. Pretoria. Water Research Commission.

Jacangelo, J., S. Adham, and J. M. Lainé. 1997. Membrane Filtration for Microbial Remova. Denver, CO: American Water Works Association.

Jacangelo, J. G., N. L. Patania, R. R. Trussell, C. N. Haas, and C. Gerba. 2002. Inactivation of waterborne emerging pathogens by selected disinfectants, American Water Works Association. Denver, Colorado. 145.

Jaques, R. S., G. M. Antonz, R. C. Cooper, and B. Sheikh. 1999. Pathogen removal effectiveness of a full-scale recycling plant. In Proceedings of the Water Environment Federation Technical Exhibition and Conference 1999. Alexandria, VA: Water Environment Federation.

Jeffrey, P. 2002. Public attitudes to in-house water recycling in England and Wales. Water and Environmental Management Journal 16: 214-217.

Jiang, S. C. 2006. Human adenoviruses in water: Occurrence and health implications: A critical review. Environmental Science & Technology 40: 7132-7140.

Jiménez, B. 2007. Helminth ova control in sludge: A review. Water Science and Technology 56: 147-155.

Jiménez, B., and T. Asano. 2008. Water reclamation and reuse around the world. In B. Jimenez and T. Asano, eds., Water Reuse: An International Survey of Current Practice, Issues and Needs. London: IWA Publishing, pp. 3-26.

Jiménez, B., and A. Chavez. 2004. Quality assessment of an aquifer recharged with wastewater for its potential use as drinking source: "El Mezquital Valley" case. Water Science and Technology 50(2): 269-276.

Jiménez, M.S., M.T. Gómez, E. Bolea, F. Laborda, and J. Castillo. 2011. An approach to the natural and engineered nanoparticles analysis in the environment by inductively coupled plasma mass spectrometry. International Journal of Mass Spectrometry. 307: 99-104.

Johnson, A. C., H. R. Aerni, A. Gerritsen, M. Gibert, W. Giger, K. Hylland, M. Jurgens, T. Nakari, A. Pickering, M. J. F. Suter, A. Svenson, and F. E. Wettstein. 2005. Comparing steroid estrogen, and nonylphenol content across a range of European sewage plants with different treatment and management practices. Water Research 39: 47-58.

Johnson, T. 2007. Battling Seawater Intrusion in the Central & West Coast Basins. Water Replenishment District Technical Bulletin 13(Fall). Available online at: http://www.wrd.org/engineering/seawater-intrusion-los-angeles.php. Accessed January 3, 2011.

Jones G. P., M. I. McCormick, M. Srinivasan, and J. V. Eage. 2004. Coral decline threatens fish biodiversity in marine reserves. Proceedings of the National Academy of Sciences, USA. 101:8251-8253.

Joo, S. H., and W. A. Mitch. 2007. Nitrile, aldehyde, and halonitroalkane formation during chlorination/chloramination of primary amines. Environmental Science & Technology 41(4): 1288-1296.

Juhna, T., and E. Melin. 2006. Ozonation and Biofiltration in Water Treatment: Operational Status and Optimization Issues. Techneau, WP5.3, Sixth Framework Programme, European Commission. Available at http://www.techneau.org/fileadmin/files/Publications/Publications/Deliverables/D5.3.1B-OBF.pdf.

Kadlec R. H., and R. L. Knight. 1996. Treatment Wetlands. Boca Raton, FL: CRC Press, p. 893.

Kaegi, R., A. Voegelin, B. Sinnet, S. Zuleeg, H. Hagendorfer, M. Burkhardt, and H. Siegris. 2011. Behavior of metallic silver nanoparticles in a pilot wastewater treatment plant. Environmental Science & Technology 45(9): 3902-3908.

Kahneman, D., P. Slovic, and A. Tversky, eds. 1982. Judgment under Uncertainty: Heuristics and Biases. New York: Cambridge University Press.

Kannan, K., L. Tao, E. Sinclair, S. D. Pastva, D. J. Jude, and J. P. Giesy. 2005. Perfluorinated compounds in aquatic organisms at various trophic levels in a great lakes food chain. Archives of Environmental Contamination and Toxicology 48: 559-566.

Kaplan, D., and A. Kaplan. 1985. Biodegrdation of N-nitrosodimethyamine in aqueous and soil systems Applied and Environmental Microbiology 50(4): 1077-1086.

Katayama, H., E. Haramoto, K. Oguma, H. Yamashita, A. Tajima, H. Nakajima, and S. Ohyaki. 2008. One-year monthly quantitative survey of noroviruses, enteroviruses, and adenoviruses in wastewater collected from six plants in Japan. Water Research 42: 1441-1448.

Keel, M. K., A. M. Ruiz, A. T. Fisk, W. K. Rumbeiha, A. K. Davis, and J. C.Maerz. 2010. Soft-tissue mineralization of bullfrog larvae (Rana catesbeiana) at a wastewater treatment facility. Journal of Veterinary Diagnostic Investigation 22(4): 655-660. Available at http://www.ncbi.nlm.nih.gov/pubmed/20622246. Accessed on November 17, 2011.

Kelly, R. T., I. D. Henriques, and N. G. Love. 2004. Chemical inhibition of nitrification in activated sludge. Biotechnology and Bioengineering 85(6): 683-694.

Kenny, J., N. Barber, S. Hutson, K. Linsey, J. Lovelace, and M. Maupin. 2009. Estimated Use of Water in the United States in 2005. USGS Circular 1344. Reston, VA: U.S. Geological Survey.

Khan, S. J. 2010. Quantitative Chemical Exposure Assessment for Water Recycling Schemes. Waterlines Report Series No. 27. National Water Commission of Australia. Available at http://www.palisade.com/downloads/pdf/papers/Waterlines_Quantitative_Chemical_Exposure.pdf.

Khan, S. and D. Roser. 2007. Risk assessment and health effect studies of indirect potable reuse schemes. Centre for Water and Waste Technology. School of Civil and Environmental Engineering. University of New South Wales, Australia. NSW, Australia.

Kidd, K. A., P. J. Blanchfield, K. H. Mills, V. P. Palace, R. E. Evans, J. M. Lazorchak, and R. W. Flick. 2007. Collapse of a fish population after exposure to a synthetic estrogen. Proceedings of the National Academy of Sciences of the United States of America. 104(21): 8897-8901.

Kidson, R., B. Haddad, and H. Zheng. 2009. Improving water supply through portfolio management: Case study from Southern California. Proceedings: 4th WEAS International Conference in Water Resources, Hydraulics, and Hydrology. University of Cambridge, February 24-26. WSEAS Press, pp. 157-162. Available at http://www.wseas.us/books/2009/cambridge/WHH.pdf.

Kim, S. B., and M. Y. Corpcioglu. 2002. contaminant transport in dual-porosity media with dissolved organic matter and bacteria present as mobile colloids. Journal of Contaminant Hydrology 59(3-4): 267-289.

Kim, Y., K. Choi, J. Jung, S. Park, P. Kim, and J. Park. 2007. Aquatic toxicity of acetaminophen, carbamazepine, cimetidine, diltiazem and six major sulfonamides, and their potential ecological risks in korea. Environment International 33(3): 370-375.

Kim, Y. M., D. Park, D. S. Lee, K. A. Jung, and J. M. Park. 2009. Sudden failure of biological nitrogen and carbon removal in the full-scale pre-denitrification process treating cokes wastewater. Bioresource Technology 100(19): 4340-4347.

Kiser, M. A., P. Westerhoff, T. Benn, Y. Wang, J. Pérez-Rivera, and K. Hristovski. 2009. Titanium nanomaterial removal and release from wastewater treatment plants. Environmental Science & Technology 43: 6757-6763.

Kohut, K. D., and S. A. Andrews. 2003. Polyelectrolyte age and N-nitrosodimethylamine formation in drinking water treatment. Water Quality Research Journal of Canada 38(4): 719-735.

Kolpin, D., E. T. Furlong, M. T. Meyer, E. M. Thurman, S. D. Zaugg, L. B. Barber, and H. T. Buxton. 2002. Pharmaceuticals, hormones, and other organic wastewater contaminants in U.S. streams, 1999-2000—A national reconnaissance. Environmental Science & Technology 35: 1202-1211.

Kramer, V. J., M. A. Etterson, M. Hecker, C. A. Murphy, G. Roesijadi, D. J. Spade, J. A. Spromberg, M. Wang, and G. T. Ankley. 2011. Adverse outcome pathways and ecological risk assessment: Bridging to population-level effects. Environmental Toxicology and Chemistry 30(1): 64-76.

Krasner, S., H. Weinberg, S. Richardson, S. Pastor, R. Chinn, M. Sclimenti, G. Onstad, and A. Thurston. 2006. Occurrence of a new generation of disinfection byproducts. Environmental Science & Technology 40: 7175-7185.

Krasner, S., P. Westerhoff, B. Chen, G. Amy, S. Nam, Z. Chowdhury, S. Sinha, and B. Rittman. 2008. Contribution of Wastewater to DBP Formation. Final Report. Denver, CO: AWWA Research Foundation.

Krasner, S. W., P. Westerhoff, B. Chen, B. E. Rittman, and G. Amy. 2009. Occurrence of disinfection byproducts in United States wastewater treatment plant effluents. Environmental Science & Technology 43(21): 8320-8325.

Krauss, M., P. Longrée, F. Dorusch, C. Ort, and J. Hollender. 2009. Occurrence and removal of N-nitrosamines in wastewater plants. Water Research 43: 4381-4391.

Kroes, R., A. G. Renwick, M. Cheeseman, J. Kleiner, I. Mangelsdorf, A. Piersma, B. Schilter, J. Schlatter, F. van Schothorst, J. G. Vos, and G. Wurtzen. 2004. Structure-based thresholds of toxicological concern: Guidance for application to substances present at low levels in the diet. Food and Chemical Toxicology 42: 65-83. Available at http://www.ilsi.org/Europe/Publications/2004Stru_Base.pdf. Accessed on November 17, 2011.

Krueger, A. 2007. Chula Vista center connected to pipes carrying treated sewage. Available at SignOnSanDiego.com, August 22. San Diego, California.

Kühn, W., and U. Müller. 2000. Riverbank filtration: An overview. Journal—American Water Works Association 92(12): 60-69.

Kumamoto, H., and E. J. Henley. 1996. Probabilistic Risk Assessment and Management for Engineers and Scientists, 2nd Ed. New York: Institute of Electrical and Electronics Engineers, Inc.

Kumar, A. Y., and M. V. Reddy. 2009. Assessment of seasonal effects of municipal sewage pollution on the water quality of an urban canal—a case study of the Buckingham canal at Kalpakkam (India): NO_3, PO_4, SO_4, BOD, COD and DO. Environmental Monitoring and Assessment 157(1-4): 223-234.

Küster, A., A. C. Alfredo, B. I. Escher, K. Duis, K. Fenner, J. Garric, T. H. Hutchinson, D. R. Lapen, A. Pery, J. Rombke, J. Snape, T. Ternes, E. Topp, A. Wehrman, and T. Knacker. 2010. Environmental risk assessment of human pharmaceuticals in the European Union: A case study with the β-blocker atenolol. Integrated Environmental Assessment and Management 6: 514-523.

LACSD (Los Angeles County Sanitation Districts). 2008. Final Project Report for Montebello Forebay Attenuation and Dilution Studies. Prepared by the Lawrence Berkeley National Laboratory and Kennedy/Jenks/Todd LLC. Oakland, CA.

Lahnsteiner, J., and G. Lempert. 2005. Water Management in Windhoek/Namibia. In Proceedings of the IWA Conference, Wastewater Reclamation & Reuse for Sustainability, November 8-11, Jeju, Korea.

Länge, R., and D. Dietrich. 2002. Environmental risk assessment of pharmaceutical drug substances—conceptual considerations. Toxicology Letters 131(1-2): 97-104.

Länge, R., T. H. Hutchinson, C. P. Croudace, F. Siegmund, H. Schweinfurth, P. Hampe, G. H. Panter, and J. P. Sumpter. 2001. Effects of the synthetic estrogen 17 alpha-ethinylestradiol on the life-cycle of the fathead minnow (Pimephales promelas). Environmental Toxicology and Chemistry 20(6): 1216-1227.

Laposata, M. M., and W. A. Dunson. 1998. Effects of boron and nitrate on hatching success of amphibian eggs. Archives of Environmental Contaminants and Toxicology 35: 615-619.

Lauer, W. C., and S. E. Rogers. 1996. The demonstration of direct potable reuse: Denver's pioneer project. In AWWA/WEF 1996 Water Reuse Conference Proceedings. Denver, CO: American Water Works Association, pp. 269-289.

Lauer, W. C., F. J. Johns, G. W. Wolfe, B. A. Meyers, L. W. Condie, and J. F. Borzelleca. 1990. Comprehensive health effects testing program for Denver's potable reuse demonstration project. Journal of Toxicology and Environmental Health 30: 305-321.

Law, I. B. 2003. Advanced reuse—from Windhoek to Singapore and beyond. Water 30(5): 31-36.

Laws, B., E. Dickenson, T. Johnson, S. Snyder, and J. E. Drewes. 2011. Attenuation of contaminants of emerging concern during surface spreading aquifer recharge. Science of the Total Environment 409(6): 1087-1094.

LeChevallier, M. W., T. M. Evans, and R. J. Seidler. 1981. Effect of turbidity on chlorination efficiency and bacterial persistence in drinking water. Applied and Environmental Microbiology 42: 159-167.

Lehmacher, A., J. Bockemuhl, and S. Aleksic. 1995. Nationwide outbreak of human salmonellosis in Germany due to contaminated paprika and paprika-powdered potato chips. Epidemiology and Infection. 115:501-511.

LeClech, P., V. Chen, and T. Fane. 2006. Fouling in membrane bioreactors used in wastewater treatment. Journal of Membrane Science 284: 17-53.

Lee, B. C., M. Kamata, Y. Akatsuka, M. Takeda, K. Ohno, T. Kamei, and Y. Magara. 2004. Effects of chlorine on the decrease of estrogenic chemicals. Water Research 38: 733-739.

Lee, C., J. Yoon, and U. von Gunten. 2007. Oxidative degradation of *N*-nitrosodimethylamine by conventional ozonation and the advanced oxidation process ozone/hydrogen peroxide. Water Research 41(3): 581-590.

Lenz, K., V. Beck, and M. Fuerhacker. 2004. Behaviour of bisphenol A (BPA), 4-nonylphenol (4-HP) and 4-nonylphenol ethoxylates (4-NP1EO, 4-NP2EO) in oxidative water treatment processes. Water Science and Technology 50(5): 141-147.

Levantesi, C., R. La Mantia, C. Masciopinto, U. Böckelmann, M. Neus Ayuso-Gabella, M. Salgot, V. Tandoi, E. Van Houtte, T. Wintgens, and E. Grohmann. 2010. Quantification of pathogenic microorganisms and microbial indicators in three wastewater reclamation and managed aquifer recharge facilities in Europe. Science of the Total Environment 408(21): 4923-4930.

Li, X., L. W. Y. Yueng, M. Xu, S. Taniyasu, P. K. S. Lam, N. Yamashita, and J. Dai. 2008. Perfluorooctane sulfonate (PFOS) and other fluorochemicals in fish blood collected near the outfall of wastewater treatment plant (WWTP) in Beijing. Environmental Pollution 156(3): 1298-1303.

Liang, J. L., E. J. Dziuban, G. F. Craun, V. Hill, M. R. Moore, R. J. Gelting, R. L. Calderon, M. J. Beach, and S. L. Roy. 2006. Surveillance for waterborne disease and outbreaks associated with drinking water and water not intended for drinking—United States, 2003-2004. Morbidity and Mortality Weekly Report 55: 31-58.

Lightbody, A. F., M. E. Avener, and H. M. Nepf. 2008. Observations of short-circuiting flow paths within a free-surface wetland in Augusta, Georgia, USA. Limnology and Oceanography 53(3): 1040-1053.

Lim, Y. A. L., W. I. W. Hafiz, and V. Nissapatom. 2007. Reduction of *Cryptosporidium* and *Giardia* by sewage treatment processes. Tropical Biomedicine 24: 95-104.

Lin, A. Y. C., J. F. Debroux, J. A. Cunningham, and M. Reinhard. 2003. Comparison of rhodamine WT and bromide in the determination of hydraulic characteristics of constructed wetlands. Ecological Engineering 20(1): 75-88.

Lin, C., G. Eshel, I. Negev and A. Banin. 2008. Long-term accumulation and material balance of organic matter in the soil an effluent infiltration basin. Geoderma 148: 35-42.

Livingston, E. 2008. Village of Cloudcroft, NM PUReWater Project. Presentation at 2008 Water Smart Innovations Conference and Exposition, October 8-10, Las Vegas, NV. Available at http://www.watersmartinnovations.com/posters-sessions/2008/PDFs/1000-%20Eddie%20Livingston-%20Village%20of%20Cloudcroft,%20NM%20PURe%20Water%20Project.pdf.

Lopez-Ramirez, J. A., M. D. Coello Oviedo, and J. M. Quiroga Alonso. 2006. Comparative studies of reverse osmosis membranes for wastewater reclamation. Desalination 191: 137-147.

Loveland, J. P., J. N. Ryan, G. L. Amy, R. W. Harvey. 1996. The reversibility of virus attachment to mineral surfaces. Colloids and Surfaces A: Physicochemical and Engineering Aspects 107: 205-221.

Loyo-Rosales, J. E., C. P. Rice, and A. Torrents. 2009. Fate and distribution of the octyl- and nonylphenol ethoxylates and some carboxylated transformation products in the Back River, Maryland. Journal of Environmental Monitoring 12: 614-621.

Lozier, J. C., M. Kitis, J. H. Kim, B. Mi, and B. J. Mariñas. 2006. Evaluation of biologic and nonbiologic methods for assessing virus removal by and integrity of pressure membrane systems. In K. J. Howe, ed., Membrane Treatment for Drinking Water & Reuse Applications: A Compendium of Peer-Reviewed Papers. Denver, CO: American Water Works Association, pp. 399-420.

Lu, G. P., J. K. Fellman, C. H. Edwards, D. S. Mattinson, and J. Navazio. 2003. Quantitative determination of geosmin in red beets (*Beta vulgaris* L.) using headspace solid-phase microextraction. Journal of Agricultural and Food Chemistry 51(4): 1021-1025.

Lundie, S., G. M. Peters, and P. C. Beavis. 2004. Life cycle assessment for sustainable metropolitan water systems planning. Environmental Science & Technology 38: 3465-3473.

Lytle, D. A., and M. R. Schock. 2005. Formation of Pb(IV) Oxides in Chlorinated Water. Journal—American Water Works Association 97(11): 102-114.

Lytle, D. A., and V. L. Snoeyink. 2004. The effect of oxidant on the properties of Fe(III) particles and suspensions formed from the oxidation of Fe(II). Journal—American Water Works Association 96(8): 112-124.

Mac Kenzie, W. R., N. J. Hoxie, M. E. Proctor, M. S. Gradus, K. A. Blair, D. E. Peterson, J. J. Kazmierczak, D. G. Addiss, K. R. Fox, J. B. Rose, and J. P. Davis. 1994. A massive outbreak in Milwaukee of *Cryptosporidium* infection transmitted through the public water supply. New England Journal of Medicine 331: 161-167.

Macpherson, L., and P. Slovic. 2011. Talking About Water: Vocabulary and Images that Support Informed Decisions about Water Recycling and Desalination. WRF-07-03. Alexandria, VA: WateReuse Foundation.

Maeng, S. K., E. Ameda, S. Sharma, G. Grutzmacher, and G. L. Amy. 2010. Organic micropollutant removal from wastewater effluent-impacted drinking water sources during bank filtration and artificial recharge. Water Research 44: 4003-4014.

Maeng, S. K., S. K. Sharma, K. Lekkerkerker-Teunissen, and G.L. Amy. 2011. Occurrence and fate of bulk organic matter and pharmaceutically active compounds in managed aquifer recharge: A review. Water Research 45: 3015-3033.

Magee, P., R. Montesano, and R. Preussman. 1976. Nitroso Compounds and Related Carcinogens. In C. Searle, ed., Chemical Carcinogens. ACS Monograph 173. Washington, DC: American Chemical Society.

Mankad, A., and S. Tapsuwan. 2011. Review of socio-economic drivers of community acceptance and adoption of decentralised water systems. Journal of Environmental Management 92(3): 380-391.

Marella, R. L. 2009. Water Withdrawals, Use, and Trends in Florida, 2005. USGS Scientific Investigations Report 2009-5125. U.S. Geological Survey. 49 pp. Available at http://pubs.usgs.gov/sir/2009/5125/.

Marks, J., and M. Zadoroznyj. 2005. Managing sustainable urban water reuse: Structural context and cultures of trust. Society & Natural Resources 18(6): 557-572.

Marks, J. S., B. Martin, and M. Zadoroznyj. 2006. Acceptance of water recycling in Australia: National baseline data. Water Journal of the Australian Water Association 33(2): 151.

Marks, J., B. Martin, and M. Zadoroznyj. 2008. How australians order acceptance of recycled water: National baseline data. Journal of Sociology 44(1): 83-99.

Martin, J. W., S. A. Mabury, S. K. Solomon, and D. C. G. Muir. 2003. Bioconcentration and tissue distribution of perfluorinated acid in rainbow trout (Oncorhynchus mykiss). Environmental Toxicology and Chemistry 22: 196-204.

Massmann G., J. Greskowiak, U. Dunnbier, S. Zuehlke, A. Knappe, and A. Pekdeger. 2006. The impact of variable temperatures on the redox conditions and the behaviour of pharmaceutical residuesduring artificial recharge. Journal of Hydrology 328(1-2): 141-156.

Matamoros, V., and J. M. Bayona. 2006. Elimination of pharmaceuticals and personal care products in subsurface flow constructed wetlands. Environmental Science & Technology 40(18): 5811-5816.

Matamoros, V., J. Garcia, and J. M. Bayona. 2005. Behavior of selected pharmaceuticals in subsurface flow constructed wetlands: A pilot-scale study. Environmental Science & Technology 39(14): 5449-5454.

Matthews, E. J., N. L. Kruhlak, R. D Benz, and J. F. Contrera. 2004. Assessment of the health effects of chemicals in humans: I. QSAR estimation of the maximum recommended therapeutic dose (MRTD) and no effect level (NOEL) of organic chemicals based on clinical trial data. Current Drug Discovery Technologies 1(1): 61-76.

Mathiason, C. K., S. A. Hays, J. Powers, J. Hayes-Klug, J. Langenberg, S. J. Dahmes, D. A. Osborn, K. V. Miller, R. J. Warren, G. L. Mason, and E. A. Hoover. 2009. Infectious prions in pre-clinical deer and transmission of chronic wasting disease solely by environmental exposure. PLoS ONE 4(6): e5916.

Maunder, M. J., P. Matthiessen, J. P. Sumpter, and T. G. Pottinger. 2007. Impaired reproduction in three-spined sticklebacks exposed to ethinyl estradiol as juveniles. Biology of Reproduction 77(6): 999-1006.

Mawhinney, D. B., R. B. Young, B. J. Vanderford, T. Borch, and S. A. Snyder. 2011. Artificial sweetener sucralose in U.S. drinking water systems. Environmental Science & Technology 45: 8716-8722.

McDowell-Boyer, L. M., J. R. Hunt, and N. Sitar. 1986. Particle transport through porous media. Water Resources Research 22(13): 1901-1921.

MCES (Metropolitan Council of Environmental Services). 2007. Recycling Treated Municipal Wastewater for Industrial Water Use. St. Paul, Minnesota. Available at http://www.metrocouncil.org/planning/environment/RTMWIWU/RecyclingWastewaterReport.pdf. Accessed November 9, 2011.

McGauhey, P. H. 1968. Engineering Management of Water Quality. New York: McGraw-Hill.

Medema, G. J., M. H. A. Juhasz-Hoterman, and J. A. Luitjen. 2000. Removal of micro-organisms by bank filtration in a gravel-sand soil. In W. Jülich and J. Schubert, eds., Proceedings of International Riverbank Filtration Conference, November 2-4, 2000, Düsseldorf, Germany. Amsterdam: Internationale Arbeitsgemeinschaft der Wasserwerke im Rheineinzugsgebiet.

Medsker, L., D. Jenkins, and J. Thomas. 1968. Odorous compounds in natural waters. An earthy-smelling compound associated with blue-green algae and actinomycetes. Environmental Science & Technology 2(6): 461-464.

Medsker, L., D. Jenkins, J. Thomas, and C. Koch. 1969. Odorous compound in natural waters—2-exo-hydroxy-2-methylbornane, the major odorous compound produced by several actinomycetes. Environmental Science & Technology 3(5): 476-477.

Melosi, M. V. 2000. The Sanitary City: Urban Infrastructure in America from Colonial Times to the Present. Baltimore, MD: Johns Hopkins University Press.

Mena, K. D., and C. P. Gerba. 2008. Waterborne adenovirus. Reviews of Environmental Contamination and Toxicology 198: 133-167.

Menegaki, A., R. Mellon, A. Vrentzou, G. Koumakis, and K. Tsagarakis. 2009. What's in a name: Framing treated wastewater as recycled water increases willingness to use and willingness to pay. Journal of Economic Psychology 30(3): 285-292.

Metzler, D. F., R. L. Culp, H. A. Stoltenberg, R. L. Woodward, G. Walton, S. L. Chang, N. A. Clarke, C. M. Palmer, and F. M. Middleton. 1958. Emergency use of reclaimed water for potable supply at Chanute, Kansas. Journal—American Water Works Association. 50: 1021-1060.

Minear, R. A., and G. L. Amy, eds. 1996. Disinfection By-products in Water Treatment: The Chemistry of Their Formation and Control, 1st Ed. Boca Raton, FL: CRC Press.

Mitch, W., J. Sharp, R. Trussell, R. Valentine, L. Alvarez-Cohen, and D. Sedlak. 2003. N-Nitrosodimethylamine (NDMA) as a drinking water contaminant: A review. Environmental Engineering Science 20(5): 389-404.

Mitch, W. A., G. L. Oelker, E. L. Hawley, R. A. Deeb and D. L. Sedlak. 2005. Minimization of NDMA formation during chlorine disinfection of municipal wastewater by application of pre-formed chloramines. Environmental Engineering Science. 22(6): 882-890.

Montgomery-Brown, J., J. E. Drewes, P. Fox, and M. Reinhard. 2003. Behavior of alkylphenol polyethoxylate metabolites during soil aquifer treatment. Water Research 37: 3672-3681.

Morgan, J. J. 2004. What Water Quality Is: Matrices of Sources, Uses, Standards, Occurrence and Technology. Abel Wolman Distinguished Lecture, National Academies, Washington, D.C. Available at http://dels.nationalacademies.org/resources/staticassets/wstb/miscellaneous/2004_What_Water_Quality_Is.pdf.

Moss, K., and B. Boethling. 1999 EPA's new chemicals program PBT chemical category. Paper presented at American Chemical Society Division of Environmental Chemistry Symposium, Anaheim, CA, March 21-25, 1999. Preprints of Extended Abstracts 39(1): 141-143.

Mujeriego, R., J. Compte, T. Cazurra, and M. Gullon. 2008. The water reclamation and reuse project of El Prat de Llobregat, Barcelona, Spain. Water Science and Technology 57(4): 567-574.

Munro, I. C. 1990. Safety assessment procedures for indirect food additives: An overview. Report of a workshop. Regulatory Toxicology and Pharmacology 12: 2-12.

Munro, I. C., R. A. Ford, E. Kennepohl, and J. G. Sprenger. 1996. Correlation of structural class with no-observed-effect levels: A proposal for establishing a threshold of concern. Food and Chemical Toxicology 34: 829-867.

NACMC (National Advisory Committee on Microbiological Criteria for Foods). 1997. Hazard Analysis and Critical Control Point Principles and Application Guidelines. Available at http://www.fda.gov/food/foodsafety/HazardAnalysisCriticalControlPointsHACCP/ucm114868.htm.

Najm, I., and R. R. Trussell. 2001. NDMA formation in water and wastewater. Journal—American Water Works Association 93(2): 93-99.

Nalinakumari, B. W. Cha, and P. Fox. 2010. Effects of primary substrate concentration on NDMA transport during simulated aquifer recharge. American Society of Civil Engineers. Journal of Environmental Engineering 36(4): 363-370.

Namkung, E., and B. E. Rittman. 1986. Soluble microbial products (SMP) formation kinetics by biofilms. Water Research 20(6): 795-806.

NAS (National Academy of Sciences). 1975. Principles for Evaluating Chemicals in the Environment. Washington, DC: NAS.

Nataro, J. P. and M. M. Levine. 1994. *Escerichia coli* diseases in humans. In *Escherichia coli* in Domestic animals and Humans. CAB International, Wallingford: 285-333.NDWAC (National Drinking Water Advisory Council). 2004. Report on the CCL Classification Process to the U.S. Environmental Protection Agency. May 19, 2004. Available at http://www.epa.gov/safewater/ndwac/pdfs/report_ccl_ndwac_07-06-04.pdf. Accessed September 18, 2010.

National Drinking Water Advisory Council Report (NDWAC). 2004. National Drinking Water Advisory Council Report on the CCL Classification Process to the U.S. Environmental Protection Agency. Available online at http://www.epa.gov/safewater/ndwac/pdfs/report_ccl_ndwac_07-06-04.pdf. Accessed April 30, 2012.

Neisess, L., R. Baird, S. Carr, J. Gute, J. Strand, and C. Young. 2003. Can *N*-nitrosomethylamine formation be affected by polymer use during advanced wastewater treatment? Proceedings of the 2003 WateReuse Research Symposium.

Nellor, M. H., and R. Larson. 2010. Assessment of Approaches to Achieve Nationally Consistent Reclaimed Water Standards. Alexandria, VA: WateReuse Research Foundation.

Nellor, M. H., R. B. Baird, and J. R. Smyth. 1985. Health effects of indirect potable water reuse. Journal—American Water Works Association 77: 88-96.

Nemeroff, C., and P. Rozin. 1994. The contagion concept in adult thinking in the United States: Transmission of germs and interpersonal influence. Ethos 22: 157-186.

Neubauer, C. P., G. B. Hall, E. F. Lowe, C. P. Robison, R. B. Hupalo, and L. W. Keenan. 2008. Minimum flows and levels method of the St. Johns River Water Management District, Florida, USA. Environmental Management 42(6): 1101-1114.

NEWater Expert Panel. 2002. Singapore Water Reclamation Study: Expert Panel Review and Findings. Available at http://www.pub.gov.sg/water/newater/NEWaterOverview/Documents/review.pdf.

NHMRC (National Health and Medical Research Council). 2004. Australian Drinking Water Guidelines, Canberra, Australia.

Nichols, T. A., B. Pulford, A. C. Wyckoff, C. Meyerett, B. Michel, K. Gertig, E. A. Hoover, J. E. Jewell, G. C. Telling, and M. D. Zabel. 2009. Detection of protease-resistant cervid prion protein in water from a CWD-endemic area. Prion 3(3): 171-183.

NJDEP (New Jersey Department of Environmental Protection). 2011. Request for Authorization Checklist: Discharge to Surface Water, Category ABR—General Reclaimed Water for Beneficial Reuse Permit Authorization. Available at http://www.state.nj.us/dep/dwq/forms_surfacewater.htm.

Nordgren, J., A. Matussek, A. Mattsson, L. Svensson, and P. E. Lindgren. 2009. Prevalence of norovirus and factors influencing virus concentrations during one year in a full-scale wastewater treatment plant. Water Research 43: 1117-1125.

Norman, W., and C. MacDonald. 2004. Getting to the bottom of "triple bottom line." Business Ethics Quarterly 14(2): 243-262.

Notermans, S., G. Gallhoff, M. H. Zwietering, and G. C. Mead. 1994. The HACCP concept: Specification of criteria using quantitative risk assessment. Food Microbiology 11(5): 397-408.

NRC (National Research Council). 1977. Chemical contaminants: Safety and risk assessment. In Drinking Water and Health, Volume I. Washington, DC: National Academy Press, pp. 19-62.

NRC. 1982. Quality Criteria for Water Reuse. Washington, DC: National Academy Press.

NRC. 1983. Risk Assessment in the Federal Government: Managing the Process. Washington, DC: National Academy Press.

NRC. 1984. The Potomac Estuary Experimental Water Treatment Plant: A Review of the U.S. Army Corps of Engineers, Evaluation of the Operation, Maintenance and Performance of the Estuary Experimental Water Treatment Plant. Washington, DC: National Academy Press.

NRC. 1989. Irrigation-Induced Water Quality Problems. Washington, DC: National Academy Press.

NRC. 1994a. Ground Water Recharge Using Waters of Impaired Quality. Washington, DC: National Academy Press.

NRC. 1994b. Public health issues. In Ground Water Recharge Using Waters of Impaired Quality. Washington, DC: National Academy Press, pp. 132-178. Available at http://www.nap.edu/catalog.php?record_id=4780#toc.

NRC. 1994c. Science and Judgment in Risk Assessment. Washington, DC: National Academy Press.

NRC. 1996. Use of Reclaimed Water and Sludge in Food Crop Production. Washington, DC: National Academy Press.

NRC. 1998. Issues in Potable Reuse: The Viability of Augmenting Drinking Water Supplies with Reclaimed Water. Washington, DC: National Academy Press.

NRC. 1999a. Identifying Future Drinking Water Contaminants. Washington, DC: National Academy Press.

NRC. 1999b. Setting Priorities for Drinking Water Contaminants. Washington, DC: National Academy Press.

NRC. 2000. Watershed Management for Potable Water Supply: Assessing the New York City Strategy. Washington, DC: National Academy Press.

NRC. 2001. Classifying Drinking Water Contaminants for Regulatory Consideration. Washington, DC: National Academy Press.

NRC. 2004. Confronting the Nation's Water Problems: The Role of Research. Washington, DC: National Academies Press.

NRC. 2006. Drinking Water Distribution Systems: Assessing and Reducing Risks. Washington, DC: National Academies Press.

NRC. 2007. Toxicity Testing in the 21st Century: A Vision and a Strategy. Washington, DC: National Academies Press.

NRC. 2008a. Characterizing the Potential Human Toxicity of Low Doses of Pharmaceuticals in Drinking Water: Are New Risk Assessment Methods or Approaches Required? Sixth Workshop of the Standing Committee on Risk Analysis Issues and Reviews.

NRC. 2008b. Desalination: A National Perspective. Washington, DC: National Academies Press.

NRC. 2008c. Prospects for Managed Underground Storage of Recoverable Water. Washington, DC: National Academies Press.

NRC. 2009a. Nanotechnology and Oncology: Workshop Summary. Washington, DC: National Academies Press.

NRC. 2009b. Science and Decisions: Advancing Risk Assessment. Washington, DC: National Academies Press. Available at http://www.nap.edu/catalog.php?record_id=12209.

NRC. 2009c. Urban Stormwater Management in the United States. Washington, DC: National Academies Press.

NRC. 2010. Progress Toward Restoring the Everglades: The Third Biennial Review—2010. Washington, DC: National Academies Press.

NRC. 2011a. Adapting to the Impacts of Climate Change. Washington, DC: National Academies Press.

NRC. 2011b. Review of Federal Strategy for Nanotechnology-Related Environmental, Health, and Safety Research. Washington, DC: National Academies Press.

NRMMC/EPHC/NHMC (Natural Resource Management Ministerial Council, Environment Protection and Heritage Council, and National Health and Medical Research Council). 2006. National Guidelines for Water Recycling: Managing Health and Environmental Risks (Phase 1). NRMMC/EPHC/AHMC: Canberra, Australia: Available at http://www.ephc.gov.au/sites/default/files/WQ_AGWR_GL__Managing_Health_Environmental_Risks_Phase1_Final_200611.pdf.

NRMMC/EPHC/NHMRC (Natural Resource Management Ministerial Council, Environment Protection and Heritage Council, and National Health and Medical Research Council). 2008. Australian Guidelines for Water Recycling: Managing Health an Environmental Risks (Phase 2). Augmentation of Drinking Water Supplies. http://www.ephc.gov.au/sites/default/files/WQ_AGWR_GL__ADWS_Corrected_Final_%20200809.pdf.

O'Brien, N., and E. Cummins. 2010. Ranking initial environmental and human health risk resulting from environmentally relevant nanomaterials. Journal of Environmental Science and Health, Part A: Toxic/Hazardous Substances and Environmental Engineering 45(8): 992-1007.

OCWD (Orange County Water District). 2009. OCWD Response to the Final Report of the September 25-26, 2008 Meeting of the Independent Advisory Panel for the GWR System, March 19.

OCWD. 2010. Technical Memorandum on Mircobial Water Quality. Supplemental Report to the NWRI Groundwater Replenishment System Independent Advisory Panel, July 29.

Odendaal, P. E., J. L. J. van der Westhuizen, and G. J. Grobler. 1998. Wastewater Reuse in South Africa. In T. Asano, ed., Wastewater Reclamation and Reuse. Lancaster, PA: Technomic, pp. 1163-1192.

Odjadjare, E. O., and A. I. Okoh. 2010. Physicochemical quality of an urban municipal wastewater effluent and its impact on the receiving environment. Environmental Monitoring Assessment 170: 393-394.

OECD (Organisation for Economic Co-operations and Development) 1996. Test No. 305: Bioconcentration: Flow through Fish Test. In OECD Guidelins for the Testing of Chemicals. Paris. June 14.

OECD. 2011. OECD guidelines for the testing of chemical, Section 2: Effects on Biotic Systems. Available at http://www.oecd-ilibrary.org/content/serial/20745761.

Okun, D. A. 1996. A history of nonpotable urban water reuse. In 1996 Water Reuse Conference Proceedings, February 25-28, San Diego, CA. Denver, CO: American Water Works Association, pp. 41-54.

Olivieri, A., D. Eisenberg, R. Cooper, J. Korte, F. Maitski, and K. Thompson. 1987. Advanced Wastewater Treatment System Reliability Evaluation. Prepared for the City of San Diego. Berkeley, CA: Western Consortium for Public Health, May.

Olivieri, A. W., D. M. Eisenberg, R. C. Cooper, G. Tchobanoglous, and P. Gagliardo. 1996. Recycled water—A source of potable water: City of San Diego health effects study. Water Science and Technology 33: 285-296.

Olivieri, A., D. Eisenberg, J. Soller, J. Eisenberg, R. Cooper, G.Tchobanoglous, R. Trussell, and P. Gagliardo. 1999. Estimation of pathogen removal in an advanced water treatment facility using Monte Carlo simulation. Journal of Water Science and Technology 40(4-5): 223-233.

Ong, C.N. and H. Seah. 2003. Singapore Water Reclamation Study. PowerPoint presentation. National University of Singapore and Public Utilities Board, Singapore.

Panguluri, S., G. Meiners, J. Hall, and J. G. Szabo. 2009. Distribution System Water Quality Monitoring: Sensor Technology Evaluation Methodology and Results, EPA/600/R-09/076. Washington, DC: U.S. Environmental Protection Agency.

Patè-Cornell, M. E. 1996. Uncertainties in risk analysis: Six levels of treatment. Reliability Engineering & System Safety 54(2-3): 95-111.

Patterson, B., M. Shackleton, A. Furness, E. Bekele, J. Pearce, K. Linge, F. Busetti, T. Spadek, and S. Toze. 2011. Behaviour and fate of nine recycled water trace organics during managed aquifer recharge in an aerobic aquifer. Journal of Contamonatn Hydrology 122: 53-62.

Payment, P., J. Siemiatycki, L. Richardson, G. Renaud, E. Franco, and M. Prevost. 1997. A prospective epidemiological study of gastrointestinal health effects due to the consumption of drinking water. International Journal of Environmental Health Research 7: 5-31.

Pehlivanogllu-Mantas, E., E. Hawley, R. Deeb, and D. Sedlak. 2006. Formation of nitrosodimethylamine (NDMA) during chlorine disinfection of wastewater effluents prior to use in irrigation systems. Water Research 40: 341-347.

Pendergast, M., and E. Hoek. 2011. A review of water treatment membrane nanotechnologies. Energy Environmental Science 4: 1946.

Pera-Titus, M., and J. Llorens. 2007. Characterization of meso- and macroporous ceramic membranes in terms of flux measurement: A moment-based analysis. Journal of Membrane Science 302: 218-234.

Pereira, V. J., K. G. Linden, and H. S. Weinberg. 2007. UV degradation kinetics and modeling of pharmaceutical compounds in laboratory grade and surface water via direct and indirect photolysis at 254 nm. Environmental Science & Technology 41(5): 1682-1688.

Pinkston, K. E., and D. L. Sedlak. 2004. Transformation of aromatic, ether- and amine-containing pharmaceuticals during chlorine disinfection. Environmental Science & Technology 38: 4019-4025.

Pisarenko, A. N., B. D. Stanford, O. Quinones, G. E. Pacey, G. Gordon, and S. A. Snyder. 2010. Rapid analysis of perchlorate, chlorate and bromate ions in concentrated sodium hypochlorite solutions. Analytica Chimica Acta 659(1-2): 216-223.

Plewa, M. J., E. D. Wagner, S. D. Richardson, A. D. Thruston, Jr., Y. T. Woo, and B. McKague. 2004. Chemical and biological characterization of newly discovered iodoacid drinking water disinfection byproducts. Environmental Science & Technology 38(18): 4713-4722.

Plumlee, M. H., J. Larabee, and M. Reinhard. 2008. Perfluorochemicals in water reuse. Chemosphere 72: 541-1547.

Po, M., J. Kaercher, and B. Nancarrow. 2003. Literature Review of Factors Influencing Public Perceptions of Water Reuse. CSIRO Land and Water Technical Report 54/03, December. Perth, Australia: Commonwealth Scientific and Industrial Research Organisation

Poiger, T., H. R. Buser, and M. D. Müller. 2001. Photodegradation of the pharmaceutical drug diclofenac in a lake: Pathway, field measurements, and mathematical modeling. Environmental Toxicology and Chemistry. 20: 256-263

Portney, P. 1992. Trouble in Happyville. Journal of Policy Analysis and Management 11(1): 131-132.

Postel, S., and B. Richter. 2003. Rivers for Life: Managing Water for People and Nature. Washington, DC: Island Press.

Powell, R. L., R. A. Kimerle, G. T. Coyle, and G. R. Best. 1997. Ecological risk assessment of a wetland exposed to boron. Environmental Toxicology and Chemistry 16: 2409-2414.

Powelson, D. K., and C. P. Gerba. 1994. Virus removal from sewage effluents during saturated and unsaturared flow through soil columns. Water Research 28(10): 2175-2181.

Powelson, D. K., J. R. Simpson, and C. P. Gerba. 1990. Virus transport and survival in saturated and unsaturated flow through soil columns. Journal of Environmental Quality 19: 396-401.

Puglisi, E., M. J. Cahill, P. A. Lessard, E. Capri, A. J. Sinskey, J. A. Archer, and P. Boccazzi. 2010. Transcriptional response of Rhodococcus aetherivorans I24 to polychlorinated biphenyl-contaminated sediments. Microbiology Ecology. Available online at http://www.springerlink.com/content/j26010727m552935/. Accessed on November 17, 2011.

Purdom, C. E., P. A. Hardiman, V. V. J. Bye, N. C. Eno, C. R. Tyler, and J. P. Sumpter. 1994. Estrogenic Effects of Effluents from Sewage Treatment Works. Chemistry and Ecology. 8(4): 275-285.

Pusch, D., D. Y. Oh, S. Wolf, R. Dumke, U. Schroter-Bobsin, M. Hohne, I. Roske, and E. Schreier. 2005. Detection of enteric viruses and bacterial indicators in German environmental waters. Archives of Virology 150: 929-947.

Quanrud, D. M., B. Arnold, K. Lansey, C. Begay, W. Ela, and A. J. Gandolfi. 2003. Fate of effluent organic matter during soil aquifer treatment: Biodegradability, chlorine reactivity and genotoxicity. Journal of Water and Health 1(1): 33-44.

Radcliffe, J. C. 2008. Australian Water Recycling Today—The Big Issues. Presentation at the National Water Recycling and Reuse Conference, May 27. Available at http://www.nwc.gov.au/__data/assets/pdf_file/0004/11488/Aus-Water-Recycling-PRES-2705081.pdf.

Radcliffe, J. C. 2010. Evolution of water recycling in Australian cities since 2003. Water Science and Technology 62(4): 792-802.

Ram, N. M. 1986. Environmental significance of trace organic contaminants in drinking water suppliers. In N. M. Ram, E. J. Calabrese, and R. F. Christman, eds. Organic Carcinogens in Drinking Water: Detection, Treatment, and Risk Assessment. New York: John Wiley & Sons, pp. 3-31.

Ramirez, A. J., R. A. Brain, S. Usenko, M. A. Mottaleb, J. G. O'Donnell, L. L. Stahl, J. B. Wathen, B. D. Snyder, J. L. Pitt, P. Perez-Hurtado, L. L. Dobbind. B. W. Brooks, and C. K. Chambliss. 2009. Occurrence of pharmaceuticals and personal care products in fish: Results of a national pilot study in the United States. Environmental Toxicology and Chemistry 28(12): 2587-2597.

Rasmussen, N. C. 1981. The application of probabilistic risk assessment techniques to energy technologies. Annual Review of Energy 6: 123-138.

Rauch-Williams, T., and J. E. Drewes. 2006. Using soil biomass as an indicator for the biological removal of effluent-derived organic carbon during soil infiltration. Water Research 40: 961-968.

Ray, C., D. K. Borah, D. Soong, and G. S. Roadcap. 1998. Agricultural chemicals: Impacts on riparian wells during floods. Journal—American Water Works Association 90(7): 90-100.

Ray, C., T. Grischek, J. Schubert, J. Wang, and T. Speth. 2002a. A perspective of riverbank filtration. Journal—American Water Works Association 94(4): 149-160.

Ray, C, G, Melin, and R. B. Linsky. 2002b. Riverbank Filtration: Improving Source-Water Quality. Dordrecht, The Netherlands: Kluwer Academic.

Ray, C., T. Grischek, S. Hubbs, J.E. Drewes, D. Haas, and C. Darnault. 2008. Riverbank filtration for drinking water supply. Encyclopedia of Hydrological Sciences. DOI: 10.1002/0470848944.hsa305.

Redshaw, C. H., M. P. Cooke, H. M. Talbot, S. McGrath, and S. J. Rowland. 2008. Low biodegradability of fluoxetine HCl, diazepam and their human metabolites in sewage sludge-amended soil. Journal of Soils and Sediments 8: 217-230.

Regli, S., J. B. Rose, C. N. Haas, and C. P. Gerba. 1991. Modeling risk from giardia and viruses in drinking water. Journal—American Water Works Association 83(11): 76-84.

Reid, D. 1991. Paris Sewers and Sewermen: Realities and Representations. Cambridge, MA: Harvard University Press.

Reinke, E. A. 1934. Experience with Sewage Farming in the Southeast United States. Pasadena, CA: California Sewage Works Association and Public Health Engineering Section of the American Public Health Association.

Revankar, S. N., A. D. Bhatt, N. D. Desai, H. V. Bolar, S. P. Sane, and H. P. Tipnis. 1999. Comparative bioavailability study of a conventional and two controlled release oral formulations of Tegretol (carbamazepine)—200 mg. Journal of Association of Physicians of India 47(9): 886-900.

Reynolds, K. A., K. D. Mena, and C. P. Gerba. 2008. Risk of waterborne illness via drinking water in the United States. Reviews of Environmental Contamination and Toxicology. 192: 117-158.

Richardson, S. D., A. D. Thruston, Jr., T. W. Collette, K. S. Patterson, B. W. Lykins, Jr., G. Majetich, and Y. Zhang. 1994. Multispectral Identification of Chlorine Dioxide Disinfection Byproducts in Drinking Water. Environmental Science & Technoogy 28: 592-599.

Richardson, S. D., F. Fasano, J. J. Ellington, F. G. Crumley, K. M. Buettner, J. J. Evans, B. C. Blount, L. K. Silva, T. J. Waite, G.W. Luther, A. B. McKague, R. J. Miltner, E. D. Wagner, and M. J. Plewa. 2008. Occurrence and mammalian cell toxicity of iodinated disinfection byproducts in drinking water. Environmental Science & Technology 42(3): 955-961.

Robertson, L. J., L. Hermansen, and B. K. Gjerde. 2006. Occurrence of Cryptosporidium oocysts and Giardia cysts in sewage in Norway. Applications in Environmental Microbiology 72: 5297-5303.

Robins, F. W. 1946. Ancient Rome. In The Story of Water Supply. Oxford: Oxford University Press, pp. 57-65.

Rodgers-Gray, T. P., S. Jobling, C. Kelly, S. Morris, G. Brighty, M. J. Waldock, J. P. Sumpter, and C. R. Tyler. 2001. Exposure of juvenile roach (Rutilus rutilus) to treated sewage effluent induces dose-dependent and persistent disruption in gonadal duct development. Environmental Science & Technology 35(3): 462-470.

Rodriguez, C., A. Cook, Van Buyner, B. Devine, and P. Weinstein. 2007a. Screening health risk assesment of micropollutants for indirect potable reuse schemes: A three-tiered approach. Water Science and Technology 56(11): 35-42.

Rodriguez, C., P. Weinstein, A. Cook, B. Devine, and P. V. Buynder. 2007b. A proposed approach for the assessment of chemicals in indirect potable reuse schemes. Journal of Toxicology and Environmental Health 70(19): 1654-1663.

Rodriguez, C., A. Cook, B. Devine, P. Van Buynder, R. Lugg, K. Linge, and P. Weinstein. 2008. Dioxins, furans, and PCBs in recycled water for indirect potable reuse. International Journal of Environmental Research and Public Health 5(5): 356-367. Available at http://www.mdpi.com/1660-4601/5/5/356/pdf. Accessed November 17, 2011.

Rodriguez, C., P. van Buynder, R. Lugg, P. Blair, B. Devine, A. Cook, and P. Weinstein. 2009. Indirect potable reuse: A sustainable water supply alternative. International Journal of Environmental Research and Public Health 6(3): 1174-1209. Available at http://www.mdpi.com/1660-4601/6/3/1174/pdf. Accessed November 17, 2011.

Rodriguez, R. A., I. L. Pepper, and C. P. Gerba. 2009. Application of PCR-based methods to assess the infectivity of enteric viruses in environmental samples. Applications in Environmental Microbiology 75: 297-307.

Rohnke, A. T., and R. H. Yahner. 2008. Long-term effects of wastewater irrigation on habitat and a bird community in central Pennsylvania. Wilson Journal of Ornithology 120(1): 146-152.

Rommelmann, D. W., S. J. Duranceau, M. W. Stahl, C. Kaminkar, and R. M. Gonzales. 2004. Industrial Water Quality Requirements for Reclaimed Water. Report prepared for the American Water Works Association Research Foundation, Denver, Colorado.

Rosario-Ortiz, F. L., E. C. Wert, and S. A. Snyder. 2010. Evaluation of UV/H2O2 treatment for the oxidation of pharmaceuticals in wastewater. Water Research 44(5): 1440-1448.

Rose, J. B., C. N. Haas, and S. Regli. 1991. Risk assessment and the control of waterborne giardiasis. American Journal of Public Health 81: 709-713.

Rose, J. B., L. J. Dickson, S. R. Farrah, and R. P. Carnahan. 1996. Removal of pathogenic and indicator microorganisms by a full scale water reclamation facility. Water Research 30: 2785-2797.

Roseberry, A. M., and D. E. Burmaster. 1992. Lognormal distributions for water intake by children and adults. Risk Analysis 12(1): 99-104.

Rostad, C. E., B. S. Martin, L. B. Barber, and J. A. Leenheer. 2000. Effect of a constructed wetland on disinfection byproducts: Removal processes and production of precursors. Environmental Science & Technology 34(13): 2703-2710.

Roy, L., E. Scallan, and M. J. Beach. 2006. The rate of acute gastrointestinal illness in developed countries. Journal of Water and Health. 4(2).

Rozin, P., and E. Royzman. 2001. Negativity bias, negativity dominance, and contagion. Personal and Social Psychology Review 5: 296-320.

Ruetten, J., J. Birkhoff, K. Darr, J. E. Drewes, J. Flatt, D. Kelso, M. McDaniel, and D. Noble. 2004. Best Practices for Developing Indirect Potable Reuse Projects: Phase 1 Report. Alexandria, VA: WateReuse Foundation.

Ruiz, A. M., J. C. Maerz, A. K. Davis, M. K. Keel, A. R. Ferreira, M. J. Conroy, L. A. Morris, and A. T. Fisk. 2010. Patterns of development and abnormalities among tadpoles in a constructed wetland receiving treated wastewater. Environmental Science & Technology 44(13): 4862-4868.

Rule, K. L., V. R. Ebbett, and P. J. Vikesland. 2005. Formation of chloroform and chlorinated organics by free-chlorine-mediated oxidation of triclosan. Environmental Science & Technology 39(9): 3176-3185.

Rulis, A. 1987. Safety assurance margins for food additives currently in use. Regulatory Toxicology and Pharmacology 7: 160-168.

Rulis, A. 1989. The Food And Drug Administration's food additive petition review process. Food Drug Cosmetic Law Journal 45: 533-544.

Russell, S., and C. Lux. 2006. Water Recycling & the Community. Public Responses and Consultation Strategies: A Literature Review and Discussion. Oz-AQUAREC WP5. Wollongong, NSW, Australia: University of Wollongong.

Russell, S., and C. Lux. 2009. Getting over yuck: Moving from psychological to cultural and sociotechnical analyses of responses to water recycling. Water Policy 11(1): 21-35.

Rygaard, M. P. J. Binning, and H. J. Albrechtsen. 2011. Increasing urban water self-sufficiency: New era, new challenges. Journal of Environmental Management 92: 185-194.

Sanches, S., M. T. Crespo, and V. J. Pereira. 2010. Drinking water treatment of priority pesticides using low pressure UV photolysis and advanced oxidation processes. Water Research 44(6): 1809-1818.

Sanchez, C. A., R. I. Krieger, N. Khandaker, R. C. Moore, K. C. Holts, and L. L. Neidel. 2005. Accumulation and perchlorate exposure potential of lettuce produced in the lower Colorado River region. Journal of Agricultural and Food Chemistry 53(13): 5479-5486.

Savage, N., and M. Diallo. 2005. Nanomaterials and water purification: Opportunities and challenges. Journal of Nanoparticle Research 7: 331-342.

SCCP (Scientific Committee on Consumer Products). 2008. Use of the Threshold of Toxicological Concern (TTC) Approach for the Safety Assessment of Chemical Substances. SCCP/1171/08.

Schempp, A., and J. Austin. 2007. Water Right Impairment in Reclamation and Reuse: How Other Western States Can Inform Washington Law. Washington, DC: Environmental Law Institute.

Schets, F. M., J. H. van Wijnen, J. F. Schijven, H. Schoon, and A. M. de Roda Husman. 2008. Monitoring of Waterborne Pathogens in Surface Waters in Amsterdam, The Netherlands, and the Potential Health Risk Associated with Exposure to *Cryptosporidium* and *Giardia* in These Waters. Applied and Environmental Microbiology 74(7): 2069-2078.

Schijven, J. F., W. Hoogenboezem, P. J. Nobel, G. J. Medema, and A. Stakelbeek. 1998. Reduction of FRNA-bacteriophages and faecal indicator bacteria by dune infiltration and estimation of sticking efficiencies. Water Science and Technology 38(12): 127-131.

Schijven, J. F., W. Hoogenboezem, S. M. Hassanizadeh, and J. H. Peters. 1999. Modelling removal of bacteriophages MS2 and PRD1 by dune recharge at Castricum, Netherlands. Water Resources Research 35(4): 1101-1111.

Schijven, J. F., G. Medema, A. J. Vogelaar, and M. Hassanizadeh. 2000. Removal of microorganisms by deep well injection. Journal of Contaminant Hydrology 44: 301-327.

Schijven, J., P. Berger, and I. Miettinen. 2002. Removal of pathogens, surrogates, indicators, and toxins using riverbank filtration. In C. Ray, G. Melin, and R. B. Linsky, eds., Riverbank Filtration Improving Source-Water Quality, Dodrecht, The Netherlands: Kluwer Academic.

Schlenk, D., D. Hinton, and G. Woodside. 2006. Online Methods for Evaluating the Safety of Reclaimed Water. Report No. 01-HHE-4a. Alexandria, VA: Water Environment Research Foundation.

Schlindwein, A. D., C. Rigotto, C. M. O. Simoes, and C. R. M. Barardi. 2010. Detection of enteric viruses in sewage sludge and treated wastewater effluent. Water Science and Technology 61: 537-544.

Schmidt, C., F. Lange, and H. J. Brauch. 2004. Assessing the impact of different redox conditions and residence times on the fate of organic micropollutants during riverbank filtration. Proceedings of the 4th International Symposium on Pharmaceutically Active Compounds and Endocrine Disrupting Chemicals, Minneapolis, Minnesota, Oct. 13-15, 2004.

Schreiber, R., U. Gündel, S. Franz, A. Küster, B. Rechenberg, and R. Altenburger. 2011. Using the fish plasma model for comparative hazard identification for pharmaceuticals in the environment by extrapolation from human therapeutic data. Regulatory Toxicology and Pharmacology 61: 261-275.

Scown, T. M., R. van Aerle, and C. R. Tyler. 2010. Review: Do engineered nanoparticles pose a significant threat to the aquatic environment? Critical Reviews in Toxicology 40(7): 653-670.

Sedgwick, W. 1914. Principles of Sanitary Science and Public Health. London: Macmillan.

Sedlak, D., R. Deeb, E. Hawley, W. Mitch, T. Durbin, S. Mowbray, and S. Carr. 2005. Sources and fate of nitrosodimethylamine and its precursors in municipal wastewater treatment plants. Water Environment Research 77(1): 32-39.

Sedlak, D. L., and U. von Gunten. 2011. The chlorine dilemma. Science 331(6013): 42-43.

Servais, P., G. Billen, and M. C. Hascoet. 1987. Determination of the biodegradable fraction of dissolved organic matter in waters. Water Research 21: 445.

Sharp, J. O., Wood, T. K., & Alvarez-Cohen, L. 2005. Aerobic biodegradation of N-nitrosodimethylamine (NDMA) by axenic bacterial strains. Biotechnology and Bioengineering 89(5): 608-618.

Sharp, J. O., C. Sales, J. LeBlanc, J. Liu, T. Wood, L. Eltis, W. Mohn, and L. Alvarez-Cohen. 2007. An inducible propane monooxygenase is responsible for *N*-nitrosodimethylamine degradation by *Rhodococcus* sp. strain RHA1. Applied and Environmental Microbiology 73: 6930-6938.

Sharpless, C., and K. Linden. 2003. Experimental and model comparisons of low- and medium-pressure Hg lamps for the direct and H_2O_2 assisted UV photodegradation of *N*-nitrosodimethylamine in simulated drinking water. Enivronmental Science & Technology 37: 1933-1940.

Shaviv, A. 2009. Use of reclaimed wastewater for irrigation in Israel. In A. Shaviv, D. Broday, S. Cohen, A. Furman, and R. Kanwar, eds., Proceedings of the Dahlia Greidinger International Symposium, Crop Production in the 21st Century: Global Climate Change, Environmental Risk, and Water Security, pp. 300-315. Available at http://dgsymp09.technion.ac.il/Proceedings%20finalized%20for%20dist.pdf.

Shen, R., and S. A. Andrews. 2011. Demonstration of 20 pharmaceuticals and personal care products (PPCPs) as nitrosamine precursors during chloramine disinfection. Water Research 45(2): 944-952.

Shepherd, J. L., and R. L. Corsi. 1996. Chloroform in indoor air and wastewater: The role of residential washing machines. Journal of the Air & Waste Management Association 46(7): 631-642.

Shon, H. K., S. Vigneswaran, and S. A Snyder. 2006. Effluent organic matter (EfOM) in wastewater: Constituents, effects, and treatment. Critical Reviews in Environmental Science and Technology 36: 327-374.

Shuval, H. I., A. Adin, B. Fattal, E. Rawitz, and P. Yekutiel. 1986. Wastewater irrigation in developing countries: Health effects and technical solutions.Washington, DC: World Bank.

Simmons, F. J., and I. Xagoraraki. 2011. Release of infectious human enteric viruses by full-scale wastewater utilities. Water Research 45: 3590-3598.

Sinclair, M., J. O'Toole, A. Forbes, D. Carr, and K. Leder. 2010. Health status of residents of an urban dual reticulation system. International Journal of Epidemiology 39(6): 1667-1675. Available at http://ije.oxfordjournals.org/content/39/6/1667.abstract. Accessed November 10, 2011.

Singapore Public Utilities Board and Ministry of the Environment. 2002. Singapore Water Reclamation Study: Expert Panel Review and Findings. Available at http://www.pub.gov.sg/water/newater/NEWaterOverview/Documents/review.pdf.

Singer, H., S. Mueller, C. Tixier, and L. Pillonel. 2002. Triclosan: Occurrence and fate of a widely used biocide in the aquatic environment: Field measurements in wastewater treatment plants, surface waters, and lake sediments. Environmental Science & Technology 36(23): 4998-5004.

Singer, P., and D. A. Reckhow. 2010. Chemical Oxidation. Water Quality and Treatment Handbook, 6th Ed. New York: McGraw-Hill.

Slifko, T., D. Huffman, and J. Rose. 1999. A most-probable-number assay for enumeration of infectious *Cryptosporidium parvum* oocysts. Applied and Environmental Microbiology 65: 3936-3941.

Sloan, D. W., C. Wingert, and I. Cadena. 2010. Potable reuse in the Permian Basin. Presentation at the 25th WateReuse Symposium, Washington, DC, September 13. Available at http://www.watereuse.org/sites/default/files/u3/David%20Sloan.pdf.

Sloss, E. M., S. A. Geschwind, D. F. McCaffrey, and B. R. Ritz. 1996. Groundwater Recharge with Recycled Water: An Epidemiologic Assessment in Los Angeles, County, 1987-1991. Report prepared for the Water Replenishment District of Southern California by RAND, Santa Monica, CA.

Sloss, E. M., D. F. McCaffrey, R. D. Fricker, S. A. Geschwind, and B. R. Ritz. 1999. Groundwater Recharge with Recycled Water: Birth Outcomes in Los Angeles County 1982-1993. Report prepared for the Water Replenishment District of Southern California by RAND, Santa Monica, CA.

Slovic, P. 1987. Perception of risk. Science 236(4799): 280-285.

Slovic, P. 1993. Perceived risk, trust, and democracy. Risk Analysis 13: 675-682.

Slovic, P., M. Finucane, E. Peters, and D. G. MacGregor. 2002. The affect heuristic. In T. Gilovich, D. Griffin, and D. Kahneman, eds., Heuristics and Biases: The Psychology of Intuitive Judgment. New York: Cambridge University Press, pp. 397-420.

Slovic, P., M. Finucane, E. Peters, and D. MacGregor. 2004. Risk as analysis and risk as feeling: Some thoughts about affect, reason, risk, and rationality. Risk Analysis 24: 311-322.

Slye, J. L, J. H. Kennedy, D. R. Johnson, S. F. Atkinson, S. D. Dyer, M. Ciarlo, K. Stanton, H. Sanderson, A. M. Nielsen, and B. B. Price. 2011. Relationships between benthic macroinvertebrate community structure and geospatial habitat, in-stream chemistry, and surfactants in the effluent-dominated Trinity River, Texas USA. Environmental Toxicology 30(5): 1127-1138.

Snyder, S. A. 2007. Evaluation of Conventional and Advanced Treatment Processes to Remove Endocrine Disruptors and Pharmaceutically Active Compounds. Denver, CO: American Water Works Research Foundation.

Snyder, S. A., T. L. Keith, D. A. Verbrugge, E. M. Snyder, T. S. Gross, K. Kannan, and J. P. Giesy. 1999. Analytical methods for detection of selected estrogenic compounds in aqueous mixtures. Environmental Science & Technology 33(16): 2814-2820.

Snyder, S. A., T. L. Keith, C. G. Naylor, C. A. Staples, and J. P. Giesy. 2001a. Identification and quantitation method for nonylphenol and lower oligomer nonylphenol ethoxylates in fish tissues. Environmental Toxicology and Chemistry 20: 1870-1873.

Snyder, S. A., D. L. Villeneuve, E. M. Snyder, and J. P. Giesy. 2001b. Identification and quantification of estrogen receptor agonists in wastewater effluents. Environmental Science & Technology 35(18): 3620-3625.

Snyder S.A., J. Leising, P. Westerhoff, Y. Yoon, H. Mash and B. Vanderford. 2004. Biological and physical attenuation of endocrine disruptors and pharmaceuticals: implications for water reuse. Ground Water Monitoring and Remediation. 24(2): 108-188.

Snyder, S. A., B. J. Vanderford, and D. J. Rexing. 2005. Trace analysis of bromate, chlorate, iodate, and perchlorate in natural and bottled waters. Environmental Science & Technology 39(12): 4586-4593.

Snyder, S. A., S. Adham, A. M. Redding, F. S. Cannon, J. DeCarolis, J. Oppenheimer, E. C. Wert, and Y. Yoon. 2006a. Role of membranes and activated carbon in the removal of endocrine disruptors and pharmaceuticals. Desalination 202: 156-181.

Snyder, S. A., R. C. Pleus, B. J. Vanderford, and J. C. Holady. 2006b. Perchlorate and chlorate in dietary supplements and flavor enhancing ingredients. Analytica Chimica Acta 567: 23-32.

Snyder, S. A., E. C. Wert, D. J. Rexing, R. E. Zegers, and D. D. Drury. 2006c. Ozone oxidation of endocrine disruptors and pharmaceuticals in surface water and wastewater. Ozone Science and Engineering 28: 445-460.

Snyder, S., E. Wert, H. Lei, P. Westerhoff, and Y. Yoon. 2007. Removal of EDCs and Pharmaceuticals in Drinking and Reuse Treatment Processes. Denver, CO: American Water Works Research Foundation.

Snyder, S., R. Trenholm, E. M. Synder, G. M. Bruce, R. C. Pleus, and J. D. Hemming. 2008a. Toxicological Relevance of EDCs and Pharmaceuticals in Drinking Water. Denver, CO: American Water Works Research Foundation. Available at http://environmentalhealthcollaborative.org/images/91238_Toxicological_Relevance.pdf.

Snyder, S., B. Vanderford, J. E. Drewes, E. Dickenson, E. M. Snyder, G. M. Bruce, and R. C. Pleus. 2008b. State of Knowledge of Endocrine Disruptors and Pharmaceuticals in Drinking Water. Denver, CO: American Water Works Research Foundation.

Snyder, S. A., B. D. Stanford, A. N. Pisarenko. G. Gordon, and M. Asami. 2009. Hypochlorite—an Assessment of Factors That Influence the Formation of Perchlorate and Other Contaminants. Denver, CO: American Water Works Association Research Foundation. Available at http://www.awwa.org/files/GovtPublicAffairs/PDF/HypochloriteAssess.pdf. Accessed November 20, 2011.

Snyder, S. A., M. Benotti, F. Rosario-Ortiz, B. Vanderford, J. E. Drewes, and E. Dickenson. 2010a. Comparison of Chemical Composition of Reclaimed and Conventional Waters. Final Report. Alexandria, VA: WateReuse Foundation.

Snyder, S., B. D. Stanford, G. M. Bruce, R. C. Pleus, and J. E. Drewes. 2010b. Identifying Hormonally Active Compounds, Pharmaceuticals, and Personal Care Product Ingredients of Health Concern from Potential Presence in Water Intended for Indirect Potable Reuse. Report WRF-05-005-01. Alexandria, VA: WateReuse Foundation.

Sobral, P., and J. Widdows. 1997. Effects of elevated temperatures on the scope for growth and resistance to air exposure of the clam ruditapes decussatus (L.), from southern portugal. Scientia Marina. 61(2): 163-171.

Soller, J.A., and M. H. Nellor. 2011a. Development and Application of Tools to Assess and Understand the Relative Risks of Regulated Chemicals in Indirect Potable Reuse Project: the Montebello Forebay Groundwater Recharge Project. Tools to Assess and Understand the Relative Risks of Indirect Potable Reuse and Aquifer Storage & Recovery Projects. 1(a). Alexandria, VA: WaterReuse Foundation.

Soller, J.A., and M. H. Nellor. 2011b. Development and Application of Tools to Assess and Understand the Relative Risks of Regulated Chemicals in Indirect Potable Reuse Projects: the Chino Basin Groundwater Recharge Project. In Tools to Assess and Understand the Relative Risks of Indirect Potable Reuse and Aquifer Storage & Recovery Projects. 1 (b). Alexandria, VA: WaterReuse Foundation.

Soller, J. A., A. W. Olivieri, J. Crook, R. C. Cooper, G. Tchobanoglous, R. T. Parkin, R. C. Spear, and J. N. Eisenberg. 2003. Risk-based approach to evaluate the public health benefit of additional wastewater treatment. Environmental Science & Technology 37(9): 1882-1891.

Solley, W.B., R.R. Pierce, and H.A. Perlman, 1998. Estimated Use of Water in the United States in 1995. U.S. Geological Survey Circular 1200. Denver, Colorado. Song, G. X., J. Wang, C. Chiu, and P. Westerhoff. 2010. Biogenic Nanoscale Colloids in Wastewater Effluents. Environmental Science & Technology 44(21): 8216-8222.

Sontheimer, H. 1980. Experience with Riverbank Filtration Along the Rhine River. Journal—American Water Works Association 72(7): 386-390.

Sontheimer, H. 1991. Trinkwasser aus dem Rhein? Sankt Augustin, Germany: Academia Verlag. (in German).

Sontheimer, H., and W. Nissing. 1977. Änderung der Wasserbeschaffenheit bei der Bodenpassage unter besonderer Berücksichtigung der Uferfiltration am Niederrhein. Gas und Wasserfach Wasser/Abwasser 57(9): 639-645.

Soucek, D. J., T. K. Linton, C. D. Tarr, A. Dickinson, N. Wickramanayake, C. G. Delos, and L. A. Cruz. 2011. Influence of water hardness and sulfate on the acute toxicity of chloride to sensitive freshwater invertebrates. Environmental Toxicology and Chemistry 30(4): 930-938.

Sowers, A. D., K. M. Gaworecki, M. A. Mills, A. P. Roberts, and S. J. Klaine. 2009. Developmental effects of a municipal wastewater effluent on two generations of the fathead minnow, Pimephales promelas. Aquatic Toxicology 95(3): 173-181.

Sperber, W. H., and R. F. Stier. 2009. Happy 50th Birthday to HACCP: Retrospective and Prospective. FoodSafety Magazine. December. Available at http://w.foodsafetymagazine.com/article.asp?id=3481&sub=sub1.

Stalter, D., A. Magdeburg, M. Weil, T. Knacker, and J. Oelmann. 2010. Toxication or detoxication? In vivo toxicity assessment of ozonation as advanced wastewater treatment with the rainbow trout. Water Research 44: 439-448.

Stanford, B. D., R. A. Trenholm, J. C. Holady, B. J. Vanderford, and S. A. Snyder. 2010. Estrogenic activity of US drinking waters: A relative exposure comparison. Journal—American Water Works Association 102(11): 55-65.

Staples, C., E. Mihaich, J. Carbone, K. Woodburn, and G. Klecka. 2004. A weight of evidence analysis of the chronic ecotoxicity of nonylphenol ethoxylates, nonylphenol ether carboxylates, and nonylphenol. Human Ecological Risk Assessment 10: 999-1017.

Starko, K. M., E. C. Lippy, L. B. Dominguez, C. E. Haley, and H. J. Fisher. 1986. Campers' diarrhea outbreak traced to water-sewage link. Public Health Reports 101(5): 527-531.

Steinberg C. E., S. R. Sturzenbaum, and R. Menzel. 2008. Genes and environment—striking the fine balance between sophisticated biomonitoring and true functional environmental genomics. Science and the Total Environment 400: 142-161.

Stenekes, N., H. Colebatch, T. Waite, and N. Ashbolt. 2006. Risk and governance in water recycling public acceptance revisited. Science Technology & Human Values 31(2): 107-134.

Stephenson, T., S. Judd, B. Jefferson, and K. Brindle. 2000. Membrane Bioreactors for Wastewater Treatment. London: IWA Publishing.

Stokes, J. R., and A. Horvath. 2006. Life cycle energy assessment of alternative water supply systems. International Journal of Life Cycle Assessment 5: 335-343.

Stokes, J. R., and A. Horvath. 2009. Energy and air emission effects of water supply. Environmental Science & Technology 43: 2680-2687.

Strachan, N., M. Doyle, F. Kasuga. O. Rotariu, and I. D. Ogden. 2005. Dose response modelling of Escherichia coli O157 incorporating data from foodborne and environmental outbreaks. International Journal of Food Microbiology 103(1): 35-47.

Sumpter, J. P., and A. C. Johnson. 2008. 10th Anniversary perspective: Reflections on endocrine disruption in the aquatic environment: from known knowns to unknown unknowns (and many things in between). Journal of Environmental Monitoring 10: 1476-1485.

Suttle, C. A. 2007. Marine viruses—major players in the global ecosystem. Nature Reviews Microbiology. 5(10): 801-812. Available at http://www.scopus.com. Accessed December 13, 2011.

Swayne, M., G. Boone, D. Bauer, and J. Lee. 1980. Wastewater in Receiving Waters at Water Supply Abstraction Points. EPA-60012-80-044. Washington, DC: U.S. Environmental Protection Agency.

SWFWMD (Southwest Florida Water Management District). 2006. Estimating Reclaimed Water Capital Costs. White Paper.

SWFWMD. 2008. Estimating Reclaimed Water Residential Distribution Capital Costs. White Paper.

Sydney Water. 2004. Media Release: Sydney Water working closely with Rouse Hill residents. Sydney, Australia: Sydney Water.

Taguchi, V., S. Jenkins, D. Wong, J. Palmentier, and E. Reiner. 1994. Determination of N-nitrosodimethylamine by isotope dilution, high resolution mass spectrometry. Canadian Journal of Applied Spectroscopy 39: 87-89.

Tanaka, H., T. Asano, E. D. Schroeder, and G. Tchobanoglous. 1998. Estimating the safety of wastewater reclamation and reuse using enteric virus monitoring data. Water Environment Research 70(1): 39-51.

Tang, C., S. Huitric, J. Kuo, R. Horvath, and J. Stahl. 2006. Disinfection alternatives to avoid NDMA formation. Proceedings of the Water Environment Federation Technical Exhibition and Conference 2006. Alexandria, VA: Water Environment Federation.

Tanji, K., S. Grattan, C. Grieve, A. Harivandi, L. Rollins, D. Shaw, B. Sheikh, and L. Wu. 2008. Salt Management Guide for Landscape Irrigation with Recycled Water in Coastal Southern California: A Comprehensive Literature Review. Alexandria, VA: WateReuse.

Tarlock, D. 2009. Law of Water Rights and Resources. Supplement 5:18. Clark Boardman. New York, NY.

Tarquin, A. 2009. High Tech Methods to Reduce Concentrate Volume Prior to Disposal. Research Report. University of Texas at El Paso, Department of Civil Engineering. Available at http://www.twdb.state.tx.us/RWPG/rpgm_rpts/0704830769_ConcentrateVolume.pdf.

Tarr, J. A. 1979. Separate vs. combined sewer problem—case study in urban technology design choice. Journal of Urban History 5: 308-339.

Tarr, J. A., J. McCurley, F. C. McMichael, and T. Yosie. 1984. Water and wastes retrospective assessment of wastewater technology in the United States, 1800-1932. Technology and Culture 25: 226-263.

Tchobanoglous, G., F. L. Burton, and H. D. Stensel. 2002. Wastewater Engineering: Treatment and Reuse. New York: McGraw-Hill.

Tchobanoglous, G., F. Burton, and H. D. Stensel. 2003. Wastewater Engineering. Treatment and Reuse. New York: McGraw-Hill.

Tchobanoglous, G., H. Leverenz, M. H. Nellor, and J. Crook. 2011. Direct: Potable Reuse: A Path Forward. Alexandria, VA: WateReuse Research Foundation.

Ternes, T. A., J. Stuber, N. Herrmann, D. McDowell, A. Ried, M. Kampmann, and B. Teiser. 2003. Ozonation: a tool for removal of pharmaceuticals, contrast media, and musk fragrances from water. Water Research 37: 1976-1982.

Teunis, P., K. Takumi, and K. Shinagawa. 2004. Dose response for infection by *Escherichia coli* O157:H7 from outbreak data. Risk Analysis 24(2): 401.

Teunis, P. F. M., C. L. Moe, P. Liu, S. E. Miller, L. Lindesmith, R. S. Baric, J. Le Pendu, and R. L. Calderon. 2008. Norwalk virus: How infectious is it? Journal of Medical Virology 80: 1468-1476.

Texas Commission on Environmental Quality. 1997. Use of reclaimed water. Texas Administrative Code Title 30, Chapter 210). Austin, TX: State of Texas. Available at http://info.sos.state.tx.us/pls/pub/readtac$ext.ViewTAC?tac_view=4&ti=30&pt=1&ch=210.

Texas Commission on Environmental Quality. 2010. Use of Reclaimed Water. Title30, Chapter 210, Part 1. Texas Administrative Code. Austin, Texas. Available online at http://info.sos.state.tx.us/pls/pub/readtac$ext.ViewTAC?tac_view=4&ti=30&pt=1&ch=210. Accessed November 30, 2011.

Thompson, K., R. C. Cooper, A. W. Olivieri, D. M. Eisenberg, L. A. Pettegrew, R. C. Cooper, and R. E. Danielson. 1992. City of San Diego study of direct potable reuse of reclaimed water: Final results. Desalination 88: 201-214.

Thorpe, K. L., R. I. Cummings, T. H. Hutchinson, M. Scholze, G. Brighty, J. P. Sumpter, and C. R. Tyler. 2003. Relative potencies and combination effects of steroidal estrogens in fish. Environmental Science & Technology 37:1142-1149.

Tibbets, J. 1995. What's in the water: The disinfection dilemma? Environmental Health Perspectives 103: 30-34.

TRA (Trinity River Authority of Texas). 2010. Trinity River Basin Master Plan. Available at http://www.trinityra.org/downloads/Master%20Plan%20Justified%20REV.pdf.

Trussell, R. R., and M. Umphres. 1978. Formation of trihalomethanes. Journal—American Water Works Association 70(11): 604.

Trussell, R., S. Adham, and J. Jacangelo. 2000. Potable Water Reuse and the Role of Membranes, in the proceedings of Membrantechnik in der Wasseraufbereitung und Abwasserbehandlung Aachen, Germany: ITV.

Tufenkji, N., J. N. Ryan, and M. Elimelech. 2002. The promise of bank filtration. Environmental Science & Technology 36(21): 422a-428a.

Turbow, D. J., N. D. Osgood, and S. C. Jiang. 2003. Evaluation of recreational health risk in coastal waters based on enterococcus densities and bathing patterns. Environmental Health Perspectives 111(4): 598-603.

Turbow, D. J., E. E. Kent, and S. C. Jiang. 2008. Web-based investigation of water associated illness in marine bathers. Environmental Research 106(1): 101-109.

TWDB. 2009. General Frequently Asked Questions. S. Kalaswad, Ed. Available at http://www.twdb.state.tx.us/innovativewater/desal/faq.asp. Accessed December 12, 2011.

UNEP (United Nations Environmental Programme). 2005. Water and Wastewater Reuse: An Environmentally Sound Approach for Sustainable Urban Water Management. Osaka, Japan: UNEP Division of Technology, Industry and Economics.

University of Florida TREEO Center. 2011. Case Histories of Selected Backflow Incidents: Backflow at a Premises with a Reclaimed Water System. University of Florida, Gainesville, Florida.

Unnevehr, L. J., and H. H. Jensen. 1996. HACCP as a regulatory innovation to improve food safety in the meat industry. American Journal of Agricultural Economics 78(3): 764-769.

U.S. Bureau of Reclamation. 2005. Quality of Water: Colorado River Basin—Progress Report No. 22. Reclamation: Managing Water in the West. Available at http://www.usbr.gov/uc/progact/salinity/pdfs/PR22.pdf. Accessed December 14, 2011.

U.S. Census. 2008. 2008 National Population Projections. Available at http://www.census.gov/population/www/projections/2008projections.html. Accessed November 16, 2010.

U.S. Census. 2010b. State Population Estimates 1970-1980, on U.S. Census website, accessed on November 16, 2010: http://www.census.gov/popest/archives/1980s/st7080ts.txt

U.S. Census. 2010c. Table of Historical National Population Estimates. Available at http://www.census.gov/statab/hist/HS-01.pdf. Accessed October 30, 2010:

U.S. Census. 2010d. Total Midyear Population for the World: 1950-2050 on U.S. Census website, accessed on October 30, 2010:http://www.census.gov/ipc/www/idb/worldpop.php.

U.S. Census. 2010e. Upper and lower UN Estimates. Available at http://www.census.gov/ipc/www/worldhis.html. Accessed October 30, 2010.

U.S. Census. 2011. U.S. Population Clock. Available at http://www.census.gov/population/www/popclockus.html. Accessed Feb. 25, 2011:

USGCRP (U.S. Global Change Research Program). 2000. Climate Change Impacts on the United States: The Potential Consequences of Climate Variability and Change. Available at http://www.globalchange.gov/publications/reports/scientific-assessments/first-national-assessment/468. Accessed on December 8, 2011..

U.S. Public Health Service. 1962. Public Health Service Drinking Water Standards. PHS Publication No. 956. Washington, DC: U.S. Department of Health, Education, and Welfare.

Urbansky, E. T., ed. 2000. Perchlorate in the Environment. Environmental Science Research. New York: Kluwer Academic/Plenum.

USACE/SFWMD (U.S. Army Corps of Engineers and South Florida Water Management District). 1999. Central and Southern Florida Project Comprehensive Review Study, Final Integrated Feasibility Report and Programmatic Environmental Impact Statement. Available at http://www.evergladesplan.org/pub/restudy_eis.cfm#mainreport. Accessed August 13, 2010.

Utah Department of Environmental Quality. 2011. Design Requirements for Wastewater Collection, Treatment and Disposal Systems. Utah Administrative Code, Rule R317-3, Salt Lake City, Utah. Available at http://www.rules.utah.gov/publicat/code/r317/r317-003.htm#T11.

Valenti Jr., T. W., P. Perez-Hurtado, C. K, Chambliss, and B. W. Brooks. 2009. Aquatic toxicity of sertraline to pimephales promelas at environmentally relevant surface water pH. Environmental Toxicology and Chemistry 28(12): 2685-2694.

Valentine, R., J. Choi, Z. Chen, S. Barrett, C. Hwang, Y. Guo, M. Wehner, S. Fotzsimmons, S. Andrews, A. Werker, C. Brubacher, and K. Kohut. 2005. Factors Affecting the Formation of NDMA in Water and Occurrence. Denver CO: American Water Works Association Research Foundation and Alexandria, VA: Water Environment Research Foundation.

van Nieuwenhuijsen, M. J., M. B. Toledano, N. E. Eaton, J. Fawell, and P. Elliott. 2000. Chlorination disinfection by-products in water and their association with adverse reproductive outcomes: A review. Occupational and Environmental Medicine 57: 73-85.

van Nieuwenhuijzen, A. F., H. Evenblij, C. A. Uijterlinde, and F. L. Schulting. 2008. Review on the state of science on membrane bioreactors for municipal wastewater treatment. Water Science and Technology 57(7): 979-986.

van Vuuren, K. 2009. Press bias and local power in the Toowoomba water referendum. Communication, Politics and Culture 42 (1): 55-73.

Vanderford, B. J., D. B. Mawhinney, F. Rosario-Ortiz, and S. A. Snyder. 2008. Real-time detection and identification of aqueous chlorine transformation products using QTOF MS. Analytical Chemistry 80(11): 4193-4199.

Veil, J. A. 2007. Use of Reclaimed Water for Power Plant Cooling. ANL/EVS/R-07/3. Argonne, IL: Argonne National Laboratory, Environmental Science Division.

Venkatesan, A. K., S. Ahmad, W. Johnson, and J. R. Batista. 2011. Systems dynamic model to forecast salinity load to the Colorado River due to urbanization within the Las Vegas Valley. Science of the Total Environment 409(13): 2616-2625.

Verdonck, F. A., T. Aldenberg, J. Jaworska, and P. A. Vanrollegham. 2003. Limitations of current risk characterization methods in probablistic environmental risk assessment. Environmental Toxicology and Chemistry 22: 2209-2213.

Verstraeten, I. M., J. D. Carr, G. V. Steele, E. M. Thurman, K. C. Bastian, and D. F. Dormedy. 1999. Surface-water/groundwater interaction: Herbicide transport into municipal collector wells. Journal of Environmental Quality 28(5): 1396-1405.

Victoria, M., F. R. Guimaraes, T. M. Fumian, F. F. Ferreira, C. B. Vieira, T. Shubo, J. P. Leite, and M. P. Miagostovich. 2010. One year monitoring of norovirus in a sewage treatment plant in Rio de Janeiro, Brazil. Journal of Water Health 8: 158-165.

von Gunten, U. 2003. Ozonation of drinking water: Part II. Disinfection and by-product formation in presence of bromide, iodide or chlorine. Water Research 37: 1469-1487.

von Gunten, U., E. Salhi, C. K. Schmidt, and W. A. Arnold. 2010. Kinetics and mechanisms of N-nitrosodimethylamine formation upon ozonation of N,N-dimethylsulfamide-containing waters: Bromide catalysis. Environmental Science & Technology 44(15): 5762-5768.

Waltman, E. L., B. J. Venables, and W. Z. Waller. 2006. Triclosan in a North Texas wastewater treatment plant and the influent and effluent of an experimental constructed wetland. Environmental Toxicology and Chemistry 25(2): 367-372.

Wang, J. Z., S. A. Hubbs, and R. Song. 2002. Evaluation of Riverbank Filtration as a Drinking Water Treatment Process. Final Report. Denver, CO: American Water Works Association Research Foundation.

Ward, R. L., D. L. Bernstein, E. C. Young, J. R. Sherwood, D. R. Knowlton, and G. M. Schiff. 1986. Human Rotavirus Studies in Volunteers: Determination of Infectious Dose and Serological Response to Infection. Journal of Infectious Diseases 154(5): 871.

Washington State Department of Health. 2007. Report from Department of Health on Related Public Health Issues. In Reclaimed Water Use Legislative Report, December. Olympia, WA: Office of Shellfish and Water Protection, Chapter 8. 8-61.

Washington State Department of Ecology. 1997. Water Reclamation and Reuse Standards 97: 23. Washington State Department of Health ad Washington State Department of Ecology. Olympia, Washington.

Weber, N. M., H. R. Sawyer, M.E. Legare, and D. N. R. Veeramachaneni. 2006. Sub-chronic exposure to dibromoacetic acid, a water disinfection by-poduct, does not affect gametogenic potential in mice. Toxicological Sciences 89 (1): 325-330.

Wehner, M. 2009. Groundwater Replenishment System Product Water Quality. Presentation to the NRC Panel, Fountain Valley, Orange County.

Weisbrod, A. V., K. B. Woodburn, A. A. Koelmans, T. F. Parkerton, A. E. McElroy, and K. Borgä. 2009. Evaluation of bioaccumulation using in vivo laboratory and field studies. Integrated Environmental Assessment and Management 5: 598-623.

Weisel, C. P., and J. Wan-Kuen. 1996. Ingestion, inhalation, and dermal exposures to chloroform and trichloroethene from tap water. Enviromental Health Perspectives 104(1): 48-51.

Weisner, M., and J.-Y. Bottero. 2007. Nanotechnogology and the environment. In Environmental Nanotechnology: Applications and Impacts of Nanomaterials. New York: McGraw-Hill.

Weiss, W. J., E. J. Bouwer, W. P. Ball, C. R. O'Melia, H. Arora, and T. F. Speth. 2002. Chapter 8: Reduction in Disinfection By-product Precursors and Pathogens During Riverbank Filtration at Three Midwestern United States Drinking-Water Utilities. In Riverbank Filtration: Improving Source Water Quality by C. Ray, G. Melin, and R.B. Linsky, pp. 147-173.

Weiss, W. J., E. J. Bouwer, R. Aboytes, M. W. LeChevallier, C. R. O'Melia, B. T. Le, and K. J. Schwab. 2005. Riverbank filtration for control of microorganisms results from field monitoring. Water Research 39: 1990-2001.

Wert, E., F. L. Rosario-Ortiz, D. D. Drury, and S. A. Snyder. 2007. Formation of oxidation byproducts from ozonation of wastewater. Water Research 41(7): 1481-1490.

226   WATER REUSE

Wert, E., J. E. Neemann, D. J. Rexing, and R. E. Zegers. 2008. Biofiltration for removal of BOM and residual ammonia following control of bromate formation. Water Research 42(1-2): 372-378.

Westerhoff, P., Y. Yoon, S. Snyder, and E. Wert. 2005. Fate of endocrine-disruptor, pharmaceutical, and personal care product chemicals during simulated drinking water treatment processes. Environmental Science & Technology 39: 6649-6663.

WHO (World Health Organization). 1989. Health Guidelines for the Use of Wastewater in Agriculture and Aquaculture. Geneva.

WHO. 2004. Guidelines for Drinking-Water Quality, 3rd Ed., Volume 1, Recommendations. Geneva, Switzerland: WHO Press.

WHO. 2005. Water Safety Plans: Managing Drinking-Water Quality from Catchment to Consumer. Available at http://www.who.int/water_sanitation_health/dwq/wsp170805.pdf.

WHO. 2006a. Guidelines for the Safe Use of Wastewater, Excreta and Greywater, Volume 2, Wastewater Use in Agriculture. Geneva, Switzerland: WHO Press.

WHO. 2006b. Guidelines for the Safe Use of Wastewater, Excreta and Greywater, Volume 3: Wastewater and Excreta Use in Aquaculture, Geneva, Switzerland: WHO Press.

WHO. 2006c. Guidelines for the Safe Use of Wastewater, Excreta and Greywater, Volume Volume 4, Excreta and Greywater Use in Aquaculture. Geneva, Switzerland: WHO Press.

WHO. 2008. Guidelines for Drinking-Water Quality, 3rd Ed. Incorporating the First and Second Addenda, Volume 1, Recommendations. Geneva, Switzerland: WHO Press.

WHO. 2009. Principles for modelling dose-response for the risk assessment of chemicals. Environmental Health Criteria 239. Available at http://whqlibdoc.who.int/publications/2009/9789241572392_eng.pdf

WHO. 2011. Guidelines for Drinking-Water Quality, 4th Ed. Available at http://www.who.int/water_sanitation_health/publications/2011/dwq_guidelines/en/.

Williams, B. L., Y. Florez, and S. Pettygrove. 2001. Inter- and intra-ethnic variation in water intake, contact, and source estimates among Tucson residents: Implications for exposure analysis. Journal of Exposure Analysis and Environmental Epidemiology 11(6): 510-521.

Wintgens, T., F. Salehi, R. Hochstrat, and T. Melin. 2008. Emerging contaminants and treatment options in water recycling for indirect potable use. Water Science and Technology 57: 99-107.

WPCF (Water Pollution Control Federation). 1989. Water Reuse Manual of Practice SM-3, 2nd Ed, Alexandria, VA: WPCF.

Wu, L., and L. Dodge. 2005. Landscape plant salt tolerance guide for recycled water irrigation. Slosson Research Endowment for Ornamental Horticulture, Department of Plant Sciences, University of California, Davis. Available at ucce.ucdavis.edu/files/filelibrary/5505/20091.pdf. Accessed on July 1, 2011.

Wu, Q. L., and W. T. Liu. 2009. Determination of virus abundance, diversity and distribution in a municipal wastewater treatment plant. Water Research 43(4): 1101-1109.

Xu, X., and C. P. Weisel. 2003. Inhalation exposure to haloacetic acids and haloketones during showering. Environmental Science & Technology 37(3): 569-576.

Xu, X. and C. P. Weisel. 2005. Human respiratory uptake of chloroform and haloketones during showering. Journal of Exposure Analysis and Environmental Epidemiology 15(1): 6-16.

Yang, W., J. Gan, W. Liu, and R. Green. 2005. Degradation of N-nitrosodimethylamine (NDMA) in landscape soils. Journal of Environmental Quality 34(1): 336-341.

Yates, M. V., C. P. Gerba, and L. M. Kelley. 1985. Virus persistence in groundwater. Applied Environmental Microbiology 49: 778-781.

Yates, M. V., S. R. Yates, J. Wagner, and C. P. Gerba. 1987. Modeling virus survival and transport in the subsurface. Journal of Contaminant Hydrology 1: 329-345.

Yoder, J., V. Roberts, G. F. Craun, V. Hill, L. Hicks, N. T. Alexander, V. Radke, R. L. Calderon, M. C. Hlavsa, M. J. Beach, and S. L. Roy. 2008. Surveillance for waterborne disease and outbreaks associated with drinking water and water not intended for drinking—United States, 2005-2006. Morbidity and Mortality Weekly Report 57(SS09): 39-62.

You, S. J., Y. P. Tsai, and R.Y. Huanf. 2009. Effect of heavy metals on nitrification performance in different activated sludge processes. Journal of Hazardous Materials 165(1-3): 987-994.

Yuan, F., C. Hu, X. Hu, J. Qu, and M. Yang. 2009. Degradation of selected pharmaceuticals in aqueous solution with UV and UV H$_2$O$_2$. Water Research 43(6): 1766-1774.

Zhang, J., H. C. Chua, J. Zhou, and A. G. Fane. 2006. Factors affecting the membrane performance in membrane bioreactors. Journal of Membrane Science 284: 54-66.

Zhou, Q., S. McCraven, J. Garcia, M. Gasca, T. Johnson, and W. Motzer. 2009. Field evidence of biodegradation of N-Nitrosodimethylamine (NDMA) in groundwater with incidental and active recycled water recharge. Water Research 43: 793-805.

Acronyms

ABS	alkylbenzenesulfonate
ADD	average daily dose
ADI	acceptable daily intake
AGI	acute gastrointestinal infection
AOP	advanced oxidation process
APEO	alkylphenol polyethoxylate
ARR	artificial recharge recovery
ASR	aquifer storage and recovery
AWWA	American Water Works Association
BAF	bioaccumulation factor
BASINS	Better Assessment Science Integrating Point and Nonpoint Sources
BCF	bioconcentration factor
BGD	billion gallons per day
BNR	biological nutrient removal
BOD	biochemical oxygen demand
BPA	bisphenol A
CCE	carbon-chloroform extract
CCL	Contaminant Candidate List
CDC	Centers for Disease Control and Prevention
CDPH	California Department of Public Health
CEC	contaminant of emerging concern
CERP	Comprehensive Everglades Restoration Plan
CRWS	Central Regional Wastewater System
CSF	cancer slope factor
CWA	Clean Water Act
CWD	chronic wasting disease
DALYs	disability-adjusted life years
DBP	disinfection byproduct

DOC	dissolved organic carbon
DOE	Department of Energy
ED	electrodialysis
EDR	electrodialysis reversal
EE2	ethinyl estradiol
EPA	U.S. Environmental Protection Agency
ERA	ecological risk assessment
FDA	Food and Drug Administration
FIFRA	Federal Insecticide, Fungicide, and Rodenticide Act
GWRC	Global Water Research Coalition
GWRS	groundwater replenishment system
HAAs	haloacetic acids
HIV	human immunodeficiency virus
IPR	indirect potable reuse
IU	infectious unit
IWF21	Interim Water Factor 21
JECFA	Joint Expert Commission on Food Additives
KWRP	Kwinana Water Reclamation Plant
LOAEL	lowest observed adverse effect level
LOEC	lowest observed effect concentration
MBR	membrane bioreactor
MCL	maximum contaminant level
MCLG	maximum contaminant level goal
MDWASD	Miami Dade Water and Sewer Department
MF	microfiltration
MGD	million gallons per day
MIB	methylisoborneol
MMWD	Marin Municipal Water District
MOE	margin of exposure
MOS	margin of safety
NDMA	nitrosodimethylamine
NDWAC	National Drinking Water Advisory Council
NEPA	National Environmental Policy Act
NGO	nongovernmental organization
NJDEP	New Jersey Department of Environmental Protection
NOEC	no observed effect concentration
NOM	natural organic matter

NPDES	National Pollutant Discharge Elimination System
NPDWR	National Primary Drinking Water Regulation
NRC	National Research Council
NTU	nephelometric turbidity unit
O&M	operation and maintenance
OCWD	Orange County Water District
OECD	Organisation for Economic Co-operation and Development
OWD	Oceanside Water Department
PCCL	Preliminary Contaminant Candidate List
PCR	polymerase chain reaction
PEC	predicted environment concentration
PFOA	perfluorooctanoate
PFOS	perfluorooctane sulfonate
PNEC	predicted no-effect concentration
PPCP	pharmaceuticals and personal care products
QALYs	quality-adjusted life years
RBAL	Risk-Based Action Level
RBF	riverbank filtration
RfD	reference dose
RO	reverse osmosis
RQ	risk quotient
RWC	recycled water contribution
SAR	sodium absorption ratio
SAT	soil aquifer treatment
SCVWD	Santa Clara Valley Water District
SDWA	Safe Drinking Water Act
SDWRP	South District Water Reclamation Plant
SFWMD	South Florida Water Management District
SWAQ	Subcommittee on Water Availability and Quality
SWFWMD	Southwest Florida Water Management District
TDI	tolerable daily intake
TDS	total dissolved solids
TEF	toxicity equivalency factor
TEQ	toxic equivalents
THM	trihalomethane
TN	total nitrogen
TOC	total organic carbon
TOR	threshold of regulation
TOX	total organic halides
TP	total phosphorus
TSCA	Toxic Substances Control Act

TSS	total suspended solids
TTC	threshold of toxicological concern
UCMR	Unregulated Contaminant Monitoring Regulation
UIC	underground injection control
UOSA	Upper Occoquan Service Authority
USAID	U.S. Agency for International Development
USACE	U.S. Army Corps of Engineers
USDA	U.S. Department of Agriculture
USBR	U.S Bureau of Reclamation
USGS	U.S. Geological Survey
WET	whole effluent toxicity
WERF	Water Environment Research Foundation
WHO	World Health Organization
WRP	water reclamation plant
YAR	yeast androgen receptor
YES	yeast estrogen

Appendixes

Appendix A

Details in Support of the Risk Exemplar in Chapter 6

In this appendix, the committee details the data and assumptions used in the risk exemplar, described in Chapter 6.

PATHOGENS

The exemplar includes four enteric pathogens: adenovirus, norovirus, *Salmonella*, and *Cryptosporidium*. In the following discussion, each organism is briefly described and an estimated density in secondary effluent for use in the exemplar is provided. Modifications in those densities are then estimated that correspond to each of the scenarios in the exemplar. Finally, the densities are adjusted so that they are in the same form as those used in dose-response testing, and a risk of illness is estimated using quantitative microbial risk estimation methodology (Haas et al., 1999).

Pathogen Occurrence in Secondary Effluents

The information and assumptions used to estimate pathogen occurrence in undisinfected secondary wastewater effluent as a starting point for the risk exemplar is discussed in this section and summarized in Table A-1. Pathogen reduction from subsequent disinfection and treatment steps is discussed in the next section.

Adenovirus

Adenovirus is a waterborne pathogen that has been associated with recreation-related outbreaks in the United States. It causes a large spectrum of human

TABLE A-1 Estimated Pathogen Densities in Secondary Effluent

Organism	Concentration
Adenovirus	5,000 gc/L
Norovirus	10,000 gc/L
Salmonella	500 cfu/L
Cryptosporidium	17 oocysts/L

diseases from diarrhea to eye and throat infections (Jiang, 2006; Mena and Gerba, 2008). Quantitative data on adenovirus occurrence in water and wastewater are available in the current literature, because their occurrence is often used as a marker for human viral contamination in waters. The dose-response model for this virus has also been developed previously based on epidemiological studies (Haas et al., 1999); thus, it is an organism for which quantitative risk assessment is possible.

Human adenovirus occurrence data in the exemplar were collected from peer-reviewed literature, which used molecular biology–based genome quantification methods (He and Jiang, 2005; Albinana-Gimenez et al., 2006; Bofill-Mas et al., 2006; Haramoto et al., 2007; Katayama et al., 2008; Fong et al., 2010; Schlindwein et al., 2010). Reported densities vary over a wide range, between 1 and 10^5 genome copies/liter (gc/L). A density of 5×10^3 gc/L, which falls in the most frequently reported range, was chosen by the committee as a typical concentration in secondary effluent.

Although the genome-based method is sensitive at detecting viral presence, it does not provide information on viral infectivity; thus the presence of a genome is not synonymous with the presence of an infectious unit (IU). Dose-response studies were conducted using tissue culture assays for quantification of IU. There is limited quantitative information on the side-by-side data for IUs and genome copies although it is generally known that infectivity decays more rapidly than does the density of genome copies (R.A. Rodriguez et al., 2009). Based on a single report (He and Jiang, 2005), where three side-by-side polymerase chain reaction (PCR) and tissue culture assays were performed on adenovirus isolated from secondary effluent, it is estimated that the ratio between genome copies and infectious units is approximately 1,000:1. Thus, genome count densities estimated for adenovirus for each scenario were reduced by three orders of magnitude to convert to IUs during the risk estimation process.

Norovirus

Norovirus is one of the most important enteric viruses for both waterborne and foodborne outbreaks in the United States. Several recent studies have focused on occurrence of this virus in water and wastewater (Pusch et al., 2005; Haramoto et al., 2006; Katayama et al., 2008; Nordgren et al., 2009; Victoria et al., 2010). In these studies, the density of the norovirus genome varies over a wide range with densities as high as 10^7 gc/L reported in raw sewage. Based on the published literature, a density of 10^4 gc/L is estimated to be the median occurrence in secondary effluent. Once again, although the genome-based method is sensitive at detecting the presence of copies of the genome of the virus, it does not provide information on viral infectivity. Norovirus has not been successfully cultivated using conventional tissue culture methods, and so no work is available to establish the ratio between genome density and IU density.

A dose-response model for norovirus was used based on the study by Teunis et al. (2008), using the estimate for single unaggregated virus. Because norovirus has not been successfully cultivated in vitro, these studies were conducted using fresh virus and the genome count quantified by PCR. Published work has shown that the fraction of genome copies that are infectious drops rapidly in the environment (R.A. Rodriguez et al., 2009). Thus, for the purposes of this exemplar, the same 1,000:1 was applied before risk estimation.

Salmonella

Salmonella has long been a well-studied waterborne enteric pathogen. The concentration of this microorganism in raw sewage ranges between 10^2 and 10^4 cfu/100 mL (Asano et al., 2007). Taking the average of these two and assuming the same 2-log reduction during primary and secondary treatment that normally occurs for Escherichia coli produces an estimate of 5 × 10^2 cfu/L in secondary effluent for the exemplar. Again, the dose-response model for this organism has been developed previously based on epidemiological studies (Haas et al., 1999).

Cryptosporidium

Cryptosporidium is associated with both drinking water and recreational water outbreaks in the United States. The health significance of this organism has motivated a number of studies to understand its occurrence and persistence in the water environment (Rose et al., 1996; Gennaccaro et al., 2003; Robertson et al., 2006; Lim et al., 2007; Castro-Hermida et al., 2008; Chalmers et al., 2010; Fu et al., 2010). The peer reviewed literature reports a range of Cryptosporidium densities in secondary treated effluents varying with season and geographical location. Studying this literature, a density of 50 oocysts/L is estimated as typical for secondary effluents. However, most of the data on oocyst concentration were determined using the indirect fluorescent-antibody assay (IFA), which also does not directly measure IUs. A study comparing oocyst densities as determined by IFA with IU densities as determined by a focus-detection-method most-probable-number technique in cell culture (Slifko et al., 1999) found a ratio of approximately 3:1 in 18 samples of secondary effluent (Gennaccaro et al., 2003). Using this ratio, a density of 50 oocysts/L produces an estimate of 17 IUs/L in secondary effluent for the exemplar. More than one dose-response model has been developed for this organism (Haas et al., 1999).

Assumptions Concerning Fate, Transport, and Removal

The following is a brief discussion of assumptions made regarding fate, transport, and removal for the pathogens in the exemplar.

Scenario 1—de Facto Reuse

As discussed in Chapter 6, Scenario 1 represents a conventional water supply drawn from a surface water source with a 5 percent contribution from treated wastewater. For this scenario a nonnitrified secondary effluent is assumed to be disinfected with chlorine prior to discharge to bring fecal coliforms from 10^5/100 mL to 200/100 mL, a 2.7-log reduction (99.8 percent). The exemplar assumes combined chlorine is the active disinfectant. According to Butterfield and Wattie (1946), *E. coli*, the principal target of the fecal coliform measurement, are generally as or more resistant to combined chlorine than *Salmonella* spp. (*S. dysenteriae*). Accordingly, the same 2.7-log reduction was assumed for *Salmonella* spp. For adenovirus and norovirus, removal was assumed to follow the removal credit for viruses in the surface water treatment rule, which was judged to be negligible. It is also assumed that this limited disinfection has no impact in the viability of *Cryptosporidium*.

The water treatment plant has been modified to be compliant with the requirements of the Long-Term-2 Enhanced Surface Water Treatment Rule (LT2ESWTR; EPA, 2006a). Assuming no diminishment during transport in the river, the *Cryptosporidium* contribution from upstream wastewater plants in the exemplar puts the density of oocysts in the water plant's source water at approximately 0.85 oocyst/L. This classifies the supply in "Bin 2" according to LT2ESWTR, which corresponds to a requirement of 1 log of removal for *Cryptosporidium* beyond the performance of conventional treatment. Hence, additional treatment to achieve 1- and 2-log removal is required for *Cryptosporidium* and viruses, respectively.[1]

For the exemplar it is assumed that the drinking

water treatment plant uses free chlorine for primary disinfection and that it has been modified to obtain 1 log of additional inactivation of *Cryptosporidium* using UV light (required dose, 2.5 mJ/cm^2). Under the LT2ESWTR, the inactivation credit for UV at a dose of 2.5 mJ/cm^2 is 1 log *Cryptosporidium* and negligible for viruses. Thus the 2-log virus inactivation requirement must be met by free chlorine. At a low temperature of 5 °C (a conservative surface water temperature), this corresponds to a $C{\cdot}t$ of 8 mg-min/L. So the process train is conventional water treatment (coagulation, flocculation, filtration) followed by UV (3 mJ/cm^2) and chlorination (8 mg-min/L) and this train will get the full 4-log removal/inactivation credit for both *Cryptosporidium* and viruses, required by the LT2ESWTR.

In the exemplar, excluding dilution, the overall reduction in *Cryptosporidium* is assumed to correspond to the 4-log removal required by EPA, and the reductions in adenovirus and norovirus are also assumed to correspond to EPA's assumptions for 2 logs of physical removal in conventional treatment and an additional 2 logs of inactivation via chlorination (totaling 4-log removal). EPA's LT2ESWTR does not provide direct guidance on *Salmonella* spp., and so an independent analysis is required. *Salmonella* spp. are understood to be more sensitive to free chorine than are *E. coli* (Butterfield et al., 1943). According to Figure 13-5 in Crittenden et al. (2005), a $C{\cdot}t$ of approximately 1 mg-min/L is required for 2-log removal of *E. coli* at 25 °C; thus, a $C{\cdot}t$ of 8 mg-min/L will achieve a 16-log reduction of *E. coli*. For the effect of chlorine on *Salmonella* spp., the exemplar discounts this to an inactivation credit of 4 logs to account for temperature. Exposure to low levels of UV light also affects *Salmonella* spp. and to some degree adenovirus and norovirus. According to data in a recent Dutch review (Hijnen et al., 2005a), a low-pressure UV dose of 2.5 mJ/cm^2 should result in a 1.5-log inactivation of *Salmonella* spp., a 0.1-log reduction in adenovirus, and a 0.3-log reduction in norovirus. In the exemplar, the effect of UV on the *Salmonella* spp. is included, and the impact of UV on these viruses is neglected. Thus, the overall water treatment plant removal is 4 logs for *Cryptosporidium*, 5.5 logs for *Salmonella* spp., and 4 logs for adenovirus and norovirus. A summary of removal for microorganisms and their resulting densities is given in Table A-2.

---

[1] Actually, the LT2ESWTR gives conventional drinking water treatment (without disinfection) credit for the physical removal of 2 logs of *Cryptosporidium* and viruses, respectively. Where *Cryptosporidium* is concerned, this is a bit confusing because the regulation requires 4-log removal of *Cryptosporidium* for any alternative process in Bin 2, but requires only one additional log removal for conventional treatment. It appears that the 2-log credit is actually a holdover from the earlier interim enhanced surface water treatment rule, which established the 2-log credit and that EPA expects 3-log removal of *Cryptosporidium*. For the exemplar it is assumed that the drinking water treatment plant achieves 3-log *Cryptosporidium* removal and requires UV disinfection to achieve one additional log. An actual plant might make other choices from the microbial treatment toolbox to accomplish similar results.

TABLE A-2 Summary of Log (and %) Removals of Pathogens in Various Steps of Scenario 1

Process	Adenovirus	Norovirus	*Salmonella*	*Cryptosporidium*
Disinfection at wastewater treatment plant	0 (0%)	0 (0%)	2.7 (99.8%)	0 (0%)
Dilution in stream	1.3 (95%)	1.3 (95%)	1.3 (95%)	1.3 (95%)
Removal in water treatment	4.0 (99.99%)	4.0 (99.99%)	4.0 (99.99%)	3.0 (99.9%)
Removal by UV	0 (0%)	0 (0%)	1.5 (96.8%)	1.0 (90%)

Scenario 2—Soil Aquifer Treatment and Groundwater Recharge

As described in Chapter 7, in Scenario 2, a nitrified and partially denitrified secondary effluent, which has been subjected to granular media filtration, is applied to surface spreading basins with subsequent soil aquifer treatment (SAT). The effluent is not disinfected. It is assumed that the water will remain in the subsurface for 6 months with no dilution from native groundwater. While the assumption of no dilution is in contrast to hydrogeological characteristics of subsurface systems, this condition was selected to assign removal credits only to physicochemical and biological attenuation processes occurring during SAT. Subsequently, the water is abstracted from a deep well, disinfected at the wellhead, and chlorinated prior to consumption, assuming no blending occurs with other source waters in the distribution system. These assumptions describe a scenario where drinking water is consumed that originates 100 percent from reclaimed water after additional treatment using SAT.

Effect of SAT on Virus Removal. During percolation through porous media or groundwater recharge, the removal of pathogens from infiltrating reclaimed water depends primarily on three attenuation mechanisms: straining, inactivation, and attachment to aquifer grains (McDowell-Boyer et al., 1986). Subsurface systems, such as riverbank filtration and SAT have been reported as efficient treatment systems for the removal of microbial contaminants. With respect to virus removal, the field experiments conducted by

Schijven et al. (1998, 1999, 2000) are considered a benchmark for removal under relatively homogeneous and steady-state conditions in a saturated sand aquifer. During dune recharge using water that was spiked with bacteriophages (MS2 and PRD1), Schijven et al. (1999) reported a virus reduction of 3 logs within the first 2.4 m and another linear 5-log removal within the following 27 m of transport in the subsurface. Spiking tests with bacteriophages conducted by Fox et al. (2001) under field conditions suggested a 7-log removal over a distance of 100 m. During a deep-well (~300 m below surface) injection study, Schijven et al. (2000) spiked pretreated surface water with bacteriophages (MS2 and PRD1) and observed a 6-log removal within the first 8 m of travel followed by an additional 2-log removal during the subsequent 30 m of travel. These values are well within the range of virus inactivation values reported by others (Dizer et al., 1984; Yates et al., 1985; Powelson et al., 1990). Findings from these field studies demonstrated that infiltration into a relatively homogeneous sandy aquifer can achieve up to 8-log virus removal over a distance of 30 m in about 25 days. Loveland et al. (1996) revealed some of the conditions that favor removal of viruses in the subsurface and concluded that precipitated ferric, manganese, and aluminum oxyhydroxides form positively charged patches on the soil grains. These patches provide favorable attachment sites for negatively charged viruses. Powelson and Gerba (1994) also reported that virus inactivation is more efficient under unsaturated than saturated infiltration conditions. In addition, some studies reported that virus inactivation may be enhanced by microbial activity (Quanrud et al., 2003; Gupta et al., 2009) resulting in the expression of enzymes that are detrimental to other microorganisms (Yates et al., 1987). Considering that these conditions (i.e., biological activity, sequence of unsaturated to saturated conditions, presence of metal oxyhydroxides) are commonly observed in SAT systems and the retention time in the potable reuse case study of the exemplar using groundwater recharge via SAT is 6 months, a conservative removal of 6-log was assumed during SAT for both adenovirus and norovirus.

Effect of SAT on Bacteria Removal. For subsurface treatment, such as SAT and riverbank filtration, several studies have reported efficient inactivation of coliform bacteria. Havelaar et al. (1995) reported removal in excess of 5 logs for total coliform during transport of impaired river water over a 30-m distance from the Rhine River and over a 25-m distance from the Meuse River to a well. During a deep-well (~300 m below surface) injection study, Schijven et al. (2000) spiked pretreated surface water with *E. coli* and observed a 7.5-log removal within the first 8 m of travel in the subsurface. During SAT in the Dan Region Project, Israel, Ickeson-Tal et al. (2003) measured 5.3-log removal of total coliform and 4.5-log removal of fecal coliform bacteria. Total coliforms were rarely detected in riverbank-filtered waters, with 5.5- and 6.1-log reductions in average concentrations in wells relative to river water (Weiss et al., 2005). The efficient removal of fecal and total coliform bacteria during subsurface treatment and essentially their absence in groundwater abstraction wells after SAT or riverbank filtration was confirmed by various other studies (Fox et al., 2001; Hijnen et al., 2005b; Levantesi et al., 2010). Considering these field and controlled laboratory studies as well as a retention time of 6 months in the subsurface for the surface spreading groundwater recharge case of the exemplar, 6 logs of removal was assumed for bacteria (*Salmonella*) through SAT treatment in the exemplar.

Effect of SAT on *Cryptosoridium*. Under the LT2ESWTR (EPA, 2006a), immobilization of *Cryptosporidium* within granular media, often accomplished by sand or riverbank filtration can result in cost-effective removal of protozoa and other pathogens (Ray et al., 2002a,b; Tufenkji et al., 2002). By meeting certain design standards (i.e., unconsolidated, predominantly sandy aquifer with 25- or 50-ft setback from the river), EPA assigns 0.5-log or 1.0-log removal credits for *Cryptosporidium*, respectively. Log removal calculations require counts per volume of the same organism in the initial water source (e.g., reclaimed water) and groundwater wells. Given the usually low counts of *Cryptosporidium* in impaired source waters, log removal studies under ambient conditions are not practical. Bacterial spores, anaerobic clostridia spores, and aerobic endospores are resistant to inactivation in the subsurface, similar in shape to *Cryptosporidium* but smaller and sufficiently ubiquitous in both impaired surface water and groundwater that log removal can be calculated. Findings from studies investigating the

fate of bacterial spores in gravel aquifers suggest a high mobility and similar removal of *Cryptosporidium*, making bacterial spores adequate surrogate measures.

Findings from various field studies suggest that large removal of anaerobic and aerobic spores occurs during passage across the surface water–groundwater interface, and lesser removal is observed during groundwater transport away from this interface. Havelaar et al. (1995) reported 3.1-log removal of anaerobic spores during transport over a 30-m distance from the Rhine River to a well and 3.6-log removal over a 25-m distance from the Meuse River to a well. Schijven et al. (1998) measured 1.9-log removal over a 2-m distance from a canal. This finding is consistent with field monitoring results from a riverbank filtration site in Wyoming, where Gollnitz et al. (2005) observed a 2-log removal of *Cryptosporidium* in groundwater wells characterized by flow paths between 6 and 300 m. At a riverbank filtration site at the Great Miami River, Gollnitz et al. (2003) reported a 5-log removal of aerobic spores in a production well located 30 m off the river. Wang et al. (2002) reported 1.7-log removal of aerobic spores over the first 0.6-m distance and 3.8-log removal over a distance of 15.2 m at a riverbank filtration facility at the Ohio River. Less efficient removal of approximately 0.6 logs over a distance of 12 m was reported for transport solely within groundwater (Medema et al., 2000). For an injection experiment in a sandy aquifer at distances relatively far from an injection well, Schijven et al. (1998) observed negligible removal of anaerobic spores over a 30-m distance. Besides straining, inactivation might be important for the attenuation of *Cryptosporidium* during subsurface treatment. For two *Cryptosporidium* strains examined, NRC (2000) assumed a 1-log inactivation over 100 days and 180 days (corresponding to an inactivation rate coefficient of 0.023/d and 0.013/d, respectively). Considering these field and controlled laboratory studies as well as a retention time of 6 months in the surface spreading groundwater recharge case of the exemplar (much longer than is the case for any of the preceding citations), a removal credit of 6 log for *Cryptosporidium* was assumed for SAT treatment.

Effect of Wellhead Chlorination. The exemplar assumes that chlorination is provided at the wellhead in order to achieve a 4-log virus removal credit, and so this is the removal assigned to adenovirus and norovirus. At 15 °C (an approximate groundwater temperature), this would require a *C·t* of 4 mg-min/L. *Salmonella* removal is estimated using the equation in Table 13-3 in Crittenden et al. (2005), and adjusting the log inactivation by a factor of 2 for every 10 °C, results in a removal of 6 logs. No removal is assumed in the exemplar for *Cryptosporidium* via chlorine. The removals are summarized in Table A-3.

Scenario 3—Reverse Osmosis, Advanced Oxidation, and Deep-Well Injection.

Scenario 3, as discussed in Chapter 6, represents a water supply drawn from a deep well in an aquifer fed by injection of reclaimed water that received secondary treatment by chloramination, microfiltration, reverse osmosis, and high-output low-pressure ultraviolet (UV) light supplemented with hydrogen peroxide (also called advanced oxidation).

Effect of Microfiltration. Olivieri et al. (1999) showed median coliphage removals of 2 logs for microfiltration and 3 logs for ultrafiltration, but for microfiltration, removals as low as 0.1 log were observed on one occasion and removals below 1 log were observed 30 percent of the time. Consequently no virus removal was assumed in the exemplar for microfiltration. There is a great deal of literature on the removal of bacteria and protozoa via membrane filtration. This literature shows virtually complete rejection so long as the membranes remain intact (Jacangelo et al., 1997). Methods used were able to demonstrate between 4 and 5 logs for *Cryptosporidium* and 7 and 8 logs for bacteria. For the purposes of the exemplar, 99.99 percent removal is assumed for both *Salmonella* and *Cryptosporidium*. It should be cautioned that, for specific projects, these removals must be demonstrated for each membrane type and, even then, they

cannot be ensured unless monitoring demonstrates that the membranes continue to perform.

Effect of Reverse Osmosis. In principle, reverse osmosis, which is designed to remove individual ions from water, should completely reject all microorganisms. On the other hand, testing has demonstrated that these organisms can pass through these installations unless special quality control practices, beyond those normally exercised in the desalination community, are undertaken (Trussell et al., 2000). This is particularly true where viruses are concerned because these organisms have been shown to pass through flaws in the membranes themselves (Adham et al, 1998). More limited quality control on the installation of the membranes and O-rings has been shown adequate to manage the rejection of bacteria and protozoa. As a result, a removal credit of 99.99 percent is assumed for both bacteria and *Cryptosporidium* but a credit of only 97 percent is assumed for viruses because this roughly corresponds to the removal of conductivity through reverse osmosis.[2]

Effect of UV/H$_2$O$_2$. UV/H$_2$O$_2$ installations in existing projects are designed using low-pressure UV to provide 1.2-log removal of NDMA. It has been demonstrated that this corresponds to a delivered UV dose of approximately 1,200 mJ/cm^2 (Sharpless and Linden, 2003). Low doses of peroxide and chloramines (both 3 to 5 mg/L) are also present and these absorb some of the UV; nevertheless, the remaining effective UV dose is nearly 10-fold above the dose specified by EPA for 4-log removal of adenovirus or *Cryptosporidium* in the LT2ESWTR. Evidence is that *Salmonella* and norovirus are more easily removed than adenovirus (Hijnen et al., 2005a). Consequently a removal of 6 logs (99.9999 percent) is assumed for all these organisms, and this is thought to be very conservative.

Effect of Deep-Well Injection on Pathogen Removal. The lack of microbial activity and the potential absence of metal oxyhydroxides in deep aquifers recharged with reverse osmosis–treated reclaimed water will provide less favorable conditions for virus removal

TABLE A-3 Summary of Logs (and %) Pathogen Removal Assumed for Processes in Scenario 2

Process	Adenovirus	Norovirus	*Salmonella*	*Cryptosporidium*
SAT + 6 mo	6 (99.9999%)	6 (99.9999%)	6 (99.9999%)	6 (99.9999%)
Chlorination at wellhead	4 (99.99%)	4 (99.99%)	6 (99.9999%)	0 (0%)

[2] Based on data from the first 2 years of operation of the Orange County Water District's Advanced Water Purification Facility (B. Dunivan, OCWD, personal communication, 2011).

and/or inactivation. Thus, in the exemplar, no removal credit for viruses was considered for the reuse scenario using direct injection into a potable aquifer. Likewise, given the lack of a surface water–groundwater interface in direct injection projects and a rather low inactivation rate in aquifers, no removal credits for *Salmonella* or *Cryptosporidium* were assigned for the direct injection process and groundwater travel time. It is noteworthy that these are conservative assumptions, because pathogen inactivation could occur in deep aquifers receiving reverse osmosis permeate.

Effect of Wellhead Chlorination. As described under Scenario 2, wellhead chlorination was assigned a 4-log virus removal credit, and a 6-log removal for *Salmonella*. No removal is assumed in the exemplar for *Cryptosporidium* via chlorine. The removals are summarized in Table A-4.

Summary of Results on Pathogen Densities

Using the assumptions and results summarized earlier, calculations were conducted to produce an estimate of the densities of each of the four pathogens studied in the drinking water produced in each of the three scenarios. The results of these calculations are summarized in Table A-5.

Quantitative Microbial Risk Assessment

The pathogen densities shown in Table A-5 can be translated into risk of illness using the methodologies for quantitative risk assessment summarized in Chapter

5. Table A-6 summarizes the coefficients derived from the literature in order to facilitate those calculations, as well as the pertinent dose-response model equations.

Table A-7 summarizes the quantitative microbial risk assessment in three parts for the three scenarios. Table A-7A details the pathogen densities at the point of exposure (i.e., the tap). The virus densities in Table A-7A were used to compute the daily risk (based on a daily consumption of 1 L) using equations (1) or (2) as appropriate for the organism being considered. Table A-7B shows the estimated levels of excess illness that result from the drinking water from a single exposure (1-L consumption). A consumption of 1 L/d is used for consumption of unboiled water as contrasted with the consumption of 2 L/d used for total consumption (Roseberry and Burmaster, 1992).

TRACE ORGANIC CHEMICALS

For potable reuse projects, there is growing concern among stakeholders and the public about potential adverse health effects associated with the presence of trace organic chemicals in reclaimed water. Reclaimed water can contain thousands of chemicals originating from consumer products (e.g., household chemicals, personal care products, pharmaceutical residues), human waste (e.g., natural hormones), industrial and commercial discharges (e.g., solvents, heavy metals), or chemicals that are generated during water treatment (e.g., disinfection byproducts) (see Chapter 3). For the risk exemplar, 24 chemicals were selected that represent different classes of contaminants (i.e., nitrosamines, disinfection byproducts, hormones, pharmaceuticals, antimicrobials, flame retardants, and perfluorochemicals).

Chemical Occurrence in Secondary Effluents

For disinfection byproducts in secondary effluents, data were obtained from Krasner et al. (2008), which reported occurrence of unregulated and regulated disinfection byproducts for secondary wastewater treatment processes with various disinfection practices for a range of different geographical regions of the United States. These datasets were validated and augmented with results from field monitoring efforts reported by Snyder et al. (2010a) and Dickenson et al. (2011). Hormones and pharmaceutical occurrence data were adopted

TABLE A-4 Summary of Logs (and %) Pathogen Removal Assumed for Processes in Scenario 3

Process	Adenovirus	Norovirus	*Salmonella*	*Cryptosporidium*
Microfiltration (MF)	0 (0%)	0 (0%)	4 (99.99%)	4 (99.99%)
Reverse osmosis (RO)	1.5 (97%)	1.5 (97%)	4 (99.99%)	4 (99.99%)
UV/H_2O_2	6 (99.9999%)	6 (99.9999%)	6 (99.9999%)	6 (99.9999%)
Deep-well Injection + 6 mo	0 (0%)	0 (0%)	0 (0%)	0 (0%)
Chlorination at wellhead	4 (99.99%)	4 (99.99%)	6 (99.9999%)	0 (0%)

TABLE A-5 Summary of Exemplar Calculations to Establish Pathogen Levels in Drinking Water for the Three Scenarios

Scenario 1 De facto Reuse: Secondary effluent, disinfected with chlorine, diluted 95%, conventional water treatment

	2° Effluent Concentration	Removal through Disinfection	Discharge Concentration	95% dilution	Concentration in River	Die-off & Predation	WTP Influent Concentration	Removal in Conventional WTP	Removal by UV	Drinking Water Concentration
Norovirus	10,000 gc/L	0%	10,000 gc/l	95%	500 gc/L	0%	500 gc/L	99.99%	0.0%	0.050 gc/L
Adenovirus	5,000 gc/L	0%	5,000 gc/L	95%	250 gc/L	0%	250 gc/L	99.99%	0.0%	0.025 gc/L
Salmonella	500 CFU/L	99.80%	1.0 CFU/L	95%	0.1 CFU/L	0%	0.1 CFU/L	99.99%	96.8%	1.6E-07 CFU/L
Cryptosporidium	17 oocyst/L	0%	17 oocyst/L	95%	0.85 oocyst/L	0%	0.85 oocyst/L	99.9%	90%	8.5E-05 oocysts/L

Scenario 2 Secondary effluent, no disinfection, followed by SAT, 6 mo retention, no dilution, free chlorine disinfection

	2° Effluent Concentration	SAT Removal	Concentration at Wellhead	Removal by Chlorination	Drinking Water Concentration
Norovirus	10,000gc/L	99.9999%	1.0E-02 gc/L	99.99%	1.0E-06 gc/L
Adenovirus	5,000 gc/L	99.9999%	5.0E-03 gc/L	99.99%	5.0E-07 gc/L
Salmonella	500 CFU/L	99.9999%	5.0E-04 CFU/L	99.9999%	5.0E-10 CFU/L
Cryptosporidium	17 oocyst/L	99.9999%	1.7E-05 oocysts/L	0.00%	1.7E-05 oocysts/L

Scenario 3 Secondary effluent, MF, RO, UV/H202, groundwater injection, free chlorine disinfection

	2° Effluent Concentration	Removal through MF	Removal through RO	Removal through UV/H2O2	AWT Effluent Concentration	Removal through Groundwater Injection	Wellhead Concentration	Removal by Free Chlorine	Drinking Water Concentration
Norovirus	10,000 gc/L	0%	97%	99.9999%	3.0E-04 gc/L	0%	3.0E-04 gc/L	99.99%	3.0E-08 gc/L
Adenovirus	5,000 gc/L	0%	97%	99.9999%	1.5E-04 gc/L	0%	1.5E-04 gc/L	99.99%	1.5E-08 gc/L
Salmonella	500 CFU/L	99.99%	99.99%	99.9999%	5.0E-12 CFU/L	0%	5.0E-12 CFU/L	99.9999%	5.0E-18 CFU/L
Cryptosporidium	17 oocyst/L	99.99%	99.99%	99.9999%	1.7E-13 oocysts/L	0%	1.7E-13 oocysts/L	0%	1.70E-13 CFU/L

TABLE A-6 Dose-Response Parameters for Quantitative Microbial Risk Assessment

	Exponential k	Beta Poisson α	Beta Poisson N_{50}	Beta Poisson β	Dose-Response Models[a]
Norovirus[b]		0.04		0.055	$p = 1 - \left(1 + \dfrac{d}{\beta}\right)^{-a}$
Adenovirus[c]	0.4172				$p = 1 - \exp(-kd)$
Salmonella[d]		0.3126	23600		$p = 1 - \left(1 + \dfrac{\left(2^{\frac{1}{a}} - 1\right)d}{N_{50}}\right)^{-a}$
Cryptosporidium[e]	0.0042				$p = 1 - \exp(-kd)$

[a]In these equations, p and d are the single exposure risk and dose, respectively. As discussed previously, when the drinking water concentration is measured in genome count densities, the concentration is divided by 1000 to convert to infectious units.

[b]Teunis et al. (2008, Table III—pooled response for infection).

[c]From Haas et al. (1999, Table 9-15).

[d]From Haas et al. (1999, Table 9-3, Pooled Salmonella strains)

[e]*Original* Iowa strain data for *Cryptosporidium* (Haas et al., 1996).

from studies comparing the chemical composition of reclaimed and conventional waters at seven field sites in the United States (Snyder et al., 2010a; Dickenson et al. 2011), with some additional data from select pharmaceuticals adopted from Krasner et al. (2008). Other chemicals of interest, such as antimicrobials, chlorinated flame retardants, and perfluorochemicals, were adopted from field monitoring efforts using secondary treated effluents reported by Snyder et al.

TABLE A-7 Summary of Quantitative Microbial Risk Assessment of Risk Exemplar

Organism	Scenario 1 De Facto Reuse	Scenario 2 SAT	Scenario 3 MF/RO/UV
A. Pathogen Densities			
Norovirus	0.050 gc/L	1.0E-06 gc/L	3.0E-08 gc/L
Adenovirus	0.025 gc/L	5.0E-07 gc/L	1.5E-08 gc/L
Salmonella	1.6E-07 CFU/L	5.0E-10 CFU/L	5.0E-18 CFU/L
Cryptosporidium	8.5E-05 oocysts/L	1.7E-05 oocysts/L	1.7E-13 oocysts/L
B. Risk of Illness (illness/(capita*d))			
Norovirus	3.6E-05	7.3E-10	2.2E-11
Adenovirus	1.0E-05	2.1E-10	6.3E-12
Salmonella	1.7E-11	5.4E-14	0
Cryptosporidium	3.6E-07	7.1E-08	0
C. Relative Risk			
Norovirus	1	2.0E-05	6.0E-07
Adenovirus	1	2.0E-05	6.0E-07
Salmonella	1	3.2E-03	0
Cryptosporidium	1	0.2	0

(2010a), Laws et al. (2011), and Drewes et al. (2003). It has been demonstrated that disinfection processes used in the treatment of wastewater and drinking water are effective in removing a significant number of hormones and pharmaceutical compounds (Snyder et al., 2007), but disinfection processes also introduce disinfectant byproducts, and for these reasons, previously cited measurements are used in the exemplar as opposed to the model-based estimates used for microbials. Table A-8 lists the concentrations of the 24 chemicals in disinfected secondary effluent, and Table A-9 shows the concentrations for undisinfected secondary effluent.

Assumptions Concerning Fate, Transport, and Removal

Scenario 1—De Facto Reuse

For the scenario describing de facto reuse (Scenario 1), it was assumed that the surface water providing dilution of treated wastewater discharge to a drinking water source represents a pristine water quality with respect to trace organic chemical concentrations, as reported by Krasner et al. (2008). The concentration of unregulated and regulated disinfection byproducts after conventional treatment (including coagulation/flocculation, filtration, free chlorine as primary disinfectant, and residual chloramines) is adopted from Krasner et al. (2008). The effectiveness of conventional water treatment for hormones, pharmaceuticals, and other trace organic chemicals was adopted from an investigation of

TABLE A-8 Estimation of Margin of Safety for Scenario 1—Drinking Water from Surface Water Source with 5% Contribution from Wastewater Discharges

Name of Chemical	Unit	2° Effluent with Disinfect.	Surface Water[a]	Blend 95% SW 5% 2° Effluent	Drinking Water[b]	Risk Based Action Level	Margin of Safety (unitless)
Nitrosamines[c,d,e]							
N-Nitrosodimethylamine (NDMA)	ng/L	10	<2	<2	<2	0.7	>0.35
Disinfection Byproducts[f,g,h]							
Bromate	μg/L	N/A	N/A	N/A	N/A	10	N/A
Bromoform	μg/L	18	<0.5	<1.1[i]	3	80	27
Chloroform	μg/L	25	<1	<1.7[i]	5	80	16
Dibromoacetic acid	μg/L	10	<1	<1[i]	<1	60	>60
Dibromoacetonitrile	μg/L	16	<1	<1.3[i]	<1.3	70	>54
Dibromochloromethane	μg/L	<1	<1	<1[i]	<1	80	>80
Dichloroacetic acid	μg/L	31	<1	<2[i]	5	60	12
Dichloroacetonitrile	μg/L	0.3	<1	<1[i]	<1	20	>20
Haloacetic acid (HAA5)	μg/L	70	<1	<4[i]	10	60	6
Trihalomethanes THMs)	μg/L	57	<0.5	<3.1[i]	30	80	3
Pharmaceuticals[f,g,h]							
Acetaminophen	ng/L	1	<1	<1[i]	<1	350,000,000	>350,000,000
Ibuprofen	ng/L	38	<1	<2.4[i]	<2.4	280,000,000	>120,000,000
Carbamazepine	ng/L	180	10	19	19	186,900,000	10,000,000
Gemfibrozil	ng/L	305	1	16	16	140,000,000	8,600,000
Sulfamethoxazole	ng/L	30	<1	<2[i]	<2	160,000,000	>80,000,000
Meprobamate	ng/L	240	5	17	17	280,000,000	17,000,000
Primidone	ng/L	98	1	6	6	58,100,000	10,000,000
Others[c,f,g,h]							
Caffeine	ng/L	210	10	20	20	70,000,000	3,500,000
17-β Estradiol	ng/L	0.15	<0.1	<0.1[i]	<0.1	3,500,000	>35,000,000
Triclosan	ng/L	2.5	<1	<0.6[i]	<0.6	2,100,000	>3,500,000
TCEP (tris(2-chloroethyl)phosphate)	ng/L	400	<10	<25[g]	<25	2,100,000	>84,000
PFOS	ng/L	54	10	12	12	200	17

NOTES: N/A = data not available.

[a]Taken from median conc. from Krasner national occurrence survey (Krasner et al., 2008)

[b]Remaining after conventional surface water treatment (including coagulation/flocculation; filtration, free chlorine; residual chloramines); no transformation occurred in surface water.

[c]Krasner et al. (2008).

[d]Snyder et al. (2010a).

[e]Dickenson et al. (2011)

[f]Bellona et al. (2008).

[g]M. Wehner, OCWD, personal communication, 2009.

[h]Bellona and Drewes (2007).

[i]When surface water concentrations were below the detection limit, one-half the detection limit was used in the dilution calculations. (In contrast, for Scenarios 2 and 3, the detection limit is used for concentrations below the detection limit to be a more conservative assumption in the relative comparison and because secondary effluent is likely to contain higher levels of contaminants than pristine surface waters.) If the final calculated concentration was below the detection limit, less than the detection limit was reported.

five conventional drinking water plants in the United States by Snyder et al. (2010a). The removal efficiencies assumed were within the same range as reported by Snyder et al. (2008a) for conventional drinking water processes.

Scenario 2—Soil Aquifer Treatment and Groundwater Recharge

For Scenario 2 describing a potable reuse system using surface spreading leading to groundwater re-

charge, an effluent quality is assumed that mirrors the secondary effluent qualities assumed in Scenario 1, except that Scenario 2 represents a *undisinfected*, filtered, secondary wastewater effluent. The water quality after 6 months of SAT, assuming no dilution with ambient groundwater, and a final disinfection with free chlorine at the wellhead, is based on findings from field monitoring efforts at SAT and riverbank filtration installations (Drewes et al., 2003; Hoppe-Jones et al., 2010; Snyder et al., 2010a; Laws et al., 2011). The data are augmented by field monitoring results for disinfection

TABLE A-9 Estimation of Margin of Safety for Scenario 2—Drinking Water from Deep-Well Supplied by Spreading of Undisinfected, Filtered, Effluent

Name of Chemical	Unit	2° Effluent, No Disinfection	Drinking Water Conc.	Risk-Based Action Level	Margin of Safety (unitless)
Nitrosamines[a,b]					
N-Nitrosodimethylamine (NDMA)	ng/L	2.7	<2	0.7	>0.35
Disinfection Byproducts[a,b,c]					
Bromate	µg/L	N/A	N/A	10	N/A
Bromoform	µg/L	2	0.5	80	160
Chloroform	µg/L	10	1	80	80
Dibromoacetic acid	µg/L	0.5	<1	60	>60
Dibromoacetonitrile	µg/L	<1	<0.5	70	>140
Dibromochloromethane	µg/L	N/A	N/A	80	N/A
Dichloroacetic acid	µg/L	1	<1	60	>60
Dichloroacetonitrile	µg/L	<1	<1	20	>20
Haloacetic acid (HAA5)	µg/L	2	5	60	12
Trihalomethanes (THMs)	µg/L	1	5	80	16
Pharmaceuticals[a,b,c]					
Acetaminophen	ng/L	10	<1	350,000,000	>350,000,000
Ibuprofen	ng/L	50	5	280,000,000	56,000,000
Carbamazepine	ng/L	200	150	186,900,000	1,200,000
Gemfibrozil	ng/L	610	61	140,000,000	2,300,000
Sulfamethoxazole	ng/L	295	221	160,000,000	720,000
Meprobamate	ng/L	320	32	280,000,000	8,800,000
Primidone	ng/L	130	130	58,100,000	450,000
Others					
Caffeine	ng/L	280	<1	70,000,000	>70,000,000
17-β Estradiol	ng/L	1.5	<0.1	3,500,000	>35,000,000
Triclosan[a,b,c]	ng/L	25	2.5	2,100,000	840,000
TCEP (Tris(2-Chloroethyl)-phosphate)[2 a,b,c]	ng/L	400	360	2,100,000	5,800
PFOS[a,b,c,d]	ng/L	54	54	200	3.7
PFOA[a,b,c,d]	ng/L	21	21	400	19

NOTES: N/A = data not available.
[a]Bellona et al. (2008).
[b]M. Wehner, OCWD, personal communication, 2009.
[c]Bellona and Drewes (2007).
[d]Snyder et al. (2010a).

byproducts after SAT reported by Krasner et al. (2008) and Dickenson et al. (2011).

Scenario 3—Reverse Osmosis, Advanced Oxidation, and Deep-Well Injection

For the potable reuse scenario via direct injection (Scenario 3), a reclaimed water quality after microfiltration, reverse osmosis, and advanced oxidation (UV/H$_2$O$_2$) is assumed. The concentration of disinfection byproducts in this reclaimed water after advanced treatment is adopted from monitoring at full-scale installations as reported by Wehner (2009), Bellona et al. (2008), and Bellona and Drewes (2007). Hormones, pharmaceuticals, and other trace organic chemicals in this highly treated reclaimed water are adopted from Wehner (2009), Bellona and Drewes (2007), Bellona et al. (2008), and Snyder et al. (2010a). The water quality

after 6 months of retention in a potable aquifer, assuming no dilution with ambient groundwater, followed by chlorination at the point of abstraction is based on field monitoring data reported by Wehner (2009) and Snyder et al. (2010a).

The concentration levels of each of the 24 chemicals discussed above are presented in Tables A-8, A-9, and A-10 for the three scenarios for the source waters or the reclaimed water applied to the spreading or direct injection projects. Additionally, the "drinking water" column represents the final water quality delivered to customers at the end of the final treatment processes from the drinking water treatment plant (Scenario 1) or after wellhead disinfection after withdrawal from the environmental buffer (Scenarios 2 and 3). Table A-11 summarizes the concentrations of contaminants at the point of exposure for all three scenarios.

TABLE A-10 Estimation of Margin of Safety for Scenario 3—Reuse with MF/RO/UV-H$_2$O$_2$ and Groundwater Injection

Name of Chemical	Unit	2° Effluent, no disinfection	Drinking Water Conc.	Risk Based Action Level	Margin of Safety (unitless)
Nitrosamines[a,b]					
N-Nitrosodimethylamine (NDMA)	ng/L	<2	<2	0.7	> 0.35
Disinfection Byproducts[a,b,c]					
Bromate	µg/L	<5	<5	10	> 2
Bromoform	µg/L	<0.5	<0.5	80	>160
Chloroform	µg/L	20	5	80	16
Dibromoacetic acid	µg/L	<1	<1	60	> 60
Dibromoacetonitrile	µg/L	N/A	N/A	70	N/A
Dibromochloromethane	µg/L	<0.5	<0.5	80	>160
Dichloroacetic acid	µg/L	<1	<1	60	>60
Dichloroacetonitrile	µg/L	N/A	N/A	20	N/A
Haloacetic acid (HAA5)	µg/L	13	5	60	12
Trihalomethanes (THMs)	µg/L	30	10	80	8
Pharmaceuticals[a,b,c]					
Acetaminophen	ng/L	<10	<10	350,000,000	>350,000,000
Ibuprofen	ng/L	<1	<1	280,000,000	>280,000,000
Carbamazepine	ng/L	<1	<1	186,900,000	>190,000,000
Gemfibrozil	ng/L	3	<1	140,000,000	140,000,000
Sulfamethoxazole	ng/L	2	<1	160,000,000	160,000,000
Meprobamate	ng/L	0.4	<0.3	280,000,000	>930,000,000
Primidone	ng/L	<1	<1	58,100,000	>58,000,000
Others					
Caffeine	ng/L	<3	<3	70,000,000	>223,000,000
17-β Estradiol	ng/L	<0.1	<0.1	3,500,000	>35,000,000
Triclosan[a,b,c]	ng/L	3	<1	2,100,000	>2,100,000
TCEP (Tris(2-chloroethyl)-phosphate)[a,b,c,d]	ng/L	<10	<10	2,100,000	>210,000
PFOS[a,b,c,d]	ng/L	<1	<1	200	>200
PFOA[a,d,c,d]	ng/L	<5	<5	400	>80

NOTES: N/A = data not available.
[a]Bellona et al. (2008).
[b]M. Wehner, OCWD, personal communication, 2009.
[c]Bellona and Drewes (2007).
[d]Snyder et al. (2010a).

Quantitative Chemical Risk Assessment

For each of the 24 chemicals identified in the three water treatment scenarios, potential lifetime health risks were assessed by calculating margins of safety (MOSs), or the risk-based action level (RBAL) divided by the concentration of contaminant in water (see Tables A-8 to A-10). RBALs represent benchmark values for risk or existing chemical-specific action levels, such as EPA maximum contaminant levels (MCLs), EPA health advisories, World Health Organization (WHO) drinking water guidelines, or chemical-specific EPA reference doses (RfDs), Agency for Toxic Substances and Disease Registry (ASTDR) minimal risk levels (MRLs), WHO acceptable daily intakes (ADIs), Food and Drug Administration (FDA)

maximum recommended therapeutic doses (MRTDs), and National Library of Medicine/National Institute of Health maximum tolerated doses (MTDs) from which a drinking water action level can be derived (see also Chapter 5). Table A-12 shows the source of the values used for each of the 24 chemicals.

These risk-based values have undergone extensive regulatory and/or peer review and incorporate uncertainty factors to account for variability and uncertainty in the hazard database, and for nonpharmaceuticals, the values consider effects on sensitive subpopulations (e.g., children, pregnant women, the elderly). Conversion of an oral reference toxicity dose to a drinking water action level uses assumptions about daily drinking water intake, consumer body weight, and the relative source contribution of water to total human exposure. The

TABLE A-11 Summary of the Levels of the 24 Chemicals in the Drinking Water for Each Scenario

Name of Chemical	Unit	Scenario 1	Scenario 2	Scenario 3
Nitrosamines				
N-Nitrosodimethylamine (NDMA)	ng/L	<2	<2	<2
Disinfection Byproducts				
Bromate	ng/L	N/A	N/A	<5
Bromoform	µg/L	3	0.5	<0.5
Chloroform	µg/L	5	1	5
Dibromoacetic acid	µg/L	<1	<1	<1
Dibromoacetonitrile	µg/L	<1.3	<0.5	N/A
Dibromochloromethane	µg/L	<1	N/A	<0.5
Dichloroacetic acid	µg/L	5	<1	<1
Dichloroacetonitrile	µg/L	<1	<1	N/A
Haloacetic acid (HAA5)	µg/L	10	5	5
Trihalomethanes THMs)	µg/L	30	5	10
Pharmaceuticals				
Acetominophen (paracetamol)	ng/L	<1	<1	<10
Ibuprofen	ng/L	<2.4	5	<1
Carbamazepine	ng/L	19	150	<1
Gemfibrozil	ng/L	16	61	<1
Sulfamethoxazole	ng/L	<2	221	<1
Meprobamate	ng/L	17	32	<0.3
Primidone	ng/L	6	130	<1
Others				
Caffeine	ng/L	20	<1	<3
17-β Estradiol	ng/L	<0.1	<0.1	<0.1
Triclosan	ng/L	<0.6	2.5	<1
TCEP (Tris(2-chloroethyl)phosphate)	ng/L	<25	360	<10
PFOS	ng/L	12	54	<1
PFOA	ng/L	11	21	<5

dose metric is expressed as concentrations in drinking water. Although numerous contaminants present in the three scenarios have existing drinking water action levels (such as an EPA MCL), a significant number of chemicals have only oral RfDs, ADIs, or are pharmaceuticals with MRTDs, all expressed as milligrams per kilogram of body weight per day. Risk values such as RfDs and ADIs are generally based upon experimental doses from repeat-dose animal studies that have been adjusted with appropriate uncertainty factors to account for animal to human extrapolation and interhuman sensitivity, while MRTDs are generally derived from doses employed in human clinical trials. To derive RBALs for chemicals without existing drinking water action levels, the following formula was used:

$$\text{Risk Based Action Level (mg/L)} = \frac{[X] \text{ mg/kg/day} \times 70 \text{ kg} \times 0.20}{2 \text{ L/day}}$$

where

X = Oral RfD, ADI, or other reference point such as MRTD;

70 kg = Default adult body weight;[3]

0.2 = Default relative source contribution from drinking water of 20%;

2 L/d = Default daily drinking water intake for a 70-kg adult.

Y = Acceptable level in drinking water (i.e., estimated action level)

[3] WHO drinking water guidelines are based upon a default adult body weight of 60 kg, while a default adult body weight of 70 kg is used by EPA and was used by this NRC committee to estimate RBALs using FDA MRTDs.

TABLE A-12 Summary of Risk-Based Action Values and Sources

Name of Chemical	Unit	Source of Risk Value	Risk Based Action Level
Nitrosamines			
NDMA	ng/L	EPA HA (EPA, 2011)	0.7
Disinfection Byproducts			
Bromate	µg/L	EPA MCL (EPA, 2011)	10
Bromoform	µg/L	EPA MCL (EPA, 2011)	80
Chloroform	µg/L	EPA MCL (EPA, 2011)	80
DBCA	µg/L	EPA MCL (EPA, 2011)	60
DBAN	µg/L	WHO Drinking Water Guideline Value (WHO, 2008)	70
DBCM	µg/L	EPA MCL (EPA, 2011)	80
DCAA	µg/L	EPA MCL (EPA, 2011)	60
DCAN	µg/L	WHO Drinking Water Guideline Value (WHO, 2008)	20
HAA5[a]	µg/L	EPA MCL (EPA, 2011)	60
THM	µg/L	EPA MCL (EPA, 2011)	80
Pharmaceuticals			
Acetominophen	ng/L	FDA MRTD (FDA, 2011)	350,000,000
Ibuprofen	ng/L	FDA MRTD (FDA, 2011)	280,000,000
Carbamazepine	ng/L	FDA MRTD (FDA, 2011)	190,000,000
Gemfibrozil	ng/L	FDA MRTD (FDA, 2011)	140,000,000
Sulfamethoxazole	ng/L	FDA MRTD (FDA, 2011)	160,000,000
Meprobamate	ng/L	FDA MRTD (FDA, 2011)	280,000,000
Primidone	ng/L	FDA MRTD (FDA, 2011)	58,000,000
Other			
Caffeine	ng/L	FDA MRTD (FDA, 2011)	70,000,000
17-β Estradiol	ng/L	FDA MRTD (FDA, 2011)	3,500,000
Triclosan	ng/L	EPA RfD (EPA, 2008)	2,100,000
TCEP	ng/L	ASTDR MRL (ASTDR, 2009)	2,100,000
PFOS	ng/L	Provisional EPA HA (EPA, 2011)	200
PFOA	ng/L	Provisional EPA HA (EPA, 2011)	400

[a]HAA5: monochloroacetic acid (MCAA) + dichloroacetic acid (DCAA) + trichloroacetic acid (TCAA) + Monobromoacetic acid (MBAA) + dibromoacetic acid (DBAA).

Ideally, the EPA bases the relative source contribution (RSC) on data regarding exposures that occur from food, air, and other important media such as personal care products or pharmaceutical agents (Donohue and Orme-Zavaleta, 2003). When data allow exposure pathways for other selected media to be quantified, default RSC values of 20, 50, or 80 percent are possible. In the absence of any data, a default RSC of 20 percent is used (Donohue and Orme-Zavaleta, 2003). EPA also assumes a daily drinking water intake of 2 L/d for an adult (EPA, 2004).

MOSs were estimated for each of the 24 contaminants (see summary of results in Table A-13). Where compounds were not detected, the lower limit on the MOS was determined using the level of detection at the concentration in drinking water.

$$MOS = \frac{RBAL}{\text{Estimated Drinking Water Level}}$$
$$(\text{Scenario 1, 2, or 3})$$

With the exception of the chemical NDMA, the MOS values are all greater than 1, indicating that there is unlikely to be a significant health risk, even after a lifetime of exposure to these individual chemicals. The analysis does not take into account combined health effects of contaminant mixtures. Simultaneous exposure to multiple chemicals would occur in all three scenarios; thus, a consideration of mixtures would not significantly affect the relative risk comparison for purposes of the risk exemplar. NDMA was not detected in any of the scenarios, but the MOS is less than 1 because the detection limit (2 ng/L) is above EPA's health advisory level of 0.7 ng/L. The large MOS for pharmaceuticals listed in Table A-13 indicates that potential health risks

TABLE A-13   Margin of Safety for 24 Chemicals for Each Scenario

Chemical	Scenario 1	Scenario 2	Scenario 3
Nitrosamines			
NDMA	>0.4	>0.4	>0.4
Disinfection Byproducts			
Bromate	N/A	N/A	> 2
Bromoform	27	160	>160
Chloroform	16	80	16
DBCA	>60	>60	>60
DBAN	>54	>140	N/A
DBCM	>80	N/A	>160
DCAA	12	>60	>60
DCAN	>20	>20	N/A
HAA5	6	12	12
THM	2.7	16	8
Pharmaceuticals			
Acetaminophen	>350,000,000	>350,000,000	>35,000,000
Ibuprofen	>120,000,000	56,000,000	>280,000,000
Carbamazepine	10,000,000	1,200,000	>190,000,000
Gemfibrozil	8,600,000	2,300,000	>140,000,000
Sulfamethoxazole	>80,000,000	720,000	>160,000,000
Meprobamate	17,000,000	8,800,000	>930,000,000
Primidone	10,000,000	450,000	>58,000,000
Others			
Caffeine	3,500,000	>70,000,000	>23,000,000
17-β Estradiol	>35,000,000	>35,000,000	>35,000,000
Triclosan	>3,500,000	840,000	>2,100,000
TCEP	>84,000	5,800	>210,000
PFOS	17	4	>200
PFOA	36	19	>80

from exposure to pharmaceuticals in reclaimed water is small. However, RBALs for pharmaceuticals presented in Table A-12 assume that long-term exposure to pharmaceuticals will result in toxicity similar to short-term exposures, which is an admitted area of uncertainty. Additional research to evaluate the effects of long-term, low-level exposure to chemicals in reclaimed water could provide additional insight on whether these areas of uncertainty are biologically significant.

VERIFICATION

The committee performed several levels of verification on this risk exemplar exercise to ensure that the results are sound. In the compilation of the water quality data that provide a basis for the analysis, three committee members worked to gather and/or review the chemical occurrence data used and three additional members gathered and/or reviewed the microbial occurrence data. After the risk analysis calculations were completed and the assumptions documented by the committee members, the chair carefully reviewed the analysis. When the report was in review, Appendix A and the spreadsheet containing the calculations were reviewed in detail by a non-committee member with experience in risk assessment. With no oversight, other than to explain the task, this individual reviewed the values and formulas used in each cell of the spreadsheet and compared them to the information documented in Appendix A. Following this verification, a few minor errors were detected that were discussed with the committee chair and staff and subsequently corrected.

Appendix B

Computation of Average Daily Dose

The Average Daily Dose from all exposures to reclaimed water (ADD_{RW}) can be estimated using the following equation (modified from Hutcheson et al., 1990):

$$ADD_{RW} = ADD_R + ADD_D + ADD_I$$

The Average Daily Dose from ingestion of the reclaimed water (ADD_R) can be estimated using the following equation (modified from Hutcheson and Martin, 1992):

$$ADD_R = \frac{Contaminant_R \times VI \times BAF\, D_2 \times C}{BW \times AP}$$

where

$Contaminant_R$	=	Concentration of chemical in reclaimed water (mass/volume),
VI	=	Daily volume of reclaimed water ingestion (mass/volume),
BAF	=	Bioavailability adjustment factor,
D_2	=	Duration of the exposure period (time),
BW	=	Average body weight (e.g., 70 kg),
AP	=	Averaging period (time),
C	=	Appropriate units conversion factor.

The Average Daily Dose from dermal contact with reclaimed water (ADD_D) can be estimated using the following equation:

$$ADD_D = Contaminant_D \times \frac{SA \times PC \times BAF \times F \times D_1 \times D_2 \times C}{BW \times AP}$$

where

$Contaminant_R$	=	Concentration of chemical in reclaimed water (mass/volume),
SA	=	Skin surface area in contact with the surface water during the period of exposure (area),
PC	=	Permeability constant (volume/time × area),
D_1	=	Average duration of each exposure event (time/event),
D_2	=	Duration of the exposure period (time),
BAF	=	Bioavailability adjustment factor,
F	=	Number of exposure events during the exposure period divided by the number of days in the exposure period (events/time),
D_1	=	Average duration of each exposure event (time/event),
D_2	=	Duration of the exposure period (time),
BW	=	Average body weight (e.g., 70 kg),
AP	=	Averaging period (time),
C	=	Appropriate units conversion factor.

The Average Daily Dose from inhalation of contaminants in reclaimed water (ADD_I) can be estimated using the following equation:

$$ADD_I = \text{Contaminant}_I \times \frac{VR \times BAF \times F \times D_1 \times D_2 \times C}{BW \times AP}$$

where

Contaminant_R = Concentration of chemical in reclaimed water (mass/volume),

VR = Daily respiratory volume during the period of exposure (volume/time),

BAF = Bioavailability adjustment factor,

F = Number of exposure events during the exposure period divided by the number of days in the exposure period (events/time),

D_1 = Average duration of each exposure event (time/event),

D_2 = Duration of the exposure period (time),

BW = Average body weight (e.g., 70 kg),

AP = Averaging period (time),

C = Appropriate units conversion factor.

Appendix C

Survey of Water Reclamation Costs

National Research Council
Committee on Assessment of Water Reuse as an
Approach for Meeting Future Water Supply Needs

Survey of Water Reclamation Costs

The National Research Council is currently conducting a comprehensive study of the potential for water reclamation and reuse of municipal wastewater to expand and enhance the nation's available water supply alternatives. This study is considering a wide range of uses, including drinking water, non-potable urban uses, irrigation, industrial process water, groundwater recharge, and water for environmental purposes. The study is considering technical, economic, institutional, and social challenges to increased adoption of water reuse, and it will provide practical guidance to decision makers evaluating their water supply alternatives. The complete task and committee membership is attached.

The study is sponsored by the U.S. Environmental Protection Agency, the National Science Foundation, the National Water Research Institute, the Centers for Disease Control, the Water Research Foundation, Orange County Water District, Orange County Sanitation District, Los Angeles Department of Water and Power, Irvine Ranch Water District, West Basin Water District, Inland Empire Utilities Agency, Metropolitan Water District of Southern California, Los Angeles County Sanitation District, and the Monterey Regional Water Pollution Control Agency. The report from this study is anticipated in January 2011.

The committee is charged to consider how different approaches to water reclamation vary in terms of cost, and how these costs compare to the costs of other available water supply alternatives. To complete its charge, the committee determined that it needed additional information on the cost of reuse from key reuse initiatives under way, representing a variety of technologies, approaches, and geographic areas. We hope that you will take the time to fill out the attached survey of costs, as the results should be valuable to many communities across the nation considering water reuse among their future water supply alternatives. Please return your completed survey by March 3, 2010.

Please note that, per our FACA requirements, your survey responses can be made available to the public upon request.

We appreciate your assistance to this committee's efforts.

Organization/Agency: _____

Contact Person: _____
      Title: _____
      Phone: _____
      Email: _____

1. **Name of the reclaimed water project (please fill out one survey for each project if your utility has multiple reuse projects/facilities):**

2. **Rated design capacity of the project (in MGD) and estimated annual production for:**
    2.1. Non-potable reuse applications: _____
    2.2. Potable reuse applications: _____

3. **Year(s) constructed:**

4. **Treatment processes included in:**
    4.1. Column (a) for treatment required for wastewater disposal:

    4.2. Column (b) for Non-potable treatment beyond Column (a):

    4.3. Column (d) for Potable reuse treatment beyond Columns (a) and (b)

5. **Major uses of effluent** (*e.g., further treatment, irrigation, agriculture, cooling, groundwater recharge, wholesale to another entity, discharge to water bodies*):
    5.1. Wastewater disposal:

    5.2. Non-potable treatment:

    5.3. Potable reuse treatment:

6. **Please fill out the *attached* Excel spreadsheet with regard to each of the three water treatment grades listed above for each of the following:**
    6.1. Capital costs, including all subsidies, as $/Kgal of rated plant capacity.
       Please, if possible, separate these costs according to major project components (e.g., treatment system, spreading system, distribution system) and include the year constructed for each.

    6.2. Annual Operation and Maintenance Cost, in $/yr/Kgal of rated plant capacity in terms of
      6.2.1. Personnel
      6.2.2. Energy (Electricity, Natural Gas, etc.)
      6.2.3. All other operations and maintenance costs
    Note that only the yellow spreadsheet cells should be filled in. The other cells will total automatically. See attached explanation sheet for more details.

7. Please describe any subsidies to the project included in the above costs, including federal, state, or local contributions to the project or land donations:

8. What rates do you charge users (in $/kgal) for:
   8.1. Non-potable reclaimed water?

   8.2. Potable reclaimed water?

   8.3. Traditional potable supply?

9. When the decision was made to implement your water reuse project(s), what other water supply alternatives were considered? What was the cost of the alternatives considered, if any (in $/Kgal)? Please note the year that those costs estimates were determined.

10. What was the decisive factor in the selection of the alternative(s) implemented?

12. Please describe any concentrate management issues faced when implementing your reuse project, and how these were resolved. Approximately what portion of the total water reclamation cost (capital + O&M) can be attributed to concentrate management?

13. Please describe the major benefits of increased reclaimed water in your area:

14. What is the per capita water use in your service area? If data are available, please include data for the past 10 years in tabular or graphical form.

Could we follow up with you if we need clarification on any of your responses? YES _____ NO _____

Thank you for your assistance!

Additional Explanations for Excel Spreadsheet

For clarification, some additional explanations of the various data categories are described here:

Row I, Capital Costs:

The capital costs include all of the costs of capital, including subsidies. If possible, please list each major project component within the overall project (e.g., treatment processes, spreader system, ASR system, reuse-specific distribution system) and indicate year constructed. Capital costs typically do not vary during the life of the project and are treated as fixed costs, over a set period of time (the amortization period).

Row II, Operating Costs:

Operating costs include the variable costs of operation over time, including energy, personnel, and other costs, such as chemicals and routine maintenance.

Column (a), Wastewater Disposal treatment costs

Column "a" focuses on the costs of the basic wastewater treatment aspects (i.e., secondary treatment steps) of a wastewater treatment for disposal purposes. If a reclaimed water facility starts with raw wastewater, column "a" would refer to the "normal" secondary treatment costs for the project. For example, this would include costs up through the disinfection stage in a conventional activated sludge plant. If the reclaimed water facility purchases the secondary effluent from a wastewater treatment plant, these costs should be stated here (enter "0" if there is no charge for the secondary effluent).

Column (b), Non-potable treatment costs beyond secondary

Column "b" focuses on the costs of the additional treatment steps for non-potable applications following those required for wastewater disposal. In other words, all other treatment after the treatment defined in Column "a". For example, if filtration or chlorination is used to produce reclaimed water for irrigation or industrial use, but these components are not part of the secondary treatment core, that cost would be shown in Column "b".

Column (c), Total Cost for Non-Potable Reuse

Column "c" will automatically add column "a" and column "b". No information needs to be entered here.

Column (d), Potable reuse treatment costs, beyond (a) and (b)

Column "d" is reserved for additional treatment steps following the wastewater treatment costs in Column "a" and the non-potable reclamation costs listed in Column "b" to further treat the water for indirect potable reuse applications. For example, a plant might consist of a secondary core of activated sludge followed by UV disinfection as the Column "a" costs. Column "b" costs might include a filtration step followed by chlorine disinfection required to produce effluent suitable for irrigation or industrial use. Column "d" costs would include costs to take the reclaimed water and polish it further to result in a product that could be injected or put into a surface impoundment for indirect potable reuse. This might include filtration with granular activated carbon or through reverse osmosis membranes.

Column (e), Total Cost for Indirect Potable Reuse

Column "e" will automatically add column "c" and column "d". No information needs to be entered here.

Incremental Costs of Water for Reuse Survey

Name of Organization/Agency and Project Name (one spreadsheet per project): _____

	(a)	(b)	(c)	(d)	(e)
	Cost of Treatment Processes Used for Wastewater Disposal[2]	Additional Process(es) for Non-potable Reuse[3]	Total Costs for Non-Potable Reuse (b)	Additional Process(es) for Potable Reuse[4]	Total Costs Indirect Potable Reuse (b)+(d)

I. Capital Costs[1] $/kgal/yr (a) $/kgal/yr (b) $/kgal/yr (c) $/kgal/yr (d) $/kgal/yr (e)

Please list major project components separately (e.g., treatment, spreader basins, reuse distribution sys.) and year constructed.

Include Subsidies (a)   Include Subsidies (b)   Include Subsidies (d)

	(a)	(b)	(c)	(d)	(e)
			$		$
			$		$
			$		$
			$		$
			$		$
			$		$
SUB-TOTAL	$ -	$ -	$ -	$ -	$ -

II. Annual Operation & Maintenance Costs $/kgal $/kgal $/kgal $/kgal $/kgal

	(a)	(b)	(c)	(d)	(e)
Personnel			$		$
Energy (Electricity & Natural Gas)			$		$
All Other Operations and Maintenance Costs			$		$
SUB-TOTAL	$ -	$ -	$ -	$ -	$ -

III. Amortized Capital Costs plus O&M $ - $ - $ - $ - $ -

Notes:
(1) Includes all Capital Costs and subsidies.
(2) If wastewater is purchased, include purchase price in $/kgal. If there is no cost for wastewater supplied from elsewhere, enter 0.
(3) Includes advanced secondary treatment, and all polishing costs such as filtration, etc.
(4) Includes all costs such as ASR, Wells, Spreading, etc.

IV. Please state any important assumptions below

_____
_____
_____

Appendix D

Water Science and Technology Board

DONALD I. SIEGEL, *Chair*, Syracuse University, New York
LISA ALVAREZ-COHEN, University of California, Berkeley
EDWARD J. BOUWER, Johns Hopkins University, Baltimore, Maryland
YU-PING CHIN, Ohio State University, Columbus
OTTO C. DOERING, Purdue University, West Lafayette, Indiana
M. SIOBHAN FENNESSY, Kenyon College, Gambier, Ohio
BEN GRUMBLES, Clean Water America Alliance, Washington, D.C.
GEORGE R. HALLBERG, The Cadmus Group, Watertown, Massachusetts
KENNETH R. HERD, Southwest Florida Water Management District, Brooksville, Florida
GEORGE M. HORNBERGER, Vanderbilt University, Nashville, Tennessee
KIMBERLY L. JONES, Howard University, Washington, D.C.

LARRY LARSON, Association of State Floodplain Managers, Madison, Wisconsin
DAVID H. MOREAU, University of North Carolina, Chapel Hill
DENNIS D. MURPHY, University of Nevada, Reno
MARYLYNN V. YATES, University of California, Riverside

Staff

JEFFREY W. JACOBS, Director
LAURA J. EHLERS, Senior Program Officer
STEPHANIE E. JOHNSON, Senior Program Officer
LAURA E. HELSABECK, Senior Program Officer
M. JEANNE AQUILINO, Financial and Administrative Associate
ANITA A. HALL, Senior Program Associate
MICHAEL J. STOEVER, Research Associate
SARAH E. BRENNAN, Senior Program Assistant

Appendix E

Biographical Sketches of Committee Members

Rhodes R. Trussell (NAE), *Chair*, is the founder of Trussell Technologies, Inc. Previously he was the lead drinking water technologist at Montgomery Watson Harza, Inc. He is recognized worldwide as an authority in methods and criteria for water quality and the development of advanced processes for treating water or wastewater to achieve the highest standards. He has worked on the process design for dozens of treatment plants, ranging from less than 1 to more than 900 MGD in capacity and has experience with virtually every physiochemical process and most biological processes as well. He has a special interest in emerging water quality problems and reuse. Dr. Trussell is a member of the National Academy of Engineering and has served for more than 10 years on the U.S. Environmental Protection Agency's (EPA's) Science Advisory Board. He also served as chair of the Water Science and Technology Board and has been a member of numerous National Research Council (NRC) committees, including the Committee on the Evaluation of the Viability of Augmenting Potable Water Supplies with Reclaimed Water and the Committee on Indicators for Waterborne Pathogens. Dr. Trussell has a B.S., M.S., and Ph.D. in environmental engineering from the University of California, Berkeley.

Henry A. Anderson is chief medical officer and state epidemiologist for occupational and environmental health in the Wisconsin Division of Public Health and adjunct professor of population health at the University of Wisconsin Medical School. Dr. Anderson's expertise includes public health; preventive, environmental, and occupational medicine; respiratory diseases; epidemiology; human health risk assessment; and risk communication. His research interests include disease surveillance, risk assessment, health hazards of Great Lakes sport-fish consumption, arsenic in drinking water, asbestos disease, and occupational fatalities and injuries. He is certified by the American Board of Preventive Medicine with a subspecialty in occupational and environmental medicine and is a fellow of the American College of Epidemiology. Dr. Anderson is chair of the Board of Scientific Councilors of the National Institute for Occupational Safety and Health and serves on the EPA Children's Health Protection Advisory Committee. He has served on several NRC committees, including the Division on Earth and Life Studies Committee, the Committee on Toxicity Testing and Assessment of Environmental Agents, and the Committee on Enhancing Environmental Health Content in Nursing Practice. Dr. Anderson received his M.D. from the University of Wisconsin Medical School.

Edmund G. Archuleta is the manager of the El Paso Water Utilities Public Service Board, a role he has served in for nearly 20 years. He is responsible for all aspects of water, wastewater, reclaimed water service, and stormwater to the greater El Paso metropolitan area. Mr. Archuleta is a past chairman of the American Water Works Association Research Foundation, and current Trustee of the Association of Metropolitan Water Agencies. He serves as chairman of the Multi-State Salinity Coalition and, in 2006, was appointed by

President George W. Bush to the National Infrastructure Advisory Council. He is a registered professional engineer. Mr. Archuleta earned his B.S. and M.S. degrees in Civil Engineering from New Mexico State University and a Master's of Management degree from the University of New Mexico.

James Crook is an independent consultant on water and environmental issues. He is an environmental engineer with 40 years of experience in state government and consulting engineering arenas, serving public and private sectors in the United States and abroad. He has authored more than 100 publications and is an internationally recognized expert in water reclamation and reuse. Previously, he spent 15 years directing the California Department of Health Services' water reuse program and developed California's first comprehensive water reuse criteria. He was the principal author of the Guidelines for Water Reuse, published by the EPA and the U.S. Agency for International Development. His honors include selection as the American Academy of Environmental Engineers' 2002 Kappe Lecturer. He served as chair of the Committee on the Evaluation of the Viability of Augmenting Potable Water Supplies with Reclaimed Water. Dr. Crook received a B.S. in civil engineering from the University of Massachusetts at Amherst, and an M.S. and a Ph.D. in environmental engineering from the University of Cincinnati.

Jörg E. Drewes is a Professor of Civil and Environmental Engineering and Director of Research for the NSF Engineering Research Center "Reinventing the Nation's Urban Water Infrastructure (ReNUWIt)." He also serves as associate director of the Advanced Water Technology Center (AQWATEC) at the Colorado School of Mines. Dr. Drewes also holds an Adjunct Professor appointment at the University of New South Wales, Sydney and a Visiting Professor appointment at the King Abdullah University of Science and Technology, Saudi Arabia. Dr. Drewes has been actively involved in research in the area of water treatment and non-potable and potable water reuse for more than nineteen years. Dr. Drewes' research interests include water treatment and potable reuse, design and operation of managed aquifer recharge (MAR) system, monitoring strategies for bulk organic carbon and emerging trace organic chemicals in natural

and engineered systems, performance modeling and optimized operation of energy-efficient membranes, and beneficial reuse of produced water during natural gas exploration. Dr. Drewes received his B.S., M.S. and Ph.D. in Environmental Engineering from the Technical University of Berlin, Germany.

Denise D. Fort is a member of the faculty at the University of New Mexico's School of Law. She has 25 years of experience in environmental and natural resources law. She served as chair of the Western Water Policy Review Advisory Commission, a presidential commission that prepared a report on western water policy concerns. In earlier positions, she served as director of New Mexico's Environmental Improvement Division, as a staff representative to the National Governors Association, as an environmental attorney, and in other capacities concerned with environmental and natural resource matters. She has served on the Water Science and Technology Board and numerous NRC committees, including the Committee on Sustainable Underground Storage of Recoverable Water. She received her B.A. from St. John's College and her J.D. from the Catholic University of America's School of Law.

Charles N. Haas is the Betz Chair Professor of Environmental Engineering at Drexel University. His areas of research involve microbial and chemical risk assessment, industrial wastewater treatment, waste recovery, and modeling wastewater disinfection and chemical fate and transport. He was one of the first scientists to examine dose-response datasets for microbial agents spread through environmental means and to implement a quantitative risk framework. Dr. Haas is also the codirector of the Center for Advancing Microbial Risk Assessment, funded by the Department of Homeland Security and EPA. Dr. Haas has been a member of several NRC committees, including a committee to define "how clean is safe" following cleanup from a bioterrorist event and the Committee on the Evaluation of the Viability of Augmenting Potable Water Supplies with Reclaimed Water. He is also a member of the Water Science and Technology Board. Dr. Haas received his B.S. in biology and M.S. in environmental engineering from Illinois Institute of Technology and his Ph.D. in environmental engineering from the University of Illinois.

Brent M. Haddad is the founder and director of the Center for Integrated Water Research and a professor of environmental studies at the University of California, Santa Cruz. His research focuses on freshwater policy and management, including urban water management strategies, utility-stakeholder communications (including risk communication and public perception), and long-range planning, and he has published research analyzing public responses to water reuse projects. Dr. Haddad serves as a member of the Research Advisory Committee of the WateReuse Foundation, on the Editorial Board of *Environmental Management*, and as a consultant for numerous public- and private-sector clients. He received a B.A from Stanford University, an M.A in international relations from Georgetown University, an M.B.A in business and public policy, and a Ph.D. in energy and resources from University of California, Berkeley.

Duane B. Huggett is Assistant Professor in the Department of Biology at the University of North Texas. Previously, Dr. Huggett worked as a research scientist with Pfizer Global Research and Development. Dr. Hugget's research interests include environmental toxicology, bioconcentration and bioaccumulation of contaminants, and ecopharmacology and physiology. Recent research has focused on the bioconcentration and toxicology of select pharmaceuticals in fish. Dr. Huggett received his B.S. in Biological Sciences from Virginia Polytechnic and State University, and he earned his M.S. and Ph.D. in Biological Sciences and Pharmacology and Toxicology, respectively, from the University of Mississippi.

Sunny Jiang is associate professor at the University of California, Irvine in the Department of Civil and Environmental Engineering. Her research focuses on water quality microbiology, microbial ecology, and epidemiology of exposure to recreational waters, and she specializes in the application of biotechnology tools for assessment and detection of microbial pathogens in the aquatic environment. She has served on the American Water Works Association Research Foundation Project Advisory Committee and the World Health Organiza-

tion Desalination Guideline Development Committee. Dr. Jiang received her B.S. in biochemistry from Nankai University in China, and her M.S. and Ph.D. in marine science at the University of South Florida.

David L. Sedlak is Professor of Civil and Environmental Engineering at the University of California, Berkeley where he is also the co-Director of the Berkeley Water Center and the Deputy Director of the National Science Foundation's Engineering Research Center on Reinventing the Nation's Urban Water Infrastructure (ReNUWIt). His areas of research interest include analytical methods for measuring organic compounds in water, fate of chemical contaminants in water recycling systems, environmental photochemistry, and ecological engineering. He has received several notable awards including the Fulbright Senior Scholar Award in 2003, Paul Busch Award for Innovation in Water Quality Engineering in 2003 and the NSF CAREER Award in 1997. Dr. Sedlak received a B.S. degree in Environmental Science from Cornell University and a Ph.D. degree in Water Chemistry from the University of Wisconsin in Madison.

Shane A. Snyder is a professor in the College of Engineering at the University of Arizona. He is also the codirector of the Arizona Laboratory for Emerging Contaminants. Dr. Snyder's research focuses on the identification, fate, and health relevance of emerging water pollutants. Prior to this appointment, he was Research and Development Project Manager at the Southern Nevada Water Authority in Las Vegas. Dr. Snyder has served on EPA advisory committees on the Endocrine Disruptor Screening Program and the Third Contaminant Candidate List. He has also served on two California Chemicals of Emerging Concern Expert Panels. Dr. Snyder is a visiting professor at the National University of Singapore where he leads research on water reuse technologies and implications for public health. He received a bachelor's degree in chemistry from Thiel College in Pennsylvania and a dual Ph.D. in environmental toxicology and zoology from Michigan State University.

Margaret H. Whittaker is the chief toxicologist and president of ToxServices, where she serves as the project manager and technical lead of ToxServices projects. In addition to her extensive program management experience, Dr. Whittaker has extensive technical experience in hazard identification and noncancer and cancer dose-response assessment, including quantitative risk assessment (e.g., benchmark dose modeling for both carcinogens and noncarcinogens). She has worked at two of the country's leading toxicology and risk assessment consulting firms (the ENVIRON Corporation and the Weinberg Group). Dr. Whittaker has over a decade of experience evaluating health hazards and quantitating human health risks for low-level contaminants in drinking water, pharmaceuticals, medical devices, cosmetics, and food additives. She is a Diplomat of the American Board of Toxicology (D.A.B.T.). Dr. Whittaker earned a Ph.D. in toxicology from the University of Maryland, Baltimore, and an M.P.H. in environmental health from the University of Michigan.

Dale Whittington is professor of environmental sciences and engineering, city and regional planning, and public policy at the University of North Carolina, Chapel Hill. Since 1986, he has worked for the World Bank and other international agencies on the development and application of techniques for estimating the economic value of environmental resources in developing countries, with a particular focus on water and sanitation and vaccine policy issues. Dr. Whittington has published extensively on cost-benefit analysis, environmental economics, and water resources planning and policy in developing countries. His current research interests include the development of planning approaches and methods for the design of improved water and sanitation systems for the rapidly growing cities of Asia. Dr. Whittington received his A.B. at Brown University, his M.P.A. at the University of Texas, his M.Sc. at the London School of Economics and Political Science, and his Ph.D. at the University of Texas.

STAFF

Stephanie E. Johnson, *study director*, is a senior program officer with the Water Science and Technology Board. Since joining the NRC in 2002, she has served as study director for twelve studies on topics such as desalination, water security, Chesapeake Bay nutrient management, and Everglades restoration progress. She has also worked on NRC studies on contaminant source remediation, the disposal of coal combustion wastes, and coalbed methane production. Dr. Johnson received a B.A. from Vanderbilt University in chemistry and geology and an M.S. and a Ph.D. in environmental sciences from the University of Virginia.

Sarah E. Brennan is a senior program assistant with the Water Science and Technology Board. Since joining the NRC in 2010, she has worked on five projects including Everglades restoration progress, U.S. Army Corps of Engineers' water resources, and water and environmental management in the California Bay Delta. Before joining WSTB, Ms. Brennan was a Peace Corps Volunteer in Ghana, West Africa. She received her B.S. in International Development from Susquehanna University.